Capitalizing on New Needs and New Opportunities:

Government-Industry Partnerships in Biotechnology and Information Technologies

CHARLES W. WESSNER, EDITOR

Board on Science, Technology, and Economic Policy

Policy and Global Affairs

National Research Council

NATIONAL ACADEMY PRESS
Washington, D.C.

NATIONAL ACADEMY PRESS • 2101 Constitution Avenue, N.W. • Washington, D.C. 20418

International Standard Book Number 0-309-08257-9

Limited copies are available from Board on Science, Technology, and Economic Policy, National Research Council, 1055 Thomas Jefferson Street, N.W., Suite 2014, Washington, D.C. 20007; 202-334-2200.

Additional copies of this report are available from National Academy Press, 2101 Constitution Avenue, N.W., Lockbox 285, Washington, D.C. 20055; (800) 624-6242 or (202) 334-3313 (in the Washington metropolitan area); Internet, http://www.nap.edu

Printed in the United States of America

THE NATIONAL ACADEMIES

National Academy of Sciences
National Academy of Engineering
Institute of Medicine
National Research Council

The **National Academy of Sciences** is a private, nonprofit, self-perpetuating society of distinguished scholars engaged in scientific and engineering research, dedicated to the furtherance of science and technology and to their use for the general welfare. Upon the authority of the charter granted to it by the Congress in 1863, the Academy has a mandate that requires it to advise the federal government on scientific and technical matters. Dr. Bruce M. Alberts is president of the National Academy of Sciences.

The **National Academy of Engineering** was established in 1964, under the charter of the National Academy of Sciences, as a parallel organization of outstanding engineers. It is autonomous in its administration and in the selection of its members, sharing with the National Academy of Sciences the responsibility for advising the federal government. The National Academy of Engineering also sponsors engineering programs aimed at meeting national needs, encourages education and research, and recognizes the superior achievements of engineers. Dr. Wm. A. Wulf is president of the National Academy of Engineering.

The **Institute of Medicine** was established in 1970 by the National Academy of Sciences to secure the services of eminent members of appropriate professions in the examination of policy matters pertaining to the health of the public. The Institute acts under the responsibility given to the National Academy of Sciences by its congressional charter to be an adviser to the federal government and, upon its own initiative, to identify issues of medical care, research, and education. Dr. Kenneth I. Shine is president of the Institute of Medicine.

The **National Research Council** was organized by the National Academy of Sciences in 1916 to associate the broad community of science and technology with the Academy's purposes of furthering knowledge and advising the federal government. Functioning in accordance with general policies determined by the Academy, the Council has become the principal operating agency of both the National Academy of Sciences and the National Academy of Engineering in providing services to the government, the public, and the scientific and engineering communities. The Council is administered jointly by both Academies and the Institute of Medicine. Dr. Bruce M. Alberts and Dr. Wm. A. Wulf are chairman and vice chairman, respectively, of the National Research Council.

Steering Committee for Government-Industry Partnerships for the Development of New Technologies *

Gordon Moore, *Chair*
Chairman Emeritus, *retired*
Intel Corporation

M. Kathy Behrens
Managing Partner
Robertson Stephens Investment Management
and STEP Board

Michael Borrus
Managing Director
The Petkevich Group, LLC

Iain M. Cockburn
Professor of Finance and Economics
Boston University

Kenneth Flamm
Dean Rusk Chair in International Affairs
LBJ School of Public Affairs
University of Texas at Austin

James F. Gibbons
Professor of Engineering
Stanford University

W. Clark McFadden
Partner
Dewey Ballantine

Burton J. McMurtry
General Partner
Technology Venture Investors

William J. Spencer, *Vice-Chair*
Chairman Emeritus
International SEMATECH
and STEP Board

Mark B. Myers
Senior Vice-President, *retired*
Xerox Corporation
and STEP Board

Richard Nelson
George Blumenthal Professor of International and Public Affairs
Columbia University

Edward E. Penhoet
Dean, School of Public Health
University of California at Berkeley
and STEP Board

Charles Trimble
Chairman
U.S. GPS Industry Council

John P. Walker
Chairman and Chief Executive Officer
Axys Pharmaceuticals, Inc.

Patrick Windham
President, Windham Consulting
and Lecturer, Stanford University

*As of August 2001.

Project Staff*

Charles W. Wessner
Study Director

McAlister T. Clabaugh
Program Associate

David E. Dierksheide
Program Associate

Christopher S. Hayter
Program Associate

Sujai J. Shivakumar
Consultant

Contributors

Paula Stephan
Georgia State University

Grant Black
Georgia State University

Wesley M. Cohen
Carnegie Mellon University

Kenneth Flamm
University of Texas at Austin

Michael McGeary
McGeary and Smith

John Walsh
University of Illinois at Chicago

---

*As of August 2001

For the National Research Council (NRC), this project was overseen by the Board on Science, Technology and Economic Policy (STEP), a standing board of the NRC established by the National Academies of Sciences and Engineering and the Institute of Medicine in 1991. The mandate of the STEP Board is to integrate understanding of scientific, technological, and economic elements in the formulation of national policies to promote the economic well-being of the United States. A distinctive characteristic of STEP's approach is its frequent interactions with public and private-sector decision makers. STEP bridges the disciplines of business management, engineering, economics, and the social sciences to bring diverse expertise to bear on pressing public policy questions. The members of the STEP Board* and the NRC staff are listed below:

Dale Jorgenson, *Chair*
Frederic Eaton Abbe Professor
of Economics
Harvard University

M. Kathy Behrens
Managing Partner
Robertson Stephens Investment
Management

Vinton G. Cerf
Senior Vice-President
WorldCom

Bronwyn Hall
Professor of Economics
University of California at Berkeley

James Heckman
Henry Schultz Distinguished Service
Professor of Economics
University of Chicago

Ralph Landau
Consulting Professor of Economics
Stanford University

Richard Levin
President
Yale University

William J. Spencer, *Vice-Chair*
Chairman Emeritus
International SEMATECH

David T. Morgenthaler
Founding Partner
Morgenthaler

Mark B. Myers
Senior Vice-President, *retired*
Xerox Corporation

Roger Noll
Morris M. Doyle Centennial
Professor of Economics
Stanford University

Edward E. Penhoet
Dean, School of Public Health
University of California at Berkeley

William Raduchel
Chief Technology Officer
AOL Time Warner

Alan Wm. Wolff
Managing Partner
Dewey Ballantine

---

*As of August 2001.

STEP Staff*

Stephen A. Merrill
Executive Director

Philip Aspden
Senior Program Officer

Craig M. Schultz
Senior Program Officer

Camille M. Collett
Program Associate

David E. Dierksheide
Program Associate

Charles W. Wessner
Program Director

Sujai J. Shivakumar
Consultant

Adam Korobow
Consultant

McAlister T. Clabaugh
Program Associate

Christopher S. Hayter
Program Associate

---

*As of September 2001.

National Research Council
Board on Science, Technology, and Economic Policy

Sponsors

The National Research Council gratefully acknowledges
the support of the following sponsors:

National Aeronautics and Space Administration

Office of the Director, Defense Research & Engineering

National Science Foundation

U.S. Department of Energy

Office of Naval Research

National Institutes of Health

National Institute of Standards and Technology

Sandia National Laboratories

Electric Power Research Institute

International Business Machines

Kulicke and Soffa Industries

Merck and Company

Milliken Industries

Motorola

Nortel

Proctor and Gamble

Silicon Valley Group, Incorporated

Advanced Micro Devices

Contents

Preface

As we begin the twenty-first century, many believe that we are also witnessing the start of a new era—one where humankind will increasingly expand its understanding of the building blocks of life, and one which will rely on advanced information technologies to process, analyze, and share the results of such research. This era may well rest on what some call the new economy – that is, an economy where higher sustained growth rates are fed by productivity improvements made possible by the application of new knowledge and new technologies. This state of affairs depends on continued public and private sector investment in productivity-enhancing technologies. It also requires substantial and expanded investment in basic research. Increased allocations of public resources to research, though, are not sufficient; continued progress also depends on government participation in the maintenance of a policy framework that supports the development of new technologies.

Government funding of research—especially university-based research—is an essential part of this framework of support. Policies encouraging partnerships and other cooperative arrangements among universities, industry, and the government have proved, in some cases, to be effective measures to foster the development of new productivity-enhancing technologies.[1] Such policies are often related to specific government missions and procurement in sectors such as

[1]See National Research Council, *The Small Business Innovation Research Program: An Assessment of the Department of Defense Fast Track Initiative*, Washington, D.C.: National Academy Press, 2000; and National Research Council, *The Advanced Technology Program: Assessing Outcomes*, Washington, D.C.: National Academy Press, 2001.

health, transport, and defense. In other cases, limited support of promising technologies with widespread applications may be the most appropriate approach.

A TRADITION OF PARTNERSHIPS

The government's role in supporting the development of new technologies is not new. During the nineteenth century, the federal government had an enormous impact on the structure and composition of the economy. The government played an essential role in developing the U.S. railway network, and—through the 1862 Morrill Act and support for the agricultural extension service—the farm sector.[2]

This support continued into the twentieth century. In 1901, the federal government established the National Bureau of Standards to help industry. Later, the federal government provided special backing for the development of (what we now call) dual-use industries—such as aircraft frames and engines and radio—seen as important to the nation's security and commerce. The unprecedented challenges of World War II generated huge increases in the level of government procurement and support for high-technology industries.[3] Today's computing industry has its origins in the government's wartime support for a program that resulted in the creation of one of the earliest electronic digital computers, the ENIAC.[4] Following that war, the federal government began to fund basic research at universities on a significant scale, first through the Office of Naval Research and later through the National Science Foundation.[5]

During the Cold War, the government continued to emphasize technological superiority as a means of ensuring U.S. security. Government funds and cost-plus contracts helped to support enabling technologies, such as semiconductors, new materials, radar, jet engines, missiles, and computer hardware and software.[6]

---

[2]See Richard Bingham, *Industrial Policy American Style: From Hamilton to HDTV*, New York: M.E. Sharpe, 1998 for a comprehensive review.

[3]David Mowery, "Collaborative R&D: how effective is it?" *Issues in Science and Technology*, 15(1), 1998, p. 37.

[4]Kenneth Flamm, *Creating the Computer*. Washington, D.C.: The Brookings Institution, 1988, chapters 1-3.

[5]The National Science Foundation was initially seen as the agency that would fund basic scientific research at universities after World War II. However, disagreements over the degree of Executive Branch control over the NSF delayed passage of its authorizing legislation until 1950, even though the concept for the agency was first put forth in 1945 in Vannevar Bush's report, *Science: The Endless Frontier*. The Office of Naval Research bridged the gap in basic research funding during those years. For an account of the politics of the NSF's creation, see G. Paschal Zachary, *Endless Frontier: Vannevar Bush, Engineer of the American Century*, New York: The Free Press, 1997, pp. 231. See also Daniel Lee Kleinman, *Politics on the Endless Frontier: Postwar Research Policy in the United States*, Durham, N.C.: Duke University Press, 1995.

[6]For an excellent review of the role of government support in developing the computer industry and the Internet, see National Research Council, *Funding a Revolution: Government Support for Computing Research*, Washington, D.C.: National Academy Press, 1999.

In the post-Cold War period, the evolution of the American economy continues to be marked by the interaction of government-funded research and activities pursued by innovative entrepreneurs. Government support in this period has been essential to progress in areas such as microelectronics, robotics, biotechnology, and the investigation of the human genome. It has also played a critical role in the development of the Internet (whose forerunners were funded by the Defense Department and the National Science Foundation [NSF]).[7] Together, these technologies underpin the new economy.

In all, both the federal and local governments in the United States have participated actively in promoting domestic industry in an increasingly global marketplace. Indeed, the U.S. has a remarkably wide range of public-private partnerships in high technology sectors.[8] In addition to the cases mentioned above, there are public-private consortia of many types. These can be classified in a number of ways: by economic objective of the partnership—that is, to leverage the social benefits associated with federal R&D activity and/or to enhance the position of a national industry, and by other objectives, including the need to deploy industrial R&D to meet military or other government missions.[9]

The U.S. economy continues to be distinguished by the extent to which individual entrepreneurs and researchers take the lead in developing innovations and starting new businesses. In doing so, they often harvest crops sown on fields made fertile by the government's long-term research investments.[10]

Recently, new Internet-based companies and biotechnology firms have been the source of major innovations. These innovations, and the economic benefit they provide, are based on information technologies that are more powerful and less expensive to use than ever before. These technologies promise to remain a source of substantial growth in the future.

The promise of better health, and the tangible benefits it represents, have prompted federal support for biomedicine. Progress in biomedicine and drug research, the development of diagnostic tools such as magnetic resonance imaging, and the rapidly expanding understanding of the human genome give credence to this promise.

By the late part of the 1990s, this belief steadily gained momentum, resulting in major yearly increases in federal funding for biomedical research. This tremendous

---

[7]National Research Council, *Funding a Revolution: Government Support for Computing Research, op. cit.* See, particularly, chapter 7.

[8]See Chris Coburn and Dan Bergland, *Partnerships*. Columbus, OH: Battle Press, 1995.

[9]See Albert Link, "Public/Private Partnerships as a Tool to Support Industrial R&D: Experiences in the United States." Paper prepared for the working group on Innovation Policy, Paris, 1998, p. 20. Partnerships can also be differentiated by the nature of public support. Some partnerships involve a direct transfer of funds to an industry consortium. Others focus on shared use of infrastructure, such as laboratory facilities.

[10]David B. Audretsch and Roy Thurik, *Innovation, Industry, Evolution, and Employment*, Cambridge, UK: Cambridge University Press, 1999.

Box A: The Central Role of a Positive Macro-economic Policy Environment

The evolution of federal policy geared toward greater support for basic and mission-oriented research after 1945 should not obscure the fundamental importance of the macro-economic policy environment in the United States. Policies of the federal government collectively define and shape the environment in which innovation takes place. For example, federal policies affecting capital formation and corporate governance play important roles in competitive performance.[11] The range and diversity of these policies are substantial. They include government policies related to taxation, especially capital gains, fiscal and monetary matters, education and training, trade promotion and expansion, regulatory policies, e.g., for anti-trust and the environment, intellectual property protection, government procurement, and export control.[12] These policies can all directly affect the process of innovation, sometimes decisively.[13]

Institutions also play an important role. Technology development depends in part on the performance of educational institutions and the quality of scientific and engineering research carried out by public and private institutions. Similarly, the breadth and depth of U.S. capital markets affects the availability and cost of capital.[14]

Firm growth and the ability of the economy to develop, commercialize, and absorb new technologies are all directly affected by the impact of a matrix of national policies.[15] These policies in turn condition the impact of the more focused government initiatives reviewed in this volume.

[11]See National Research Council, *U.S. Industry in 2000: Studies in Competitive Performance*, Washington, D.C.: National Academy Press, 1999, p. 5.

[12]See, for example, the observations of Ed Zchau, a Member of Congress in the 1980s from Silicon Valley, in his article, "Government Policies for Innovation and Growth" in National Research Council, *The Positive Sum Strategy, Harnessing Technology for Economic Growth*, Washington, D.C.: National Academy Press, 1986, pp. 535-539.

[13]For example, intellectual property protection plays a key role in the continued development of the biotechnology industry. See Wesley M. Cohen and John Walsh, "Public Research, Patents and Implications for Industrial R&D in the Drug, Biotechnology, Semiconductor and Computer Industries" in this volume. The interaction of supportive technology policies and restrictive trade policies proved effective in some countries in the seventies and eighties. See Daniel I. Okimoto, *Between MITI and the Market: Japanese Industrial Policy for High Technology*, Stanford, CA: Stanford University Press, 1989.

[14]See Ralph Landau, "The Dynamics of Long-Term Growth: Gaining and Losing Advantage in the Chemical Industry" in National Research Council, *U.S. Industry in 2000, op. cit.*, p. 20.

[15]See Ralph Landau, "The Dynamics of Long-Term Growth: Gaining and Losing Advantage in the Chemical Industry," *op. cit.*, pp. 17-74.

research effort has until recently been spearheaded largely by the National Institutes of Health (NIH). This has raised concerns, even among the NIH leadership, that other areas of promising research, which directly contribute to the development of medical technologies, are suffering from relative neglect.[16] The authors of this study believe that sustained scientific and technological advance depends on progress across a broad spectrum of scientific and engineering disciplines.

The federal government has had a long-standing role in fostering scientific and technological progress. Yet, the scope and diversity of this effort is not always fully appreciated by the general public. While support to universities and grant-making institutions—such as that by the NSF and the NIH—is well known, the important role that agencies—such as the Department of Defense and the Department of Energy—play in providing support to diverse academic disciplines and technological developments is less widely understood.

The nation has long held the conviction that new technologies offer the best means of meeting societal challenges whether in the realms of defense, energy, or the environment.[17] The substantial federal investment in research and development reflects this conviction. However, some observers believe that the breadth of potential applications of new technologies, their greater complexity, and the rising costs and substantial risks of developing these new technologies, means that a supportive policy framework is necessary to capture their full potential for society.[18] They advocate public-private cooperation as an effective means of bringing new, welfare-enhancing, and wealth-generating technologies to the market. [19] Cooperative activities among industry, government, and universities are important elements in such a framework.

---

[16]See Harold Varmus, "The Impact of Physics on Biology and Medicine." Plenary talk, Centennial Meeting of the American Physical Society, Atlanta, March 22, 1999.

[17]See Linda R. Cohen and Roger G. Noll, *The Technology Pork Barrel*, Washington, D.C.: The Brookings Institution, 1991. The authors observe that "the government's optimism about technology knows neither programmatic, partisan, nor ideological bounds" (p.1). They cite William Ophuls' observation that American public policy has a long history of technological optimism. See William Ophuls, *Ecology and the Politics of Scarcity: Prologue to a Political Theory of the Steady State*, San Francisco: Freeman, 1977.

[18]For a discussion of the research process, see Donald Stokes, *Pasteur's Quadrant,* Washington, D.C.: Brookings Institution Press, 1997.

19 See: David Vogel, *Kindred Strangers: The Uneasy Relationship Between Politics and Business in America,* Princeton: Princeton University Press, pages 113-137, 1996. Vogel notes that arguments, both for and against government participation in the development of new technologies, largely overlook the prevailing tradition in U.S. industrial policy. He points out that, given the constraints of the American federal system and the strength of private capital markets, U.S. industrial policy focuses more on government-industry partnerships, in contrast to direct subsidies or government ownership found in other countries.

THE ROLE OF THE STEP BOARD

Since 1991, the National Research Council's Board on Science, Technology, and Economic Policy (STEP) has undertaken a program of activities to improve policymakers' understanding of the interconnections between science, technology, and economic policy and their importance to the American economy and its international competitive position. The Board's activities have corresponded with an increased recognition by policymakers of the importance of technology to economic growth. The new economic growth theory emphasizes the role of knowledge and technology creation, which is believed to be characterized by significant growth externalities.[20] A consequence of the renewed appreciation of growth externalities is recognition of the economic geography of development. With growth externalities coming about, in part from the exchange of knowledge among innovators, certain regions become centers for particular types of high-growth activities.[21]

Some economic analysis suggests that high technology is often characterized by increasing rather than decreasing returns. This justifies to some the proposition that governments can capture permanent advantage in key industries by providing relatively small but potentially decisive support to bring national industries up the learning curve and down the cost curve.[22] In part, this is why the economics literature now recognizes the relationship between technology policy and trade policy.[23] Recognition of these linkages and a corresponding ability of governments to shift comparative advantage in favor of the national economy provides the intellectual underpinning for government support for high-technology industry.[24] Another widely recognized rationale for government support for new technologies exists in cases where a technology is expected to

[20]See Paul Romer, "Endogenous Technological Change," *Journal of Political Economy,* 98(5): 71-102, 1990. See also: Gene Grossman and Elhannan Helpman, *Innovation and Growth in the Global Economy*, Cambridge, MA: MIT Press, 1993.

[21]See Paul Krugman, *Geography and Trade*, Cambridge MA: MIT Press, 1991, p. 23 points out how the British economist Alfred Marshall initially observed in his classic, *Principles of Economics*, how geographic clusters of specific economic activities arose from the exchange of "tacit" knowledge among businesses.

[22]Paul Krugman, *Rethinking International Trade,* Cambridge, MA: MIT Press, 1990.

[23]In addition to Krugman, see J.A. Brander and B.J. Spencer, "International R&D Rivalry and Industrial Strategy," *Review of Economic Studies,* 50(4): 707-722, 1983. See also: A.K. Dixit and A.S. Kyle, "The Use of Protection and subsidies for Entry Promotion and Deterrence," *American Economic Review,* 75(1): 139-152, 1985 and P. Krugman and M. Obsfeldt, *International Economics: Theory and Policy,* 3d Edition, New York: Addison-Wesley Publishing Company, 1994.

[24]For a review of governments' efforts to capture new technologies and the industries they spawn for their national economies, see National Research Council, *Conflict and Cooperation in National Competition for High Technology Industry,* Washington, D.C.: National Academy Press, 1996, pp. 28-40. For a critique of these efforts, see: Paul Krugman, *Peddling Prosperity*, New York: W.W. Norton Press, 1994.

generate benefits beyond those that can be captured by innovating firms—a phenomenon often referred to as spillovers. There are also cases in which the cost of developing a given technology may be prohibitive for individual companies, even though expected benefits to society are substantial and widespread.[25]

PROJECT ORIGINS

The growth in government programs to support high-technology industry within national economies—and their impact on international science and technology cooperation and on the multilateral trading system—are of considerable interest worldwide. Accordingly, these topics were taken up by STEP in a study carried out in conjunction with the Hamburg Institute for Economic Research (HWWA) and the Institute for World Economics (IFW) in Kiel.[26] One of the principal recommendations for further work emerging from that study was an analysis of the principles of effective cooperation in technology development. These analyses include lessons from national and international consortia, such as assessment mechanisms and modes of cooperation that might be developed to improve national and international cooperation in high-technology products.[27] As indicated in the box below, many countries seek to nurture their new technology-based industries in order to capture benefits and anchor them in the national economy.

---

[25]See: Ishaq Nadiri, *Innovations and Technological Spillovers,* NBER Working Paper No. 4423, 1993; and Edwin Mansfield, "Academic Research and Industrial Innovation," *Research Policy*, 20(1): 1-12, 1991. See also: Council of Economic Advisors, *Supporting Research and Development to Promote Economic Growth: The Federal Government's Role,* Washington, D.C.: Executive Office of the President, 1995.

[26]This study resulted in two NRC reports: *Conflict and Cooperation, op. cit.,* and *International Friction and Cooperation in High-Technology Development and Trade,* Washington, D.C.: National Academy Press, 1997. A third report was published by the German HWWA, Georg Koopmann and Hans-Eckart Scharer, editors, *The Economics of High-Technology Competition and Cooperation in Global Markets,* Baden-Baden, Germany: HWWA (Institute for Economic Research), 1996.

[27]The NRC Report, *Conflict and Cooperation, op. cit.,* recommends further analytical work concerning principles for effective cooperation in technological development (See Recommendation 24, p. 8). David Mowery of the University of California at Berkeley has recently noted the rapid expansion of collaborative activities and emphasized the need for comprehensive assessment. David Mowery, "Collaborative R&D: How effective is it?" *op. cit.,* p. 44. See also David Mowery, "Using Cooperative Research and Development Arrangements as S&T Indicators: What do We Have and What Would We Like?" in National Science Foundation, Division of Science Resources Studies, *Workshop on Strategic Research Partnerships, 13 Strategic Research Partnerships: Proceedings from an NSF Workshop*, NSF 01-336, Project Officers, John E. Jankowski, Albert N. Link, Nicholas S. Vonortas (Arlington, VA, 2001), pp. 93-132.

BOX B. The Benefits of High Technology Industry

Government policymakers have increasingly focused their attention on high technology industry and the new technologies and entrepreneurial activities that generate them. Many believe that this policy focus is justified. A growing body of economic literature argues that the composition of the economy matters and that high technology industries bring special benefits to national economies.[28] The attributed benefits rest on a set of interlocking observations:

- High technology firms are associated with rapid rates of innovation. Such firms, in turn tend to gain market share, create new product markets, and use resources more productively than traditional industries do.[29]
- Similarly, high technology firms perform larger amounts of R&D than traditional firms do. In fact, they are distinguished by the high percentage of revenue devoted to research—10 percent of revenues on research, in contrast to 3 percent for more traditional industries.[30]
- High technology firms create positive spillover effects that are often locally concentrated. Spillovers benefit other commercial sectors by generating new products and processes that can lead to productivity gains. A substantial literature in economics underscores the potential for high returns from technological innovation; with private innovators obtaining rates of return in the 20-30 percent range and spillover (or social return) averaging about 50 percent.[31]
- High technology products are a major source of growth in the major industrialized countries. Sectors such as aerospace, biotechnology, and information systems contribute to the growing global market for high technology manufactured goods.
- High technology firms are associated with high value-added manufacturing and with the creation of high-wage employment.[32]
- Finally, many high technology industries contribute to core government missions, including national defense, environmental protection, and energy development.[33]

—National Research Council, *Conflict and Cooperation,* 1996.

---

[28]For an analysis of the role of new information technologies in the recent trends in high productivity growth, often described as the "New Economy," see Council of Economic Advisors, *Economic Report of the President,* H.Doc. 107-2, Washington, D.C.: USGPO, January 2001.

[29]*Ibid.*

30Lawrence M. Rausch, *Asia's New High-Tech Competitors,* NSF 95-309, Arlington, VA: National Science Foundation, 1995.

[31]See Ishaq Nadiri, "Innovations and Technological Spillovers," *op. cit.* See also, Council of Economic Advisors, *Supporting Research and Development to Promote Economic Growth, op. cit.*, and Analee Saxaninan, *Regional Advantage: Culture and Competition in Silicon Valley and Route 128,* Cambridge, MA: Harvard University Press, 1994.

[32]See Laura Tyson, *Who's Bashing Whom? Trade Conflict in High Technology Industries,* Washington, D.C.: Institute for International Economics, 1992.

[33]For example, see Flat Panel Display Task Force, *Building U.S. Capabilities in Flat Panel Displays: Final Report*, Washington, D.C.: U.S. Department of Defense, October 1994.

PROJECT PARAMETERS

To advance our understanding of the operation and performance of partnerships, the STEP Board has undertaken a major study of programs relying on public-private collaboration for the development of new technologies. The project's multidisciplinary Steering Committee[34] includes members from academia, high-technology industries, venture capital firms, and the realm of public policy. The committee's principal tasks are to provide overall direction and relevant expertise to assess the issues raised by the project. Rather than address general questions of principle regarding the appropriateness of government involvement in partnerships, the Committee's charge is to take a pragmatic approach in addressing such issues as the rationale and organizing principles of government-industry cooperation, current practices, sectoral differences, means of evaluation, the experience of foreign-based partnerships, and the roles of government laboratories, universities, and other non-profit research organizations.

The Committee's analysis has included a significant but necessarily limited portion of the variety of cooperative activity that takes place between the government and the private sector.[35] The selection of specific programs to review is conditioned by the Committee's desire to carry out an analysis of current partnerships directly relevant to contemporary policy making. The Committee also recognizes the importance of placing each of the studies in the broader context of U.S. technology policy, which continues to employ a wide variety of ad hoc mechanisms developed through the government's decentralized decision-making and management process.

The Committee's desire to ensure that its deliberations and analysis are directly relevant to current policy making has allowed it to be responsive to Executive Branch and Congressional requests for examinations of various policies and programs of current policy relevance. This includes the White House and State Department request for an evaluation of opportunities for greater transatlantic cooperation—a result of the signature of the U.S.-E.U. Agreement on Science and Technology Cooperation. It also includes the request by the Defense Department's Under Secretary for Technology and Acquisitions to review the SBIR Fast Track initiative at the Department of Defense. Also included in the Committee portfolio of activities is the assessment of the Advanced Technology Program, requested by NIST in compliance with Senate Report 105-235, and the subject of two reports.[36] These intermediate reports on these programs and top-

---

[34]For a list of Committee members, see the front matter of this volume.

[35]For example, DARPA's programs and contributions have not been reviewed. For an indication of the scope of cooperative activity, see C. Coburn and D. Berglund, *Partnerships: A Compendium of State and Federal Cooperative Technology Programs,* Columbus, OH: Battelle Press, 1995; and the RaDiUS database, www.rand.org/services/radius/.

[36]See Senate Report 105-235, Departments of Commerce, Justice, and State, the Judiciary, and Related Agencies Appropriation Bill, 1999, and the Report from the Committee on Appropriations to accompany Bill S. 2260, which included the Commerce Department FY1999 Appropriations Bill.

ics have contributed to national policymaking and will contribute to the Committee's final report.[37]

To meet its objective of policy-relevant analysis, the Committee has focused on the assessment of current and proposed programs, drawing on the experience of previous U.S. initiatives, foreign practices, and emerging areas (e.g., bioinformatics) resulting from federal investments in advanced technologies.[38] A summary of the partnerships reviewed by the study is included in Box C.

SUPPORT FOR ANALYSIS OF COOPERATIVE PROGRAMS

There is broad support for this type of objective analysis among federal agencies and the private sector. Among these are the U.S. Department of Defense, the U.S. Department of Energy, the National Science Foundation, the National Institutes of Health (especially the National Cancer Institute and the National Institutes of General Medical Sciences), the National Aeronautics and Space Administration, the Office of Naval Research, and the National Institute of Standards and Technology. Sandia National Laboratories and the Electric Power Research Institute have also contributed. Support has also come from a diverse group of private corporations. The sponsors are listed in the front of this report.

WORKSHOP, CONFERENCE, COMMISSIONED RESEARCH, AND DISCUSSIONS

At its scoping workshop on this topic, the Project Steering Committee decided to focus its attention on the emerging needs, synergies, and opportunities

---

[37]The 2001 Senate Appropriations language cites extensively the STEP Board report called for in 1999. Recent work with recommendations and findings includes *The Advanced Technology Program: Assessing Outcomes,* Washington, D.C.: National Academy Press, 2001 and *The Small Business Innovation Research Program: An Assessment of the Department of Defense Fast Track Initiative*, Washington, D.C.: National Academy Press, 2000. Other volumes in this series include *The Small Business Innovation Research Program: Challenges and Opportunities,* Washington, D.C.: National Academy Press, 1999, Industry-*Laboratory Partnerships: A Review of the Sandia Science and Technology Park Initiative*, Washington, D.C.: National Academy Press, 1999;and *The Advanced Technology Program: Challenges and Opportunities*, Washington, D.C.: National Academy Press, 1999. The international component of the project was addressed with the conference and report on *New Vistas in Transatlantic Science and Technology Cooperation*, Washington, D.C.: National Academy Press, 1999.

[38]The Committee has focused its attention on the "best practices" rather than the practices of less successful partnerships—although it is certainly true that much can be learned from failures as well as successes. For an analysis of lessons that might be learned from comparing the experience of a less successful and a successful partnership, see John B. Horrigan, "Cooperating Competitors: A Comparison of MCC and SEMATECH." Monograph, Washington, D.C.: National Research Council, 1997.

Box C. Partnerships Reviewed by the Government-Industry Partnerships Study

The NRC study of Government-Industry Partnerships for the Development of New Technologies reviewed a wide range of partnerships. The analysis can be divided into four primary areas: (1) current U.S. partnership programs, (2) potential U.S. partnership programs, (3) industry-national laboratory partnerships, and (4) international collaboration and benchmarking. The analysis of current U.S. partnerships has focused on the Small Business Innovation Research Program, the Advanced Technology Program, and partnerships in Biotechnology and Computing. The review of potential partnerships for specific technologies, based on the project's extensive generic partnership analysis, has focused on needs in biotechnology, computing, and opportunities for solid-state lighting. The industry-laboratory analysis reviewed the potential of science and technology parks at Sandia National Laboratories and the NASA Ames Research Center. International collaboration and benchmarking studies have included outlining new opportunities resulting from the U.S. – E.U. Science and Technology Agreement and a review of regional and national programs to support the semiconductor industry, focusing on Japan, Europe, Taiwan, and the United States.

between the fields of biotechnology and computing. Special attention, therefore, is directed to the differences and similarities of government support for technology development in biotechnology and information technology, the different uses of intellectual property in these sectors, and the need for investments across different disciplines. This report also presents the Committee's findings and recommendations concerning the special needs and opportunities in Biotechnology and Information Technology.

The Steering Committee's fact-finding initiatives include a major workshop, a conference, and the commissioned research presented in this report. The Committee's deliberations have also benefited from the rich and diverse experience of its members as it developed the recommendations and findings of this report. The responsibility for these findings and recommendations rests with the Steering Committee and not with the individual conference presenters or researchers.

ACKNOWLEDGEMENTS

On October 25 and 26, 1999, the Committee convened a conference on *Government-Industry Partnerships in Biotechnology and Computing*. Conference highlights included remarks from Representative Sherwood Boehlert of the U.S. Congress and NASA Administrator Daniel Goldin. Other senior participants included Tom Kalil, then of the White House National Economic Council,

Rita Colwell of the National Science Foundation, and Marvin Cassman, Director of the National Institute of General Medical Sciences. A complete list of participants is included in Annex B of this volume. The Proceedings section of this volume contains summaries of their presentations and discussions.

Recognition is also due to outside contributors. Grant Black and Paula E. Stephan of Georgia State University, Wesley M. Cohen of Carnegie Mellon, John Walsh of the University of Illinois at Chicago, Kenneth Flamm of the University of Texas at Austin, and Michael McGeary of McGeary and Smith have contributed original research to this report.

Given the quality and the number of presentations, summarizing the papers and conference proceedings has been a challenge. We have made every effort to capture the main points made during the presentations and the ensuing discussions. We apologize for any inadvertent errors or omissions in our summary of the proceedings.

A number of individuals deserve recognition for their contributions to the preparation of this report. Among the STEP staff, John Horrigan contributed to the development of the 25-26 October 1999 conference, and helped to prepare the initial draft of this volume. Sujai Shivakumar contributed to the preparation of the report for publication and played a key role in the review process. The preparation of the report has also benefited from the unstinting efforts of David Dierksheide, McAlister Clabaugh, and more recently, Christopher Hayter. Without their collective efforts, this report would not be possible. Special thanks are also due to Marilyn Baker for her facilitation of the NRC review.

NRC REVIEW

This report has been reviewed in draft form by individuals chosen for their diverse perspectives and technical expertise, in accordance with procedures approved by the NRC's Report Review Committee. The purpose of this independent review is to provide candid and critical comments that will assist the NRC in making its published report as sound as possible. Further, it is to ensure that the report meets institutional standards for objectivity, evidence, and responsiveness to the study charge. The review comments and draft manuscript remain confidential to protect the integrity of the deliberative process.

We thank the following individuals for their review of this report: Dr. Maryann Feldman, Johns Hopkins University; Dr. Robert Archibald, College of William and Mary; Dr. Philip Auerswald, Harvard University; Dr. Mary L. Good, Venture Capital Investors, LLC; Mr. George Scalise, President of the Semiconductor Industry Association; Dr. Lewis Edelheit, General Electric; and Dr. William J. Rutter, Chiron, *retired*. Although the reviewers listed above have provided many constructive comments and suggestions, they were not asked to endorse the conclusions or recommendations nor did they see the final draft of the report before its release.

Maureen Henderson and Gerry Dinneen have overseen the Academies review process for this report. Appointed by the National Research Council, they were responsible for making certain that an independent examination of this report was carried out in accordance with institutional procedures and that all review comments were carefully considered. Responsibility for the final content of this report rests entirely with the authoring committee and the National Research Council.

STRUCTURE

Following this preface, Part I of this report presents an introductory overview of the conference presentations and the papers. Part II elaborates on the new needs and opportunities in Biotechnology and Information Technology. Part III presents the Findings and Recommendations, which are the collective responsibility of the Steering Committee. Part IV summarizes the Conference Proceedings, setting out the views of the conference participants. Finally, Part V presents four related studies, which, though subject to NRC editing, remain the responsibility of the authors.

This report's goal is to advance our understanding of the new needs and opportunities arising in biotechnology and information technology and to ensure that a strengthened national commitment in biotechnology is not compromised by inadequate investments in the disciplines and technologies required to make that commitment a reality.

Gordon M. Moore | William Spencer | Charles W. Wessner

Executive Summary

BACKGROUND

Government-industry cooperation to achieve national goals continues to make important contributions to the growth of the U.S. economy. New technologies are widely seen as essential to sustain continued U.S. economic growth and a rising standard of living. To unlock the potential of new technologies, substantial private and public investment in research and development is often required, especially to bring promising new technologies forward to the marketplace. The U.S. government has played and continues to play an important supporting role in the development of new technologies, often through cooperative arrangements or partnerships with industry. In some cases, individual partnerships, as in semiconductors, have made major contributions to the resurgence of a key American industry. In other cases, the close interaction of a variety of federal programs over a number of years have contributed to the development of whole new industries, as in the cases of computing and biotechnology.

Despite the importance of these programs, and their interaction with the private sector, there has been little systematic analysis of the operation and impact of this form of government-industry collaboration. To improve policymakers' understanding of the performance of partnerships, the National Research Council's Board on Science, Technology, and Economic Policy (STEP) undertook a major review of programs relying on public-private collaboration. The project on Government-Industry Partnerships for the Development of New Technologies was formed under the direction of the STEP Board and is directed by a distinguished Steering Committee. Chaired by Gordon Moore, Chairman Emeritus of Intel Corporation, the Committee includes members from academia, high-technology industries, venture capital firms, and the realm of public policy.[1] Recognizing that partnerships are an integral part of the U.S. innovation system, the Committee has taken a pragmatic approach: It has focused its work on the operation and assessments of government-industry partnerships, rather than the broad questions of principle concerning the desirability of government-industry cooperation.

While the Committee's analysis has focused on a variety of current and recent programs, the study has addressed only a limited portion of the cooperative activity that takes place between the government and the private sector.[2] The selection of specific programs to review is conditioned by the Committee's desire to carry out an analysis of current partnerships directly relevant to contemporary policy making. The Committee also recognizes the importance of placing each of the studies in the broader context of U.S. technology policy, which continues to employ a wide variety of ad hoc mechanisms developed through the government's decentralized decision-making and management processes.

In the course of the Committee's analysis, it became apparent that there are substantial differences in the mechanisms and levels of federal support to different sectors of the American economy, notably in biotechnology and computing. At the same time, the Committee had a strong interest in the growing synergies between biotechnology and information technology and the emerging gaps in funding and training in related disciplines. To address these issues, the Committee organized a series of meetings, including a major workshop and a conference in 1999. The conference brought together academic experts, entrepreneurs, government officials, and others with knowledge and experience in government support for biotechnology and information technology.

The conference focused on the nature and implications of emerging trends of the federal research portfolio in biotechnology and information technology, particularly, unplanned shifts in the allocation across sectors of federal funding over the past decade. It further examined historical perspectives on partnerships in this sector, as well as new needs and emergent opportunities in biotechnology and information technology. Finally, it considered steps necessary to ensure that the nation maximizes its return on its investments in research. The conference deliberations—summarized in this volume—are supplemented by a series of commissioned research papers. The commissioned analyses address the most recent trends in federal funding, discuss the different impacts of the intellectual

---

[1]For a list of Committee members, see the front matter of this report.

[2]For example, DARPA's programs and contributions have not been reviewed. For an indication of the scope of cooperative activity, see C. Coburn and D. Berglund, *Partnerships: A Compendium of State and Federal Cooperative Technology Programs,* Columbus, OH: Battelle Press, 1995; and the RaDiUS database, www.rand.org/services/radius/.

property regimes, identify emergent needs in biotechnology and information technology, reveal decreases in federal support for computing and to supporting disciplines, and add empirical support to the Committee recommendations.

In sum, the Committee's assessment in this report of new opportunities and new needs in biotechnology and information technology should be seen as an integral part of its overall multi-year review of government-industry partnership programs in the U.S. and abroad. This report thus contributes to the Committee's broader assessment by providing a comparative perspective on the federal role in these important and increasingly interdependent technology sectors.

REPORT FOCUS

To address these issues, this report presents the proceedings of the conference, commissioned research on central issues, and the findings and recommendations of the Committee concerning emerging needs and opportunities for partnerships in biotechnology and information technology and related fields. Specifically, the report focuses on:

- The consequences of post-Cold War adjustments to the federal research and development (R&D) budget in the physical sciences and engineering on one hand, and on biomedical research, on the other.[3]
- The shift in the allocation of federal research funding in recent years. Funding has increased substantially for a few important and promising fields—e.g., biomedicine and computer science—but has decreased in real terms for research in the physical sciences and much of engineering. [4]
- The development of new technologies, which increasingly require collaboration across disciplinary boundaries.[5] Furthermore, it reveals that research, now increasingly multidisciplinary, calls new attention to the way it is funded.
- The implications for research, the development of new technologies, and the impact on U.S. competitiveness given the multidisciplinary nature of research and the "imbalance" of funding across complementary disciplines.
- The disjuncture between the incentives faced by young academics to

[3]See comments in this volume by Congressman Sherwood Boehlert, "Biomedical research is consuming too much of the federal research budget."

[4]See Michael McGeary, "Recent Trends in the Federal Funding of Research and Development Related to Health and Information Technology," in this volume.

[5]See, for example, comments in this volume by Mark Boguski of the National Center for Biotechnology Information (NIH), "One measure of the growth of scientific information is the growth in the MEDLINE database at the NIH's National Library of Medicine. The data base contains over 10 million articles, and this is growing by 40,000 articles per year—these are peer reviewed articles."

specialize in particular disciplines as against the industry need for knowledge workers able to work across disciplinary boundaries.[6]
- The disjuncture between exponentially accumulating information and linear research frameworks geared to generate knowledge.
- The highly localized nature of processes of technological innovation, even as information technologies have facilitated communication and collaboration across distances.[7]
- The role of small firms, particularly in the biotechnology sector, in enhancing the accumulation of both basic knowledge and the technology and in the advancement of early stage product development for the industry.
- The different tools of federal government support for the development of new technologies.[8]
- Issues related to industrial competitiveness of the U.S., particularly in regard to declining federal investments in key sectors of the nation's R&D enterprise.[9]
- The need for policymakers to possess better information on the magnitudes, distribution, and mechanisms for federal support of R&D.[10]

The Committee's core findings and recommendations from this study are summarized below. The "Findings and Recommendations" section of this report presents them in full.

FINDINGS AND RECOMMENDATIONS

In light of the key role that the computing and biotechnology sectors will play in America's economic well-being and security in the twenty-first century, the Committee finds that there is, in essence, insufficient understanding of needs and opportunities inherent in the increasingly multidisciplinary nature of research. This is evidenced in:

---

[6]See, for example, comments in this volume by Marvin Cassman, "There will always be only a few people sufficiently trained in physics and biology to perform experiments in both fields at a high level. It is more important … to have an adequate number of people conversant in the language of the two disciplines so that a meaningful collaboration can occur. Training such people is feasible, though it will take time to produce such trained individuals."

[7]See, for example, remarks by Edward Penhoet and Wesley M. Cohen in this volume.

[8]See, for instance, remarks in this volume by Gordon Moore, who notes that "flexibility and timing are critical elements to successful technological development. A government contract takes a long time to negotiate, and often the goal is obsolete by the time the contract is completed."

[9]In this regard, see comments by Kenneth Flamm, Gordon Moore, and Wesley M. Cohen in this volume.

[10]See, in particular, remarks by William Bonvillian, of the office of Senator Joseph Lieberman in this volume.

- Increased federal funding, in recent years, in the area of biomedicine, combined with
- Reduced federal funding (in real terms) for research and training in the physical sciences and much of engineering. [11]

Importantly, these changes in funding for science and engineering have not been the product of national debate or conscious policy. Rather, they reflect the overall impact of separate efforts by individual agencies to come to terms with post-Cold War R&D expenditure reductions.[12]

Support for innovation requires an understanding of the complementarities and synergies in biotechnology and computing research. Investments across seemingly unrelated disciplines are necessary to meet the research challenges of the future.

- While current U.S. investments in research—especially in biomedicine—show great promise, capturing their full potential requires complementary research investments in information technologies, where the federal support has recently waned.[13]
- Partnerships across disciplines and between universities and companies, although often challenging to implement, are increasingly required to capture the full potential of current investments in biotechnology.[14]
- Regular and more timely review by industry, the scientific community, and the federal government can draw out the implications of the recent decline in federal funding for non-defense research in the physical sciences and engineering, and address possible solutions.[15]

[11]For recent analysis of federal research funding, see National Research Council, *Trends in Federal Support of Research and Graduate Education*, Washington, D.C.: National Academy Press, 2001.

[12]*Ibid.* p. 4.

[13]*Ibid.* pp. 2-5. See also the research paper prepared for this volume by Kenneth Flamm, "The Federal Partnership with U.S. Industry in U.S. Computer Research: History and Recent Concerns."

[14]See presentations by Marvin Cassman, "Exploring the Biotechnology Revolution," and Edward Penhoet, "The View from the Biotechnology Industry," in this volume.

[15]The recent study by the STEP Board, cited above, and conducted in parallel with this research, recommends, that (A) "the White House Office of Science and Technology (OSTP) and the Office of Management and Budget (OMB), with assistance from federal agencies and appropriate advisory bodies, should evaluate the federal research portfolio, with an initial focus on fields related to industrial performance and other national priorities and a recent history of declining funding"; (B) "Congress should conduct its own evaluation of the federal research portfolio through the budget, appropriations, or authorization committees"; and (C) "for the longer term, the Executive branch and Congress should sponsor the following types of studies: (1) in-depth qualitative case studies of selected fields, taking into account not only finding trends across federal agencies and non-federal supporters and international comparisons but also subtler differences in the foci, time horizons, and other research characteristics that are obscured by quantitative data; (2) studies of agency research portfolios and decision making to understand the reasons for shifts in funding by field and the extent

New technological opportunities exist in a variety of related research areas involving biomedicine, and advanced computing, and require expanded cross-disciplinary instruction and research. Strategic supplementary investments in R&D and appropriate patent protections—needed to realize these opportunities—point to a unique federal role.[16]

- The history of the U.S. biotechnology and computing sectors illustrates the value of such strategic federal support.
- The U.S. venture capital market, while large and well developed, is not a substitute for long-term government support of scientific and technological research.[17]
- Federal patent policy has a significant impact on innovation in biotechnology and computing.[18]

In all, the Committee recommends the consideration of a variety of measures to address specific gaps and to realize historic opportunities. These recommendations include:

- Developing an alert system to identify critical needs in important disciplines.
- Fostering interdisciplinary competence in fields such as bioinformatics.
- Increasing support and training for interdisciplinary graduate students.
- Exploring unresolved questions about research partnerships.
- Reviewing the impact of patents on technological progress.

The Findings of this report are elaborated in the Summary of Findings (Part III) below and are further documented by four original research papers commissioned for this report (Part VI).

This report's goal is to advance our understanding of the new needs and opportunities arising in the Biotechnology and Information Technology fields and to ensure that a strengthened national commitment in biotechnology is not compromised by inadequate investments in the disciplines and technologies required to make that commitment a reality.

---

to which the health of individual fields and interrelationships among fields are taken into account; and (3) studies of methodologies for allocating federal research funding according to national rather than merely departmental criteria and priorities."

[16]Glue grants, to support initiatives—such as the Berkeley Health Initiative—can help address this need. See the presentation by Edward Penhoet and the analysis by Paula Stephan and Grant Black in this volume.

[17]See statements by prominent venture capitalists, including David Morgenthaler, in National Research Council, *The Advanced Technology Program: Assessing Outcomes*, Washington, D.C.: National Academy Press, 2001, pp. 108-112.

[18]See the paper by Wesley Cohen and John Walsh in this volume.

I

INTRODUCTION and OVERVIEW

Introduction and Overview

The biotechnology revolution is likely to be emblematic of the 21st century just as the information revolution has characterized the latter part of the 20th century. The momentum of the information revolution is not spent, however, because computing and communications will continue to change our economy and society as electronic devices become smaller and more pervasive, and as information networks expand in capacity.

The biotechnology revolution, though presently in its early stages, will grow in importance, in part, because information technology (IT) will enable scientists to take full advantage of advances in areas such as genomics. At the same time, biological systems may provide useful models for advances in computing. Indeed, the information and biotechnology revolutions are not likely to continue as separate phenomena. Rather, biotechnology and computing will interact increasingly with each other and, in this way, contribute to the pace of advance in each field. Federal partnerships with other principal actors—namely, industry, universities, and laboratories—are likely to play a key role in the continued advance of both biotechnology and information technology.

CONCERNS ABOUT THE FEDERAL ROLE

The federal government's role remains significant although non-federal entities increased their share of national funding for R&D from 60 to 74 percent between 1990 and 2000. Federal funding still supports a substantial component, 27 percent, of the nation's total research expenditures.[1] Importantly, federal

[1]See National Research Council, *Trends in Federal Support of Research and Graduate Education*, Washington, D.C.: National Academy Press, 2001, p. 4.

expenditures constitute 49 percent of basic research spending. In addition, federal funding for research is often more stable and based on a longer time horizon than funding from other sources. Shifts in the composition of federal research support are therefore important in their own right and for the impact these shifts may have on the future development of biotechnology and computing—two of our most innovative industries and a source of substantial growth in the economy.

There are two main concerns. The first is the absolute amount of federal investment and the second is the allocation of the funds available. Concerning the first point, there is a *growing concern that the United States is not investing enough and broadly enough in research and development.* In the private sector, the demise of large industrial laboratories, such as IBM's Yorktown facility and Bell Laboratories, has reduced the amount of basic research conducted by private companies.[2] Although private sector R&D has steadily increased in the United States in recent years, almost all of it has been product-oriented rather than geared towards basic research.[3] In addition, the increase in corporate spending on research is concentrated in sectors such as the pharmaceutical industry and information technology.

The second area of concern regards the allocation of federal research funds, specifically, the *unplanned shifts in the level of federal support within the U.S. public R&D portfolio.* As highlighted in an earlier report by the STEP Board,[4] the United States has experienced a largely unplanned shift in the allocation of public R&D (Figure 1).[5] The end of the Cold War and a political consensus to reduce the federal budget deficit resulted in reductions in federal R&D funding in real terms. For example, the decline of the defense budget means that military support has fallen for research in physics, chemistry, mathematics, and most fields of engineering. The previous STEP study, noted above, showed that in 1997—even though the overall level of federal research spending was nearly the same as it had been in 1993—several agencies were spending substantially less

[2]See Richard Rosenbloom and William Spencer, *Engines of Innovation: U.S. Industrial Research at the End of an Era*. Boston: Harvard Business School Press, 1996.

[3]See Charles F. Larson, "The Boom in Industry Research," *Issues in Science and Technology*, Summer 2000, p. 27. With the exception of pharmaceuticals, only a small fraction, (for example, less than four percent in computers and semiconductors) of corporate R&D is classified as basic research. National Research Council, *Trends in Federal Support of Research and Graduate Education,* Washington, D.C.: National Academy Press, 2001, p. 4.

[4]See McGeary, M., and S.A. Merrill. "Recent Trends in Federal Spending on Scientific and Engineering Research: Impacts on Research Fields and Graduate Training," Appendix A in National Research Council, *Securing America's Industrial Strength*. Washington, D.C.: National Academy Press, 1999.

[5]See Stephen A. Merrill and Michael McGeary, "Who's Balancing the Federal Research Portfolio and How?" *Science*, vol. 285, September 10, 1999, p. 1679-1680. For a more recent analysis, see National Research Council, *Trends in Federal Support of Research and Graduate Education*, *op. cit.*

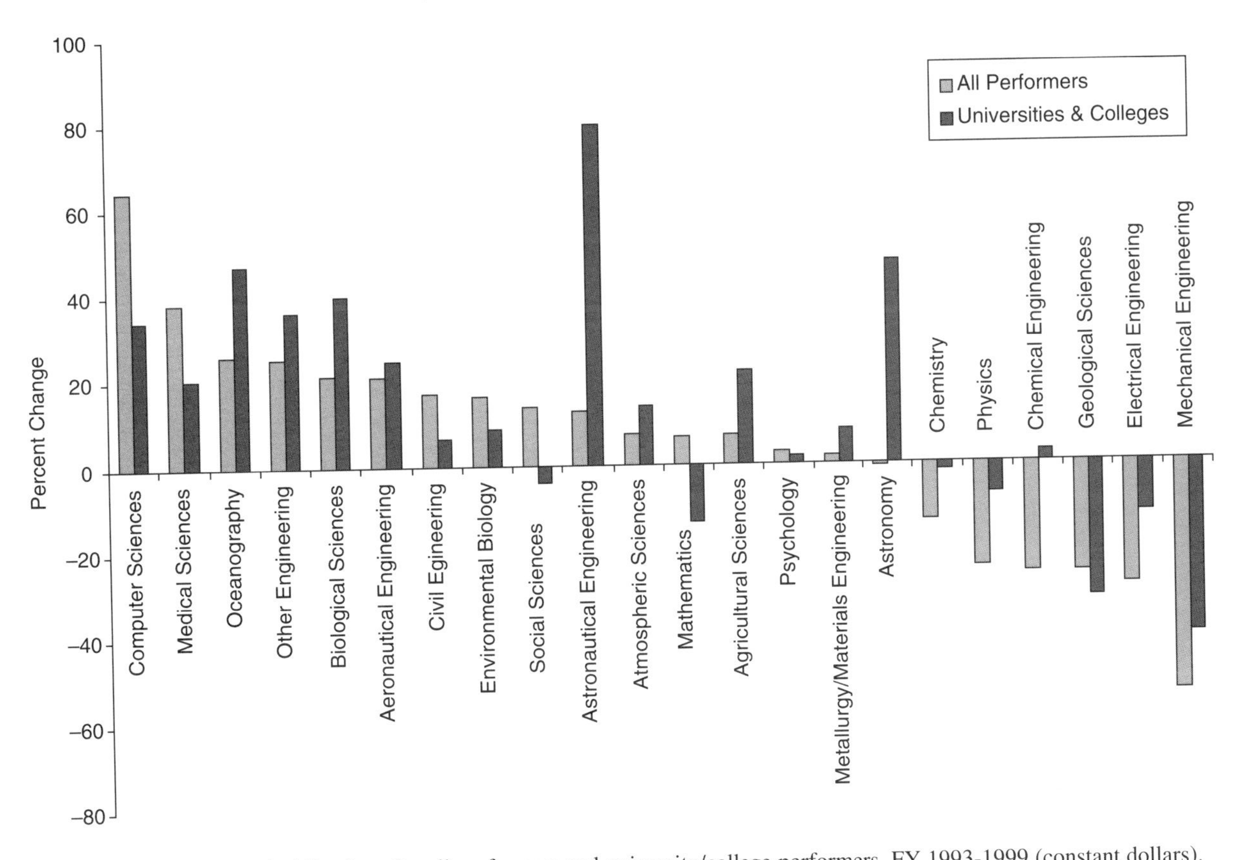

FIGURE 1 Changes in federal obligations for all performers and university/college performers, FY 1993-1999 (constant dollars).

on research than in 1993: The Department of Defense had dropped 27.5 percent, the Department of the Interior was down by 13.3 percent, and the Department of Energy had declined by 5.6 percent.[6] Declines in funding for the departments of Defense and Energy are significant because, traditionally, these agencies have provided the majority of federal funding for research in electrical engineering, mechanical engineering, materials engineering, physics, and computer science.[7]

After five years of stagnation, federal funding for R&D did recover in FY1998. In 1999, total expenditures were up 11.7 percent over the 1993 level. These changes were driven mainly by the increases to the NIH appropriations. Breakthroughs in biotechnology, and the promise of effective new medical treatments, have resulted in a substantial increase in funding for the NIH, which is slated for further increases by the current administration.[8]

Notwithstanding this change in overall R&D funding, the most recent analysis shows that even with the increase in federal research funding after 1997, the shift in the composition of federal support is largely unchanged. In 1999, the life sciences had 46 percent of federal funding for research, compared with 40 percent in 1993.[9] This difference in funding trends between the physical sciences and engineering on the one hand and the life sciences on the other hand, is disturbing insofar as progress in one field seems to depend increasingly on progress in others. Growing imbalances in federal investment in research across disciplines may therefore have major consequences for our ability to exploit fully the existing public investments in the biomedical sciences and for continued United States leadership in biotechnology innovation and commercial applications.

The policy community has recently begun to recognize the implications of these differential trends in funding. Indeed, one of the early attempts to focus policy makers' attention to address these concerns was the Committee's October 1999 conference on Government-Industry Partnerships in Biotechnology and Information Technologies. This report summarizes the proceedings of that conference and includes several subsequent studies commissioned by the Committee for Government-Industry Partnerships for the Development of New Technologies. An overview of the conference and the papers are presented below. The Findings and Recommendations of the Committee, intended to address the issues raised in the conference and commissioned analyses, are presented in Part III of this volume.

---

[6]See McGeary, M., and S.A. Merrill. "Recent Trends in Federal Spending on Scientific and Engineering Research: Impacts on Research Fields and Graduate Training," *op. cit.*, Table A-1.

[7]See Michael McGeary, "Recent Trends in the Federal Funding of Research and Development Related to Health and Information Technology," in this volume, p.3.

[8]See National Research Council, *Trends in Federal Support of Research and Graduate Education, op. cit.*, p. 2

[9]*Ibid.*

OVERVIEW OF THE CONFERENCE

The conference on Government-Industry Partnerships in Biotechnology and Computing took place at the National Academy of Sciences in Washington, D.C. Panel presentations and discussion centered on four related topics. These were (1) the nature and emerging trends of the federal research portfolio in biotechnology and information technology, (2) historical perspectives on federal support for these sectors, (3) new needs and emergent opportunities in biotechnology and information technology, and (4) how the United States might best meet these needs and capture potential benefits. Key points of speakers are summarized below:

A Supportive Policy Framework for Semiconductors and Biotechnology

Gordon Moore, Chairman Emeritus of Intel Corporation, in his review of the semiconductor industry, noted how a positive policy framework from the federal government has helped that industry to achieve its current position of competitive strength. This support, he noted, has evolved from policies that protected intellectual property rights to the support of industry research consortia to continuing partnerships between national laboratories and industry in emergent technologies.

In reviewing federal support to the biotechnology industry, Edward Penhoet of the University of California, observed that it has been mainly in the form of basic research support—much of it in university settings with funding from the National Institutes of Health. As biotechnology continues to mature, he noted, it needs greater multidisciplinary collaboration to reach its full capability.[10] In turn, the nature of federal funding will have to evolve if it is to accommodate the changing nature of the industry.

Taking a historical perspective, Kenneth Flamm of the University of Texas noted that the federal government role in supporting information technology evolved from being the main customer to providing long-term basic research to supporting near-term targeted projects. This supporting role would continue to evolve, he predicted, even as the industry continued to mature. Somewhat surprisingly, Dr. Flamm reported that the level of public support for computing has declined in recent years.[11]

---

[10]A field is "interdisciplinary" if it deals with issues that are at the periphery of two or more standard disciplines. At the same time, research is "multidisciplinary" if it attacks issues at the periphery of two or more standard disciplines by, for example, building teams of researchers from several disciplines.

[11]See Kenneth Flamm, "The Federal Partnership with U.S. Industry in U.S. Computer Research," in this volume.

In tracing the development of medical research and the biotechnology industry, Leon Rosenberg of Princeton University noted that universities play a critical research role and thus have to be accounted for in any government partnership framework. He further noted that medical and biotechnology research enjoys broad-based political support and is therefore well funded—a theme resonated by William Bonvillian of Senator Joseph Lieberman's office. Indeed, both speakers concurred that biotechnology research is expected to become increasingly multidisciplinary. To adjust to this trend they suggested, the distribution of funding to support biomedical research will increasingly have to be more broad based—that is, across a variety of seemingly unrelated disciplines.

In exploring new needs and emergent opportunities in biotechnology, Marvin Cassman of the NIH noted that the marriage of biology and information technology will usher an incredible pace of change in the advance of knowledge. Using the data being generated in genomics and other fields, scientists will be able to understand biological systems more completely. To this end, skills in bioinformatics and related information technology will be increasingly called for. Yet, as Rita Colwell of the National Science Foundation observed, interdisciplinary training—-needed for bioinformatics—is often limited by constraints posed by traditional departmental boundaries. As a result, she concluded, there is an inadequate supply of such scientists with the needed interdisciplinary skills.

Dr. Paula Stephan confirmed that there is great demand for information technology professionals, especially in the biotechnology industry, but that there are problems in meeting this demand. In explaining this supply problem, Dr. Stephan offered that universities respond more to research funding than to industrial demand. She also noted that traditional departmental structure in universities often inhibit interdisciplinary training. In addition, the operation of the market—where jobs in information technology traditionally pay more than do jobs in biology—also reduces incentives for information technology professionals to acquire training in biology.

Meanwhile, research in biotechnology holds promise for applications in a variety of areas. As publicized at the conference, crosscutting technologies—such as nanotechnology and functional genomics—are expected to yield useful applications ranging from tailor-made medication to biosensors that detect environmental dangers. In his luncheon address, Daniel Goldin of NASA outlined the enormous potential of biology and information technology, and explained how synergies between the two hold the key to future space exploration.

In order to capture the important complementarities between biological research and the physical sciences, policy shifts will be required to support and promote continued U.S. leadership in biotechnology and information technology. As Congressman Sherwood Boehlert pointed out, the question is not one of *if*, but more of *how* and *at what levels* government should support biotechnology and information technology research.

Indeed, as several speakers at the conference pointed out, the existing policy framework has, in some respects, quite different impacts on the evolving needs of the information technology and biotechnology sectors. For example, Wesley Cohen of Carnegie Mellon University noted that appropriate patenting rules are especially important in the biotechnology sector since these are often the principle assets of a start-up venture.[12] By contrast, firms in the information technology sector rely more on trade secrecy and long lead times to protect intellectual property. Robert Blackburn of Chiron Corporation explained that the courts have been applying patent law doctrine, developed for the synthetic chemical industry, to biotechnology. However, unlike synthetic chemicals, which are marketable products, biotechnologies are research tools. This makes patents difficult to enforce, he noted, leading in some cases to friction between university and industry researchers. Thus, current applications of patent laws can sometimes act as a disincentive to drug development. In a similar vein, Maryann Feldman of Johns Hopkins University noted that existing patenting rules—as per the Bayh-Dole University and Small Business Patent Act—yield only modest benefits to universities in terms of licensing revenue while, in some cases, even discouraging university-industry collaborations.

Differences also exist in the nature of direct government support to the information technology and biotechnology sectors. While there is a long history of government-industry partnerships in information technology, government support of biotechnology has, to date, relied largely on funding for basic and clinical research, with industry concentrating its resources on the development of commercial applications, albeit under the supervision and review of the Food and Drug Administration.

Yet, as the gap between fundamental research and application narrows in biotechnology, and as the field becomes more multidisciplinary, the nature and distribution of government support—as well as the policy framework—will need to evolve. The conference presented two government-industry-university consortia—the Mouse Sequencing Consortium and the Nucleotide Polymorphism Consortium—as illustrations of possibilities in this sphere.

OVERVIEW OF THE PAPERS

Following the conference, four research papers were commissioned to deepen the analysis of points raised at the conference. Two deal with macro trends—one on government-industry partnerships in information technology, and the other on the composition of the present portfolio of federally supported research. The remaining two take up complementary micro-level examinations of the rele-

[12]See Wesley Cohen and John Walsh, "Public Research, Patents, and Implications for Industrial R&D in the Drug, Biotechnology, Semiconductor, and Computing Industries," in this volume. A summary of their argument is presented in the next section of this chapter.

vance of current patent policy to the biotechnology and information technology sectors, and the need to address critical requirements in bioinformatics. Together, the four papers provide empirical support and greater depth of analysis to key aspects of the conference deliberations.

Declining Support for Information Technology

The paper by Kenneth Flamm examines the history and current concerns in federal support and government-industry partnerships in U.S. computer research.[13] He notes that large-scale R&D investment in computers and computer architecture has dropped in both absolute and relative terms over the 1990s. Further, he finds that longer-term investment in fundamental and basic research for computers is falling in comparison with historical levels.

This is cause for significant concern, argues Dr. Flamm, since computer technology, though pervasive, is by no means fully mature. Indeed, a growing body of economic literature links improvements in the technology industries with the recent healthy performance of the U.S. economy, and points out that continued technical advance in information technology yields substantial payoffs in biotechnology and other areas.[14] To maintain the commitments of the industry and to ensure continued welfare-enhancing gains in productivity, the federal government should continue and expand its investments in the research that supports this critical industry.

In his analysis, Michael McGeary documents recent trends in federal funding of research related to health and information technology.[15] Drawing on the recent assessment by the STEP Board,[16] he describes the substantial shift in the composition of the federal research portfolio that has taken place over the 1990s. He notes that in 1999, the life sciences had 46 percent of federal funding for research, compared with 40 percent in 1993, while funding for physical science and engineering decreased from 37 to 31 percent of the federal research portfolio over the same period. Dr. McGeary's paper raises the question of whether, consequently, "the federal research portfolio has become 'imbalanced.'" While it is normal that the allocation of the nation's research funding should evolve over time in response to new scientific opportunities, the sustained shifts in the distribution of this research portfolio allocation have important implications for future advances in both information technologies and biotechnology.

---

[13]See Kenneth Flamm's paper, "The Federal Partnership with U.S. Industry in U.S. Computer Research: History and Recent Concerns," in this volume.

[14]See Jorgenson, D. and Kevin J. Stiroh, "Raising the Speed Limit: U.S. Economic Growth in the Information Age." *Brookings Papers on Economic Activity* 1, Washington, D.C., 2000.

[15]See Michael McGeary, *op. cit.*

[16]See National Research Council, *Trends in Federal Support of Research and Graduate Education*, Washington, D.C.: National Academy Press, 2001.

Defining Roles of Patents

Wesley Cohen and John Walsh study patterns of publicly funded research and patent policy, and note their implications for industrial R&D in the pharmaceutical, biotechnology, and semiconductor and computer industries.[17] In particular, they assess the impact of the Bayh-Dole patent and trademark act of 1984. This act seeks to exploit what was believed to be a stock of innovations in universities and other research centers by creating incentives to commercialize them. In particular, Bayh-Dole permits these organizations to obtain patent rights on federally sponsored research.

Based on surveys of R&D managers, Cohen and Walsh find that government support for research is important for the pharmaceutical, biotechnology, and semiconductor industries. Among these, however, patent protection is effective in stimulating research in the pharmaceutical and biotechnology industries, but less powerful in the semiconductor industry. This underscores the importance of understanding the differential effect of the current public policy framework on different industries.[18] In the semiconductor industry, firms tend to protect their inventions through a combination of lead-time advantage, secrecy, and the exploitation of complementary capabilities rather than rely on patents.

Bioinformatics

Finally, Paula Stephan and Grant Black examine emerging opportunities and emerging gaps in bioinformatics.[19] Bioinformatics, they note, is a highly interdisciplinary field, requiring understanding of mathematics, computer science, and biology. There is today a critical need for individuals with skills in bioinformatics to maintain the pace of research in biology.

Stephan and Black note that while the demand for individuals trained in bioinformatics is strong and professional salaries are high, too few students are being trained in this area. They suggest that the funding signals—which encourage faculty research, and hence the demand for doctoral and postdoctoral students—are inadequate. Moreover, the traditional organization and incentives of the relevant disciplines and university structures tend to discourage cooperation. Stephan and Black suggest, among other remedial measures, that federal research funds should be targeted to promote the institutional and pragmatic changes necessary to produce a greater supply of students trained in bioinformatics.

[17] See Wesley Cohen and John Walsh, *op. cit.*

[18] See National Research Council, *Trends in Federal Support of Research and Graduate Education, op. cit.*

[19] See Paula Stephan and Grant Black, "Bioinformatics: Emerging Opportunities and Emerging Gaps," in this volume.

A VIEW TO THE FUTURE

Despite some recent turbulence, the United States continues to enjoy, at this writing, an unprecedented period of economic growth even taking into account the recent cyclical downturn. This sustained growth, historically low unemployment, and—since 1995—substantial increases in productivity, combined with the great promise of new technologies, provides the conceptual basis for those who believe the U.S. is developing a "New Economy," that is one with higher long term growth in productivity.

Despite the recent cyclical downturn, our nation's scientists, engineers, and policymakers have reason to be proud of this economic performance. Yet, many of the new technologies that drive growth today are based on innovations that are a number of years, even decades, old. The Committee is concerned that recent downward trends in funding for basic research in the public and private sectors put at risk the investments that are necessary to nurture and sustain such growth.

It is for this reason that they believe that government policymakers and industrialists should devote more attention to the investments that must be made today to underpin future productivity and prosperity. Longer-term, sustained, and increasing support for a diversified R&D portfolio is key to sustained economic growth and increased national welfare. Effective government-industry partnerships are a key component of this productivity-enhancing policy framework. With this review of the opportunities and challenges posed by partnerships in the biotechnology and computing industries, the Committee hopes to encourage greater public understanding of the importance of the national research effort and of the need for well-conceived partnership programs to help bring the fruits of this research to all Americans.

II

ISSUES IN BIOTECHNOLOGY and INFORMATION TECHNOLOGY

Issues in Biotechnology and Information Technology

This analysis is based on the belief that biotechnology and information technology are the two most innovative industries in the United States today, and that, while each is distinct in character, they display important complementarities. Indeed, the links and synergies in their innovation processes are such that progress in one sector increasingly depends in many ways on progress in the other. Innovation, which maintains U.S. technological leadership and helps to sustain economic growth, depends in turn on a policy framework that encourages basic, applied, and multidisciplinary research. Continued U.S. technology leadership will depend on the ability of the government, universities, and industry to collaborate effectively in the development of appropriate policies, especially for these innovative sectors.

It is for this reason that issues in biotechnology and information technology have been integral to the STEP Board's broader program of study of Government Industry Partnerships for the Development of New Technologies since the inception of that effort. This section reviews the new needs and opportunities in biotechnology and information technology as discussed by experts from industry, academia, and government. We begin by noting that the U.S. government has historically supported biotechnology and information technology in different ways. While this has led to significant technological progress, this success has necessarily created new issues requiring policymakers' attention. The second section looks at some future technologies that rely on complementarities between biotechnology and information technology and identifies some of the challenges that must be overcome if we are to realize the full potential of these technologies.

GOVERNMENT SUPPORT OF BIOTECHNOLOGY AND INFORMATION TECHNOLOGY

Government-industry and university-industry partnerships have a long history in the information technology sector.[1] In contrast, the biomedical sector has seen a greater division of labor, with government funding basic research and industry concentrating on applying these research results to develop new pharmaceuticals and other products.[2] Even so, as the distance from fundamental research to applications has shortened in biomedical research—already a well-established trend in the information technology sector—the demand for partnerships and other forms of close collaboration has grown. Partnerships have also been a mechanism for promoting multidisciplinary research—another strong trend underlying research in both sectors. Finally, differing uses of patent protection provides another perspective for reviewing the impact of public policy on the biotechnology and information technology sectors. In the sections below, the roles played by partnerships, funding, and patents in each sector are highlighted by various participants at the conference.

Partnerships in IT

In his opening remarks, Gordon Moore gave a personal tour of the evolution of government support to the semiconductor and computing industries. He emphasized that the government aid to these industries was through general support of science and technology, as well as through serving as an early market for innovative devices. He also underscored the importance of the government's policy on intellectual property protection for the growth of the semiconductor industry. As an example, he cited the government's requirement that AT&T license its semiconductor technology as part of a 1956 antitrust decree, observing that this was a key catalyst for the growth of the industry. He noted that the government still plays an important role in the computer industry through the purchase of high-performance computers.

---

[1]See Kenneth Flamm, "The Federal Partnership with U.S. Industry in U.S. Computer Research: History and Recent Concerns," in this volume, for a brief history. For a comprehensive review of government support for this sector, see National Research Council, *Funding a Revolution: Government Support for Computing Research*, Washington, D.C.: National Academy Press, 1999.

[2]For an account of the innovations and issues in biotechnology, see Fredrick B. Rudolph and Larry V. McIntire, Eds. *Biotechnology: Science, Engineering, and Ethical Challenges for the 21st Century.* Washington, D.C.: Joseph Henry Press, 1996. For an account of the evolution of the pharmaceutical industry, see Rebecca Henderson, Luigi Orsenigo, and Gary P. Oisano, "The Pharmaceutical Industry and the Revolution in Molecular Biology: Interactions among Scientific, Institutional, and Organizational Change." In David C. Mowery and Richard R. Nelson, eds., *Sources of Industrial Leadership: Studies of Seven Industries.* New York: Cambridge University Press, 1999, pp. 267-311.

Direct federal support of R&D, Dr. Moore noted, did not play as large a role in the semiconductor industry's early evolution as in computers, although the government contributed substantially as a reliable early buyer of high-end semiconductor devices.[3] As the U.S. semiconductor industry matured, it adopted an innovative means of addressing the need for industry-related research by establishing the Semiconductor Research Corporation in 1982. This program initiated cooperative research efforts between the industry and selected universities.[4]

In response to severe qualitative and pricing competition from Japan, the industry campaigned and secured agreement to create a consortium to redress its competitive position. Called SEMATECH, the consortium represented a substantial and innovative effort by a fiercely competitive U.S. industry. It began in 1987 with matching funds from the federal government and industry, with each contributing $100 million a year. Reflecting the tremendous growth and commercial success of the American industry, SEMATECH's membership decided, in 1996, to "sunset" the federal contribution and fund the consortium from its own resources—albeit at a reduced level.[5] In 2000, it became International SEMATECH with the inclusion of members from Korea and Europe.[6]

Government-industry cooperation on research and development continues through a variety of mechanisms, noted Dr. Moore. For example, cooperative R&D agreements have proved effective mechanisms for encouraging cooperative research and development with the National Laboratories. Research and development in priming and lithography technologies—such as that in Extreme Ultraviolet Lithography (EUV)—are likely to prove critical for the future growth of the world's semiconductor industries. EUV promises a means of continuing the remarkable pace of advance in the capacity of computer chips, as captured in

---

[3]It is important to distinguish between the role of federal R&D funding in the fifties versus its less prominent role in the late seventies and eighties. Kenneth Flamm extensively documents the significant federal role in providing R&D support to leading firms. See Flamm, *Mismanaged Trade*, Washington, D.C.: Brookings Institution Press, 1996, pp. 32-37. Other analysts also underscore the importance of the Apollo and Minuteman programs to the growth of the American semiconductor industry, although, this was not—as Dr. Moore notes—in the form of direct R&D support for the industry. See Timothy J. Sturgeon, "How Silicon Valley Came to Be," in Kinney, Martin (ed.) *Understanding Silicon Valley: The Anatomy of an Entrepreneurial Regime*, Stanford, CA: Stanford University Press, 2000.

[4]See J. Horrigan, "Cooperating Competitors: A Comparison of MCC and SEMATECH." Monograph, Washington, D.C.: National Research Council.

[5]For a review of SEMATECH's goals and evolution until 1996, see National Research Council, *Conflict and Cooperation in National Competition for High Technology Industry*, Washington, D.C.: National Academy Press, 1996, pp. 141-151. That analysis identified three elements contributing to the resurgence of the U.S. industry.

[6]Although invited, the Japanese did not join in inter-consortia cooperation with U.S. companies. Japanese semiconductor firms, preferring to have a common language with their suppliers, formed SELETE. See interview with Mark Melliar-Smith: www.semiconductor.net/semiconductor/issues/Issues/1998/mar98/docs/ind_news1.asp.

Moore's Law. EUV Lithography is being developed under a Cooperative Research and Development Agreement (CRADA) between the Sandia and Lawrence Livermore National Laboratories and the leading SEMATECH companies, such as Intel, Motorola, AMD, and Infineon, to which industry contributes $100 million a year.[7]

Expanding on Dr. Moore's discussion, Kenneth Flamm provided evidence of recent trends in government funding for computers.[8] The computer industry benefited from significant federal support in the years following the Second World War, with the government encouraging private sector initiatives, serving as the primary market for most early machines, and developing much of the basic architecture of today's computer.[9]

As the commercial market for computers grew in the 1960s, the government's role in supporting R&D evolved, with the government continuing to push the leading edge of technology through support for advanced research and as a consumer of high-performance machines.

Recent federal support for innovation in this sector, however, has declined. Dr. Flamm reported that, since 1990, government's support for computer hardware R&D, already at historically low levels, declined substantially. Further, industry investment also fell, both in absolute terms and as a percentage of computer sales. At the same time, both federal and industrial R&D investments have shifted over time from long-term basic research to near-term targeted projects.

In concluding, Dr. Flamm argued that the information technology sector is not yet a mature industry, that innovation in the computer industry remains an important source of economic growth, and that government-industry collaboration, highly productive in the past, should be encouraged in the future.

Partnerships in the Biotechnology Industry

In contrast to semiconductors, where the government has employed a variety of agencies and mechanisms, the biotechnology industry has relied heavily

---

[7]A major advance in lithography was announced by Lawrence Livermore National Laboratory in April 2001 that will allow chip manufacturers to create circuits as small as 10 nanometers wide using extreme ultraviolet (EUV) lithography. This advance was made under a CRADA in which the government laboratories provided facilities and expertise at Sandia and Lawrence Livermore National Laboratories, while the private companies such as Intel, AMD, Motorola, and Infineon provided funding and their own expertise.

[8]Dr. Flamm's paper in this report presents this data formally For a history of federal support for computing, see Kenneth Flamm, *Creating the Computer*. Washington, D.C.: The Brookings Institution, 1988, chapters 1-3. For a more recent view, see the paper by Dr. Flamm in this volume.

[9]See Kenneth Flamm, 1988, *op. cit.*, and National Research Council, *Funding a Revolution: Government Support for Computing Research,* Washington, D.C.: National Academy Press, 1999.

on a single government agency to underwrite basic research support. In this partnership of government, industry, and universities much of the technology commercialized by the biotechnology industry has been developed in the university setting, with funding from the National Institutes of Health. This relationship, however, is changing rapidly as the industry evolves.

Whereas the first 20 years of the biotechnology industry were dominated by two technologies—recombinant DNA and techniques to produce monoclonal antibodies—new technologies now are based on a wider range of scientific disciplines. These are becoming important to both the biotechnology and pharmaceutical industries.[10] DNA chips and the growing output from the Human Genome Project are just two examples of recent developments that require new responses from government, industry, and universities.

Describing trends in bio-pharmaceutical research, Leon Rosenberg emphasized the interaction among industry, government, and university. Although the government and industry fund about 90 percent of bio-pharmaceutical research, universities conduct much of this research. Universities act as an important source of R&D output, in addition to providing trained scientists. Continued advances in bio-pharmaceutical research will depend on advances or applications of new information technologies, notes Dr. Rosenberg, and it is important that academic R&D retain its emphasis on basic research. Even with the growing role of industry in funding university research, Dr. Rosenberg affirmed that academic research has remained focused on basic research rather than on applications.

Intellectual Property Protection

The intellectual property regime, which structures the incentives to innovate and the ability to capture innovation's fruits, is another important piece of the policy puzzle. There have been a number of changes in intellectual property policy and practice in recent years.

Research by Wesley Cohen and John Walsh, presented at the conference and included in the report, shows the varying importance of patenting across the biotechnology and computing sectors. Their research looks at the impact of the 1980 Bayh-Dole patent and trademark act, which extended patent protection to

---

[10]For an account of many of the innovations and issues in biotechnology, see Frederick B. Rudolph and Larry V. McIntire, eds., *Biotechnology: Science, Engineering, and Ethical Challenges for the 21st Century*. Washington, D.C.: Joseph Henry Press, 1996. For an account of the evolution of the pharmaceutical industry, see Rebecca Henderson, Luigi Orsenigo, and Gary P. Pisano, "The Pharmaceutical Industry and the Revolution in Molecular Biology: Interactions Among Scientific, Institutional, and Organizational Change," in David C. Mowery and Richard R. Nelson, eds., *Sources of Industrial Leadership: Studies of Seven Industries*. New York: Cambridge University Press, 1999, pp. 267-311.

publicly funded research, such as that conducted in universities.[11] They found that patenting is especially important in the biotechnology sector, because it is often the sole asset of a young start-up venture. On the other hand, patenting seems to be less important in the computing and information technology sectors, because companies in these industries tend to rely on trade secrecy and long lead times to protect intellectual property. In that situation, giving universities exclusive rights to research results will have little effect and may even inhibit commercial exploitation of publicly funded research, noted Drs. Cohen and Walsh. This is because it can inhibit the free flow of information between universities and firms. They conclude that policymakers should take into account the various methods that firms in different sectors employ to protect intellectual property when considering changes in intellectual policy law.

In exploring the patenting activity of universities in the post-Bayh-Dole era, Maryann Feldman of Johns Hopkins University found that patenting activities at some universities seem to be governed more by past practices and cultures than by Bayh-Dole. For most universities—including those most actively generating patents—a few patents generate most of the license revenue. Therefore, she found that widespread patenting efforts by universities may yield only modest benefits in terms of licensing revenue. At the same time, Dr. Feldman noted that such patenting practices raise the transaction costs of university-industry collaboration and thus can have the effect of discouraging collaboration.

Dr. Feldman also found that universities and companies have developed a broad range of mechanisms for exchanging information in the post-Bayh-Dole era. In addition to exclusive licenses, arrangements include sponsored research, consulting arrangements, recruitment of students, equity arrangements, support for start-up companies, and even gifts to university endowments. As Drs. Cohen and Walsh also observed, the types of transactions depend on the sector, technology, and size of the company.

Policy Concerns

Apart from reviewing past support for the biotechnology and information technology sectors, conference participants and paper authors identified a variety of policy concerns related to the federal research portfolio and patent laws.

[11]The Bayh-Dole University and Small Business Patent Act of 1980 changed the intellectual property landscape and altered the nature of collaborative relationships between university and industry. The main premise behind Bayh-Dole was to release the "urn full of untapped innovations" with commercial potential, as contained within universities. By granting universities the ability to patent innovations from government-funded research, Bayh-Dole attempted to create incentives for universities to move such innovations into the marketplace.

Evolution in the Federal Research Portfolio

Speaking at the conference, William Bonvillian, from the staff of Senator Joseph Lieberman, noted that overall, federal support for R&D fell nearly 9 percent in real terms from 1992 to 1997. These cuts, however, were not evenly distributed, he noted, with the impact in some disciplines (including some fields related to information technology) more severe than in others. For example, from FY 1993 to 1997, in real terms, mathematics was down by 6 percent, electrical engineering by 36 percent, and physics by 29 percent.[12] Federal support for health and biomedical research has increased, while support has declined for research in the physical sciences, mathematics, and many fields of engineering, on which advances in biomedical research has depended and will depend, he noted.

Although the allocation of federal research funding among scientific fields should shift, ideally, with changes in needs and opportunities, Mr. Bonvillian argued that the recent shifts in funding were due more to organizational changes than to changes in scientific opportunity. For example, the steep decline in the defense budget for R&D—a major source of funding for the physical sciences and engineering—has played a central role in these shifts in federal funding. At the same time, the fact that many of the most important scientific advances in biotechnology and information technology come from cross-disciplinary research argues for sustained funding across a broad range of fields. He recommended that policymakers develop an "alert system" to signal when R&D funding levels for certain disciplines fall below critical levels.

The analysis by Michael McGeary (in this report)—which documents recent trends in federal funding of research related to biotechnology and IT—supports this view. The data indicate a substantial shift in the federal research portfolio over the 1990s. McGeary raises "the question of whether the federal research portfolio has become 'imbalanced.'" He finds that federal support for biomedical research has expanded substantially relative to that in most fields of the physical sciences and engineering. Noting that overall federal funding for research was flat in real terms from 1993 to 1997,[13] McGeary finds that the Administration and Congress over that period increased the budget for the National Institutes of Health—which provides more than 80 percent of the federal support for the life sciences. While this development is in itself positive, the research

---

[12] In the same period, funding for computer science research was up by 39 percent in real terms. For additional detail on trends in federal R&D, see Michael McGeary and Stephen A. Merrill, "Recent Trends in Federal Spending on Scientific and Engineering Research: Impacts on Research Fields and Graduate Training," in National Research Council, *Securing America's Industrial Strength*, Washington, D.C.: National Academy Press, 1999, pp. 53-98.

[13] Merrill and McGeary, 1999, *op. cit.*

budgets of DOD and DOE—which together provide the majority of funding for research in electrical engineering, mechanical engineering, materials engineering, physics, and computer science—have suffered substantial reductions in the same period.[14]

At the conference, Tom Kalil, formerly of the White House National Economic Council, described how the Clinton administration's effort to enact a 28 percent increase in spending on information technology research for FY 2000 was unsuccessful, despite support from House Science Committee Chairman James Sensenbrenner. While the administration won some increases for the NSF and DARPA, Congress chose not to fund research for high-end computing in the Energy Department. Mr. Kalil said the administration had hoped to raise the level of funding for information technology researches in FY 2001. Indeed the current administration has proposed a $2.6 billion increase in funding for R&D in FY2001, including major initiatives in nanotechnology and information technology (including development of a second terascale computer for civilian researchers), as well as substantial increases for NSF and NIH.[15]

Asymmetries in Political Support for Funding

In explaining this trend, Leon Rosenberg observed that the bio-pharmaceutical community has done a good job in building a constituency for research funding among policymakers in Washington. He noted that the NIH budget doubled in the 1990s, and the increase of 15 percent in FY 1999 marked the first year of a five-year effort, supported by most members of Congress, to double the 1998 NIH budget by 2003.[16] Mr. Bonvillian, in this regard, observed that the biotechnology industry has done an excellent job in taking its message to Con-

---

[14]See Michael McGarry, "Recent Trends in the Federal Funding of Research and Development Related to Health and Information Technology" in this volume. Also, see Figure 1, "Changes in federal research obligations for all performers and university/college performers, FY 1993-1999" in the Findings and Recommendations part of this volume.

[15]Information on President Bush's FY 2001 R&D proposal can be found in *Budget of the United States Government: Fiscal Year 2001*, pp. 95-106. At: http://www.w3.access.gpo.gov/usbudget/fy2001/maindown.html. In the end, the increase for R&D was $7.6 billion, including $2.5 billion for NIH. There were also increases for NSF (14 instead of 17 percent), nanotechnology initiative (55 instead of 100 percent), and IT initiative (24 instead of 36 percent). See AAAS, *Congressional Action on Research and Development in the FY 2001 Budget,* Washington, D.C.: AAAS, 2000, at http://www.aaas.org/spp/dspp/rd/ca01main.htm.

[16]NIH received increases of 14.1 percent and 14.4 percent in FY 2000 and 2001, respectively. President Bush pledged to continue the doubling effort and proposed an increase of 13.5 percent in FY 2002. DHHS Secretary Tommy Thompson recently pledged an increase of $4 billion in FY 2003 to reach the doubling target of $27 billion. See Colin MacIlwain, "Bush Favours Research at Pentagon and NIH," *Nature*, April 12, 2001, p. 731.

gress. He pointed out that their success stands in stark contrast to the more limited success of proponents for research support in information technologies.[17]

Mr. Bonvillian went on to note that this growing support for the biomedical enterprise has two interconnected explanations. First, it may be that the focus of much of the biomedical research activity of one agency—the National Institutes of Health, (which also benefited from exceptionally able leadership through the nineties)—contributed to the ability of proponents to focus policymakers' attention. Second, a series of high profile research efforts, notably the Human Genome Project, captured the public's imagination in the 1990s, much as the exploration of space did in the 1960s. This initiative has helped provide support from the general public for increased funding.

In contrast, noted Mr. Bonvillian, support for computer research is scattered among many government departments and agencies with diverse objectives, constituencies, and modes of operation. In addition, investments in health R&D are meant to lead to cures for diseases that afflict us all, including members of Congress. The fact that much of the NIH research effort is extramural and widely distributed among the nation's universities may tend to augment its already broad-based support.

Reviewing Patent Laws

The pace of innovation in the biotechnology industry has caused a strain in the patent system, noted Robert Blackburn of Chiron Corporation. He recounted that in some instances, courts have been applying patent law doctrine developed for the synthetic chemical industry to biotechnology—including patents on DNA sequences and other things found in nature. This practice is problematic when applied to biotechnology, he noted, since biotechnology directly concerns research tools, rather than consumer products. Furthermore, patents covering research tools are difficult to enforce and cause friction between university and industrial researchers. In conjunction with limited resources of the Patent Office and with only intermittent legislative guidance from Congress, the patent system can sometimes have the unintended effect of discouraging drug development. Mr. Blackburn recommended, therefore, that Congress undertake an extensive review of patent law.[18]

---

[17]A few months after the conference, several new initiatives were launched in nanotechnology and IT. The Administration proposed a near doubling of research on nanotechnology, from $270 million in FY 2000 to $495 in FY 20001, and a 36 percent increase in Information Technology research, from $1.7 billion in FY 2000 to $2.3 billion in FY 2001. The Information Technology research initiative included $823 million for a new component led by NSF, "Information Technology for the 21st Century," or Information Technology, focusing on fundamental long-term research as recommended by the President's Information Technology Advisory Committee in its report, *Information Technology Research: Investing in Our Future*, February 1999.

[18]After this conference, the STEP Board undertook a major study on *Intellectual Property Rights in a Knowledge-Based Economy*. For more information, see www.nationalacademies.org/ipr.

COMPLEMENTARITIES IN BIOTECHNOLOGY AND INFORMATION TECHNOLOGY RESEARCH

A leading theme of the conference was the importance that each field, biotechnology and IT, has for the other. Typically, we think of the critical role that information technology will play in fulfilling the promise of the genetic revolution. Just as importantly, but less well understood, biological systems and concepts have much to offer the information technology and other high-technology fields such as materials and nanotechnology. The spillover effects of scientific advances propelling the biotechnology industry help the computing industries and vice versa. This holds great promise for the economy and for society at large.

Marvin Cassman noted that the tools of molecular genetics, structural biology, genomics, and other fields are creating an enormous amount of information about humans and other living organisms. With the necessary technology now in hand, he noted, scientists can begin to understand biology as a "wiring diagram"—a complex system that is dynamic and richly interconnected. This contrasts with the discipline's traditional approach to assessing biological organisms using "parts lists." To understand an organism's complexity, biologists will have to process the huge amounts of data that will flow from research. This is where bioinformatics and the tools of information technology enter the picture.

As we see below, organizations including the Department of Defense and NASA are already interested in research on small and lightweight high-performance technologies to meet their own needs.

Frontier Technologies Based on Biotechnology and Information Technology

Space Medicine

In his luncheon keynote speech, NASA Administrator Daniel Goldin, observed that the marriage of biology and information technology is key to space exploration in the future. Travel to Mars will take many years, and sophisticated medical treatments—drawn from biotechnology and remote communication systems—will be necessary to detect, diagnose, and cure health problems that astronauts might encounter. Administering medical treatment over the immense distances of space will require remote sensing, highly sophisticated software, and advanced robotic technology. Developing such technology is, perhaps, the largest challenge facing biologists and computer scientists today.

In Mr. Goldin's view, biology has tremendous potential—through biometrics, bioinformatics, and genomics—to change electronics, computational hardware and software, sensors, instruments, control systems, and materials. Developments in biology will also call for new platform concepts and system architectures to integrate these technologies, he noted. They will be key to the development of more complex and self-correcting—that is, more intelligent—

technologies that are also smaller, lighter, faster, more adaptable in changing environments, and less power-intensive.

Even as succeeding administrations and agencies (including NASA, NIH and NSF) lay the groundwork for bridging the gap between biology and computing, others are already looking to the future. In doing so, they help to chart what policymakers must do today so that the full potential of current innovations can be more fully exploited.

Functional Genomics

Functional genomics, which Mark Boguski discussed, is an example of an approach to closing the gap between biology and computing that will push innovation in computing, while enabling biology to realize the benefits of the Human Genome Project. Functional genomics uses computing technology to model organisms (using data from the human genome) so that these organisms can be understood as a system. Rather than comparing a few genes, functional genomics will allow scientists to compare many genes—or systems of genes—across organisms in order to make inferences about their function in the body and to determine what proteins are involved. This will require, however, a great deal of computing power and the development of sophisticated new visualization and simulation technologies.

Miniaturization and Nanotechnology

Miniaturization of devices is another area that will rely on biology and computing as it offers exciting possibilities in medicine, energy efficiency, national security, and environmental protection. Al Romig of Sandia National Laboratories reported that miniaturization will proceed both by making today's components smaller and by taking micro-scale materials, including biological materials, and building up from there. Thus, he said, not only will the physical sciences help biomedical research, but also biological research will lead to advances in the physical sciences. He is confident that over the next 10 to 20 years, a growing number of materials and devices on the market will be manufactured using biological methods.

In the field of nanotechnology, for example, nanocrystals have been injected into living cells. These crystals do not damage cells. Rather, they emit light enabling scientists to explore further the properties and behavior of cells. Nanotechnology will also enable the creation of more efficient materials, which should reduce energy consumption with major benefits for the environment.

Battlefield Applications

Timothy Coffey of the Naval Research Laboratory discussed how miniaturization in military applications will contribute to national security. Nanoelectronics and microelectromechanical systems (MEMS) will enable electronic systems to carry out essential battlefield functions that humans do now at their peril. For example, micro-air vehicles are being developed to jam enemy radar systems, and biosensors will detect the presence of dangerous substances in the environment. As battlefield requirements increase, the military will look to biological systems for solutions, but developing those technologies will rely on advances in computing and electronics. In turn, this will require the confluence of a variety of research disciplines, including the biosciences, materials science, computers and information technologies, and electronics.

Other interesting developments at the intersection of biology and information technology were described by Jane Alexander of the Defense Advanced Research Project Agency (DARPA). One such example is an effort to develop an "electronic dog's nose" that can sniff out explosives. If adequately portable, it can be carried into the battlefield by soldiers. Another initiative would be to intervene at the larval stage of insects to enable them, when mature, to detect explosives. DARPA is also considering putting electronic chips on insects in order to track their hunting patterns for use in developing search algorithms for Department of Defense sensors.

Realizing Synergies in Biology and Information Technology

The combination of breakthroughs in computing technologies along with the sequencing of the human genome offers exciting possibilities. Paul Horn of IBM observed that in the future, for example, doctors will prescribe medications tailored precisely to a patient's problem and genetic makeup. To do this, new information technology will be necessary.

Government support for basic research will be indispensable in creating it. New bridges must be built between basic research and industry, Dr. Horn said, and government-funded partnership programs—such as Engineering Research, Science and Technology Centers, the Advanced Technology Program, and the Technology Reinvestment Program—have and should play a role in speeding the exchange of information between universities and industry.[19] Government can also help by providing the information infrastructure, such as the development of very high-end computers that are needed to turn the vast amount of data information from the Human Genome Project into knowledge and understanding.

Dr. Romig took a similarly positive view, noting that multidisciplinary government-university-industry partnerships are essential for progress in nanotech-

[19]See Kathleen Kingscott, "Lowering Hurdle Rates for New Technologies" in National Research Council, *The Advanced Technology Program; Assessing Outcomes,* Washington, D.C.: National Academies Press, 2001, pp. 112-116.

nology research. He predicted that the federal government will have a prominent role as a supporter of research, and as an owner of national laboratories.

Supporting Multidisciplinary Research

Scientific advance is inherently interconnected; neglecting one area of inquiry can jeopardize advancement in other fields of inquiry. Progress in solving research problems requires interaction across disciplines as well as across supporting government agencies, industry groups, academic organizations, and other nonprofit associations. Capturing the full potential of existing investments requires additional multidisciplinary research collaboration among biologists and computer scientists, as well as among those active in related disciplines, such as physics, chemistry, and electrical and chemical engineering.[20] While the research community seems well aware of these challenges, more needs to be done to educate policymakers on this issue.

A further implication is the need for interdisciplinary training. This is clearly the case in biotechnology—particularly in the new field of bioinformatics. In fact, one of the challenges to analyzing bioinformation and promoting collaboration is the shortage of bioinformatics specialists. Paula Stephan and Grant Black, in a paper in this report, provide explanations for this problem: They note that while bioinformatics is inherently multidisciplinary, institutionalized practices at universities often inhibit cross-departmental collaborations. Further, they note that information technology professionals, often in great demand in their own fields, find few incentives to undertake additional training in biology.

Encouraging more interdisciplinary training is a concern that is being addressed on various fronts. Rita Colwell, Director of the National Science Foundation (NSF), noted that the NSF recognizes the challenges facing those studying entire biological systems. The NSF now funds efforts to develop more bioinformatics tools and seeks to encourage cross-disciplinary collaboration, she noted.[21] Dr. Marvin Cassman described the initiatives of the National Institute of General Medicine Sciences, which he directs, in establishing a new Center for Bioinformatics and Computational Biology to support interdisciplinary training in conjunction with biology. The center will assume oversight of the NIH's Biomedical Information Science and Technology Initiative. Dr. Cassman also described related NIH initiatives—such as "glue grants," designed to bring biologists, physical scientists, and information technologists closer together—that

[20]The recent STEP report, *Trends in Federal Support of Research and Graduate Education*, observes that these are "all fields with less funding at the end of the 1990s than they received earlier in the decade," (page 2) *op. cit.*

[21]In July 2000, the NSF awarded the University of California at Los Angeles a $2.7 million award for integrative graduate education and research training in bioinformatics.

are intended to encourage collaboration across disciplines.[22] Dr. Penhoet described the Berkeley Health Sciences Initiative—a major effort to promote such cross-disciplinary collaboration. This initiative brings together a variety of disciplines in the same location to address research problems, such as improved gene chips and computational methods for analyzing the data that the gene chips produce.

Building bridges across disciplines is a long-term effort. To succeed, it will require leadership and innovation as well as other new partnerships between government, industry, and universities. It will also require appropriate allocation of federal R&D resources and, in some cases, new institutional frameworks to foster sustained technological.

The Future of Partnerships

In his conference presentation, Congressman Sherwood Boehlert focused on the practical aspects of allocating scarce federal R&D resources to help realize the possibilities of biotechnology and new information technologies. He said that the question is not one of whether the government should support biotechnology and information technology research. Rather, the question is one of how and at what level such support is to be realized. These were bound to be "the trickiest questions" in R&D policy, he observed. Mr. Boehlert said that he believed partnerships between government, industry, and academia would be part of the answer.

Mr. Boehlert suggested a careful examination of partnerships—especially their effects on the important role of universities in long-term research and the production of scientists and engineers—to see whether these partnerships produce commercially viable results. He also brought up the question of whether

[22]In September 2000, the National Institute of General Medical Sciences awarded the first glue grant, a mechanism to fund large-scale projects involving multiple institutions. The grantee is the Alliance for Cell Signaling, a public-private collaboration involving laboratories at five universities. The alliance is also funded by other NIH institutes, a consortium of pharmaceutical companies, and private donors. The purpose of the alliance is to study signaling pathways within cells to determine how cells communicate. The ultimate goal is to create a virtual cell, that is, a computer model, that could be used to help predict the impact of chemical compounds on cell function and behavior. The results will be in the public domain, and biotechnology and pharmaceutical companies will be able to use the results to design better drugs and diagnostic reagents. NIGMS is providing $25 million over the first five years of the 10-year, $10 million-a-year effort. See Kate Devine, "Cell Signaling Alliance Gets Under Way," *The Scientist*, 14(20): 1, October 16, 2000; Sophie Wilkinson, "Big Science Takes on Cellular Signaling," *Chemical & Engineering News*, October 16, 2000. The project is being staffed by a range of disciplines, including cell and molecular biologists, biochemists, protein chemists, microscopists, programmers, and statisticians. See Diane Gershon, "Alliance Signals a Fresh Type of Scientific Research Endeavour as the Post-Genomic Face of 'Big Biology' Gets Underway," *Nature*, April 19, 2001.

the current allocation of research support among scientific fields was appropriate. He urged members of the research community to take their views on investment in research not just to members of Congress on science committees, but to all members of the House and Senate.

Several ongoing partnerships illustrate why increased collaboration among sectors in the biomedical area is so important. The Alliance for Cellular Signaling, a government-industry-university partnership, was established in September 2000 with a glue grant from the NIH. Dr. Cassman described this as a new mechanism to increase multidisciplinary collaboration and inter-institutional sharing of resources.

Other current partnerships include the Mouse Sequencing Consortium (MSC) and the Single Nucleotide Polymorphism (SNP) Consortium. MSC, also begun in 2000, supports sequencing at several academic and nonprofit research institutions. Six NIH institutes, along with the Wellcome Trust and several pharmaceutical and biotechnology companies, fund MSC.[23] The SNP Consortium, created in 1999, seeks to map genetic variations within the human genome in order to increase the understanding of their role in causing disease. SNP was formed at the initiative of the Wellcome Trust, which provided substantial funding and convinced 10 competing companies to contribute funding.[24] In both cases, as with the Alliance for Cellular Signaling, the results are being made public. The goal is to provide wide access to basic data that not only will speed up progress in research but also disseminate information that industry can use to develop new products.

CONCLUSION

This summary provides an overview of the issues the Committee sought to address at the conference. It underscores the need for additional analysis and careful monitoring of national policy on resource allocation in the biotechnology and information technology areas. These issues are presented above in an abbreviated form as a means of providing the background and necessary context for the Committee's Findings and Recommendations as detailed in the section that immediately follows.

---

[23]Eugen Russo, "Stepping Up Mouse Sequencing," *The Scientist*, 14(22): 12, November 13, 2000. At: http://www.the-scientist.com/yr2000/nov/russo_p12_001113.html.

[24]Eugene Russo and Paul Smaglik, "Single Nucleotide Polymorphisms: Big Pharma Hedges its Bets," *The Scientist*, 13(15): 1, July 19, 1999.

III

FINDINGS AND RECOMMENDATIONS

Summary of Findings

I. Information Technology *and* Biotechnology are Key Sectors for the 21st Century. The information and biotechnology sectors are each very important to America's economy, security, and well-being.[1] Although some expect the 21st century to be dominated by the Biotechnology Revolution, the Information Revolution begun in the 20th century has not run its course. Advances in information technologies remain central to economic growth; they will also be critical to progress in the Biotechnology Revolution itself. Reaping the health benefits of sequencing the human genome depends on processing and making sense of enormous amounts of data that, in turn, will be made possible by advances in computing and networking technologies. At the same time, information technology will advance based on a better understanding and use of the principles of biological systems and mechanisms. Already, NASA has developed flight control software for the F-15 aircraft based on neural network principles.[2] Biological models also may

[1]As William Spencer points out in his Introduction to the proceedings, a substantial portion of the $225 billion in private and public R&D spending in 1998 were in the fields of biotechnology and computing.

[2]This software requires 10,000 lines of code instead of the million lines required by traditional flight control software. The aircraft performed better in test flights, because the neural network program was more responsive to changes (such as loss of aircraft control surfaces) and adapted itself to correct related software faults. See Daniel S. Goldin, "The Cornucopia of the Future" in this volume.

lead to improved computer hardware architecture and other technologies, such as MEMS and nanotechnology.[3]

II. **Multidisciplinary Approaches Are Increasingly Needed in Science and Engineering Research.** More and more, progress in research depends on multidisciplinary efforts.[4] Complex research problems require the integration of both people and new knowledge across a range of disciplines. In turn, this requires knowledge workers with interdisciplinary training in mathematics, computer science, and biology. Bioinformatics is a key example of a highly multidisciplinary field requiring workers with interdisciplinary training.[5]

Biotechnology R&D and information technology R&D each provide tools and models useful for the other. Interdependencies also exist among chemistry, physics, and structural biology, and among mathematics, computer engineering, and genomics. Further examples of these interdependencies can be found in the complementary roles of the physical sciences and engineering in nano-scale semiconductor work, and in the overall important role engineering research plays in providing new research tools and diagnostics in all of these areas.[6]

III. **Government-Industry and University-Industry Partnerships Have Often Been Effective in Supporting the Development of New Technologies.** Symposium participants noted many examples of partnerships involving the federal government, universities, and industry that have contributed to the development and strengthening of leading U.S. industries.[7] Mechanisms to

---

[3]Jane Alexander, "Biofutures for Multiple Missions" in this volume, gives a number of examples of microsystems based on biological models.

[4]See the presentations of Edward Penhoet, William Bonvillian, Marvin Cassman, and Rita Colwell in the Proceedings. These participants note that biotechnology research is expected to become increasingly multidisciplinary but must overcome particular constraints if biotechnology is to reach its full capacity and usher a rapid advance in knowledge.

[5]See Paula E. Stephan and Grant Black, "Bioinformatics: Emerging Opportunities and Emerging Gaps," in this volume. The authors find that the field of bioinformatics is highly multidisciplinary, requiring a combination of understanding and skills in mathematics, computer science, and biology. Their paper explores constraints to more robust multidisciplinary research in this important area.

[6]See the presentations of Daniel Goldin, Al Romig, Paul Horn, and Mark Boguski, in the Proceedings section, on the future technologies that require current investments in multidisciplinary research.

[7]For a description of the importance of the federal partnership with U.S. industry in the development of the computer, see the paper by Kenneth Flamm in this volume. He notes that "In the first decade after the war, most significant computer research and development projects in the United States depended—directly or indirectly—on funding from the United States government, mainly from the U.S. military. The second decade of the computer, from the mid-1950s through the mid-1960s, saw rapid growth in commercial applications and utilization of computers. By the mid-1960s, business applications accounted for a vastly larger share of computer use. However, the government

support partnerships include the Engineering Research Centers and Science and Technology Centers funded by the NSF, NIST's Advanced Technology Program, and the DOD's former Technology Reinvestment Program. They also included activities funded by Cooperative Research and Development Agreements (CRADAs).[8] SEMATECH, initially funded by government as well as industry, has played a key role in improving manufacturing technologies for the U.S. semiconductor industry.[9]

A Multidisciplinary Approach: Partnerships between universities and companies that cut across disciplines, though often challenging, are increasingly important to progress as biotechnology and information technologies become dependent on contributions from other fields. While projects led by individual investigators remain vital to general scientific and engineering advancement, solving complex problems in new areas such as bioinformatics and next-generation computing requires larger, multidisciplinary collaborations among scientists and engineering researchers. Both government agencies and industry groups can help overcome these institutional problems by funding multidisciplinary research projects focused on important complex problems.

Federal laboratories also offer important capabilities and institutional lessons for dealing with complex research problems. Historically, NIH has not directly supported industry R&D, but this is changing. In 1998, NIH laboratories entered into 166 CRADAs, and in 1999 NIH's Small Business Inno-

---

continued to dominate the market for high-performance computers, and government-funded technology projects pushed much of the continuing advance at the frontiers of information technology over this period.

From the mid-1960s through the mid-1970s, continued growth in commercial applications of vastly cheaper computing power exploded. By the middle of the decade of the seventies, the commercial market had become the dominant force driving the technological development of the U.S. computer industry. The government role, increasingly, was focused on the very high-end, most advanced computers, in funding basic research, and in bankrolling long-term/leading-edge technology projects." For a broader view, see Roger Noll and Linda Cohen, *The Technology Pork Barrel*, Washington, D.C.: The Brookings Institution, 1991. The authors observe that there are frequent failures as well as successes among federal R&D programs. They count among the successes telegraphy, hybrid seeds, aircraft, radio, radar, computers, semiconductors, and communications satellites. In short, much of the foundation for the modern economy. Noll and Cohen, *op. cit.* p. 3.

In this volume, see the presentations by Dale Jorgenson, Gordon Moore, and Kenneth Flamm in the Proceedings. In his overview of the semiconductor industry, Gordon Moore notes that government support has been effective in developing technologies—particularly those related to high-end semiconductor devices—that would not likely have otherwise emerged. In describing the role of SEMATECH, he also notes that federal support has strengthened the competitive position of the United States. See also the Preface to this report.

[8]See, for example, the presentation by Paul Horn, "Meeting the Needs: Realizing the Opportunities" in this volume.

[9]Gordon Moore, "The View from the Semiconductor Industry" in this volume.

vation Research Program awarded more than $300 million to small companies.[10] This is expected to rise to $410 million in 2001.[11]

The availability of genome sequences has also prompted research on a new scale, leading to new types of partnerships in the biotechnology, pharmaceutical, and biomedical areas. Understanding the functioning of cells at the system level, or how thousands of genes and proteins interact, requires collaborations among large numbers of researchers with interdisciplinary backgrounds, including those with sophisticated computer skills.

Multidisciplinary research, with teams from different departments and organizations is needed to attack issues at the periphery of standard disciplinary boundaries.[12] The advent of such organizations as the SNP Consortium, the Alliance for Cellular Signaling, and the Mouse Sequence Consortium—not to mention the organizational effort mounted to sequence the genome itself—illustrate the need for and possibilities of partnerships in biomedicine.[13]

IV. Federal R&D Funding and Other Innovation Policies Have Been Important in Supporting the Development of U.S. Industrial Capabilities in Computing and Biotechnology. The history of the U.S. biotechnology and computing sectors illustrates the value of sustained federal support, although federal roles and contributions have been different in the two areas:

A. Computing and Semiconductors. In computing and semiconductors, the federal government has played four important roles since World War II.[14]

- **Antitrust.** Antitrust policy played an important role in the creation of the semiconductor industry by requiring the pre-1984 AT&T to license broadly its patents in transistors.[15]

---

[10]Leon Rosenberg, "Partnerships in the Biotechnology Enterprise" in this volume, discusses the extent and importance of university-industry partnerships in the biomedical field. Some recent press reports put the amount substantially higher.

[11]Substantial funds are available to non-profit research centers, some to clinics associated with private firms. For increasing relationships between nonprofit research institutions and for-profit firms, see Chris Adams, "Laboratory Hybrids: How Adroit Scientists Aid Biotech Companies with Taxpayer Money—NIH Grants Go to Non-profits Tied to For-profit Firms set up by Researchers," *Wall Street Journal,* New York: Dow Jones and Company, January 30, 2001.

[12]See Editorial, "Post-Genomic Cultures," *Nature*, February 1, 2001, p. 545.

[13]These organizations are described in "Issues in Biotechnology and Information Technology," in this volume.

[14]See the presentation by Gordon Moore in the Proceedings section for a nuanced view of the federal role in the development of the semiconductor industry.

[15]Moore, *op. cit.*

- **Procurement.** Government procurement in the early days of both semiconductors and computers presented a valuable stimulus to technological development and industry growth; in addition to providing generous contracts, the government served as a reliable first customer, providing companies with early markets.[16]

- **R&D Funding.** Federally funded R&D was important in developing certain niche areas (e.g., gallium-arsenide chips) and certain new areas of computing (e.g., ARPANET and 3D graphics). Reflecting the semiconductor industry's dependence on research and its special needs, the federal research effort has been complemented by private support, such as the industry-created Semiconductor Research Corporation,[17] and later through the SEMATECH Consortium—the highly regarded federal-industry partnership.[18]

- **High Performance Computing.** The government has long played, and continues to play, an essential role in high-performance computing. There is usually only a small commercial market for the most powerful supercomputers; the government itself continues to act as the main purchaser of leading-edge machines for a variety of national missions. As a result, federal R&D support and procurement remain critical to progress in this field.[19]

B. Biotechnology. It is widely recognized that federally funded research has played a central role in the development of the biotechnology industry. The federal government funds 90 percent of university-based health research in the United States.[20] Many of the key discoveries and techniques of

---

[16]Kenneth Flamm, "Partnerships in the Computer Industry" in this volume. See also Kenneth Flamm, *Targeting the Computer*. Washington, DC: The Brookings Institution, 1987. p. 107. See also Martin Kenney, ed., *Understanding Silicon Valley: The Anatomy of an Entrepreneurial Region,* Stanford, CA: Stanford University Press, 2000.

[17]Moore, *op. cit.*

[18]See P. Grindley, D. C. Mowery, and B. S. Silverman, "The Design of High-Technology Consortia: Lessons from SEMATECH," In *Technological Infrastructure Policy: An International Perspective*, M. Teubal, editor, Dordrecht: Kluwer Academic Publishing, 1996. Also see John Horrigan, "Cooperating Competitors: A Comparison of MCC and SEMATECH," Monograph, Washington, D.C.: National Research Council, 1997. Also see National Research Council, *Conflict and Cooperation in National Competition for High-Technology Industry*, Washington, D.C.: National Academy Press, 1996, Supplement B, p. 141.

[19]National Research Council, *Funding a Revolution: Government Support for Computing Research*. Washington, D.C.: National Academy Press, 1999.

[20]Rosenberg, *op. cit.*

modern biotechnology have come out of university research funded by the National Institutes of Health (NIH). While the NIH continues to play a crucial role, corporate research also broadens the technology base. Indeed, other non-biological technologies, such as computing, have become increasingly important to the continued progress of the biotechnology industry.[21]

C. Patent Policies. Federal patent policy also has a significant impact on these fields. Covering, as it does, areas as diverse as biotechnology and e-commerce, current federal patent policy is sometimes a source of uncertainty and, thus, could prove to be an obstacle to innovation.[22] The federal government could help further stimulate innovation by clarifying patent rules (e.g., regarding biotechnology research tools) to reduce risks of litigation over patent issues.[23]

V. Limitations of Venture Capital. The U.S. venture capital market—the largest and best developed in the world—often plays a crucial role in the formation of new high-technology companies. Even so, the nature of the role and constraints posed by venture capital markets need to be better understood. The provision of venture capital, with its informed assessment and management oversight, is but one element of a larger innovation system. It is not a substitute for government support of long-term scientific and technological research. Indeed, the venture capital industry is not designed to support early-phase research and rarely does so. Venture capitalists do provide vital support to business efforts that show the potential of exploiting new technologies and business ideas.[24] As such, venture funding tends to

---

[21]See the presentation by Mark Boguski on "Computing and the Human Genome" in this volume. See also the presentation of Jeffrey Schloss in National Research Council, *The Advanced Technology Program: Challenges and Opportunities*, Charles W. Wessner, editor, Washington, D.C.: National Academy Press, 1999, pp. 56-59. To hasten the development of computational tools in biotechnology, the NIH established the Biomedical Information Science and Technology Initiative in April 2000. The initiative "is aimed at making optimal use of science and technology to address problems in biology and medicine." More information on the initiative is available at the NIH Web site: http://grants.nih.gov/grants/bistic/bistic2.cfm.

[22]The STEP Board is currently conducting a study on these and related issues led by Richard Levin, President of Yale University, and Mark Myers, Senior Vice President (retired) of Xerox, now with the Wharton School. A major conference was organized to explore this topic in February 2000. A report on the meeting, entitled *Intellectual Property Rights in a Knowledge-Based Economy,* is presently undergoing review. See also the National Research Council Web site on this topic: www.nationalacademies.org/ipr.

[23]Robert Blackburn, "Intellectual Property and Biotechnology" in this volume.

[24]See the statement by David Morgenthaler, past president of the National Venture Capital Association, in Panel I of the Proceedings of National Research Council, *The Advanced Technology Program: Assessing Outcomes*, Washington, D.C.: National Academy Press, 2001. Morgenthaler notes that " [the ATP] is an excellent program for developing enabling, or platform, technologies,

be concentrated, rather than evenly spread out across different sectors.[25] As a result, areas with recognized long-term potential such as biotechnology may experience periods of very low venture capital funding. In part, this may be because they do not offer the returns perceived by investors in other areas such as information technologies and new Internet businesses. Consider, for example, that in 1999, U.S. venture capitalists invested approximately $59.5 billion, up from $4.9 billion in 1993.[26] However, eighty percent of venture capital in 1999 was concentrated on Internet and related businesses, with perhaps $1.2 billion invested in biotechnology companies.[27] Government support—through funding for promising technologies with less-assured returns—can partly compensate for this concentration.

VI. **Biotechnology and Information Technology Present New Technological Challenges and Opportunities.** There are a number of research areas in the two fields in which federal R&D funding could be very valuable. These include:

A. **Lithography**. The semiconductor industry needs new lithography technologies. Within a few years, the size of features on chips will become so small that traditional optical lithography will no longer work. The Department of Energy and the national laboratories have conducted research on extreme ultraviolet (EUV) lithography. Funded originally for defense purposes, this research is having valuable spillover applications in the semiconductor industry. To capture this potential, a government-industry partnership, in the form of a CRADA led by Intel, is collaborating with DOE researchers to advance this technology.[28]

---

which can have broad applications but are long-term risky investments. Venture capitalists are not going to fund these opportunities, because they feel that they are at too early a stage of maturity. Governments can and should fund these technologies. In fact, it should do more than it is doing" (p.67). Kenneth Flamm makes the same point in his presentation.

[25] Joshua Lerner discusses this "herding tendency" in the venture capital industry. See National Research Council, *The Advanced Technology Program, Challenges and Opportunities*, Washington, D.C.: National Academy Press, 1999, Introduction *et passim*. For a broader analysis of the venture capital industry, see Paul Gompers and Joshua Lerner, *The Venture Capital Cycle*. Cambridge, MA: The MIT Press, 2000. For a discussion of the requirements of early stage financing—in particular institutional requirements—and the limitations of the venture capital industry in this regard, see Lewis M. Branscomb and Philip E. Auerswald, *Taking Technical Risks*, Cambridge, MA: The MIT Press, 2001, Chapter 5.

[26] See the Web site of the National Venture Capital Association, www.nvca.org.

[27] *Ibid.*

[28] See David Mowery, Rosemarie Ziedonis, and Greg Linden, "National Technology Policy in Global Markets," in *Innovation Policy in the Knowledge-Based Economy*, Maryann P. Feldman and Albert N. Link, eds., Boston, Dordrecht and London: Kluwer Academic Publishers, 2001.

B. High Performance Computing. The future in high performance computing may belong to machines that are based on standard microprocessors rather than expensive specialty chips and architectures. However, harnessing the computing power of large numbers of microprocessors creates special technical problems. Federal support at DOE and NASA is proving valuable in addressing these issues.[29]

C. Genomics. The genomics revolution has created a pressing need for improvements in computational biology/bioinformatics. Progress in this new field will depend, in turn, on the ability of universities and others to build and fund multidisciplinary research teams. Yet, the current university environment does little to encourage such teamwork. Along with research needs in bioinformatics, there is also a great need to train people with skills in both biology and computing. In addition to the computing aspects of bioinformatics, important opportunities exist in combining biology and semiconductors into so-called "gene chips," or DNA analysis chips.

D. Interdisciplinary Instruction and Multidisciplinary Research. Increasingly, the domains of both biotechnology and computing cover more than one disciplinary field. Given this fact, federal policies that encourage research collaboration within universities and among universities, federal laboratories, and companies may be particularly helpful in promoting future U.S. innovation.

VII. The Allocation of Federal Funding among Fields of Research Has Shifted Sharply in the Last Decade. In recent years, federal research funding has increased substantially for a few important and promising fields—e.g., biomedicine and computer science—while it has decreased in real terms for research in the physical sciences and much of engineering. Biomedical research in 1960 accounted for approximately one-quarter of federal nondefense R&D. Today it accounts for approximately half. Meanwhile, research support for most areas of the physical sciences and engineering has fallen sharply in real terms since the early 1990s. For example, federal spending on research in the life sciences increased by 28 percent between 1993 to 1999. Federal support of mathematics and computer science also increased by 45 percent over the same period, while spending on the physical sciences was 17 percent less than in 1993. Federal funding of engineering research

[29]President's Information Technology Advisory Committee, *Information Technology Research: Investing in Our Future*, February 24, 1999, http://www.ccic.gov/ac/report/. The report recommends increasing the funding base for information technology research by $1.37 billion per year by fiscal year 2004, p. 3.

was 2 percent more in 1999 than in 1993, but chemical, electrical, and mechanical engineering research support fell more than 25 percent.[30]

It is wise, indeed necessary, to invest in a broad portfolio of research, because it is impossible to predict where breakthroughs will come, or how advances in one field will benefit another.[31] Research is also becoming more multidisciplinary. The declines in government resources for the physical sciences and certain fields of engineering run counter to continued progress in these fields and to the broader goal of effective multidisciplinary research; they also run counter to the immediate needs of biomedicine.

The development of new medicines is one of the most prominent products of medical research. The development of new medical techniques and new drugs relies heavily on contributions from a variety of sciences. The traditional method of random prospecting for a few promising chemicals has largely been replaced by methods based on molecular structures, computer-based images, and chemical theory. Similarly, synthesis of promising compounds relies on new chemical methods able to generate pure preparations in a single molecule or collections of subtle variants.[32] The exploitation of these new possibilities requires the contributions of many disciplines such as mathematics, physics, and chemistry. In short, advances in health depend on a broad range of disciplines, not just the biomedical sciences.

A case in point might be the enormous progress seen in medical diagnostics in the past two decades. Medical diagnostics now rely on imaging technologies such as ultrasound, positron-emission tomography, and computer-assisted tomography. Magnetic resonance imaging, for example, considered by many to be one of the great advances in diagnosis, is the product of atomic, nuclear and high-energy physics, quantum chemistry, computer science, cryogenics, solid state physics, and applied medicine.[33]

VIII. Priority Setting or Random Disinvestment? For the most part, the shifts in federal research spending shown in Figure 1 have not been the result of a conscious national debate on priorities.[34] The R&D budgets of most agencies were cut in real terms in the 1993-1997 period in response to

[30]See Michael McGeary, "Recent Trends in Federal Funding of Research and Development Related to Health and Information Technology," in this volume.

[31]Gordon Moore, in his concluding remarks—reported in the Proceedings, Part V of this volume—characterizes science as an "interconnected whole" in which advances depend on progress in all scientific and technical disciplines.

[32]Harold Varmus, "Squeeze on Science," *Washington Post*, October 4, 2000, p. A33.

[33]*Ibid.*

[34]One presenter at the conference wondered if the shifts in research funding constituted "random disinvestments," unintended but nonetheless injurious to national progress in R&D. See William Bonvillian, "Trends in Federal Research" in this volume. In addition, the biomedical community (scientific societies, patient advocacy groups, industry, et al.) is better organized and more effective

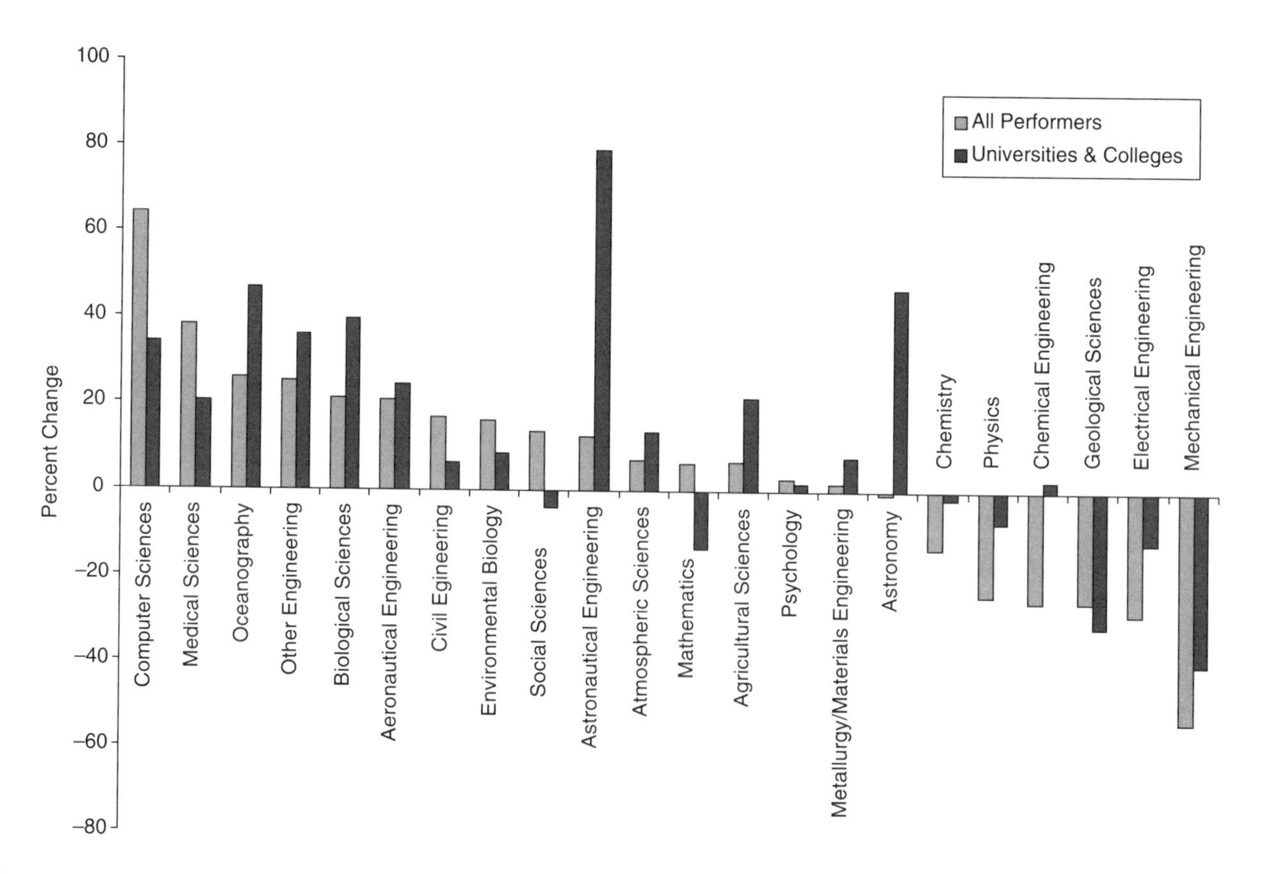

FIGURE 1 Changes in federal obligations for all performers and university/college performers, FY 1993-1999 (constant dollars).

the end of the Cold War and the national priorities in federal deficit reduction. The agencies cut research programs selectively. For example, DOD, the agency with the steepest cuts in R&D (more than 25 percent), increased funding of oceanographic research and held funding of research in computer science constant. Yet, it cut research investments in other fields, often substantially.[35] However, the net impact on the overall pattern of federal investment in research was not planned.

Agencies such as DOD and DOE reduced R&D expenditure in response to overall post-Cold War reductions. As they provided the majority of support for certain disciplines—electrical, mechanical, and materials engineering and computer science in the case of DOD, and physics in the case of DOE—changes in their priority-setting contributed to a national-level shift in federal funding from the physical sciences and engineering to the life sciences and computer science.

Indeed, shifts in agency disbursement priorities were not the result of considered government-wide reviews, either by the Executive Branch or by Congress.[36] Declining support for the physical sciences and fundamental engineering worries many observers, including many in biomedicine who know how important chemistry, physics, and mathematics have been to recent advances in molecular biology and genomics.[37] Central advances in health and communications necessarily depend on more than increased funding for biomedical research and computer science in themselves.

---

in making the political case for increased funding than their counterparts in the physical sciences. See Leon Rosenberg, op. cit., in this volume on the biomedical community's ability to make their case to policymakers. Bonvillian, *op. cit.*, makes a similar point.

[35]DOD cuts in physics and chemistry were 63 and 27 percent, respectively. Funding of electrical, chemical, mechanical, civil, and materials engineering was reduced by 40, 60, 52, 44, and 28 percent, respectively. See Michael McGeary and Stephen A. Merrill, "Recent Trends in Federal Spending on Scientific and Engineering Research: Impacts on Research Fields and Graduate Training," Appendix A in STEP, *Securing America's Industrial Strength*. Washington, D.C.: National Academy Press, 1999.

[36]Such a government-wide review has been recommended by various groups examining federal R&D priorities. See, for example, NAS [Press Committee], *Allocating Federal Funds for Science and Technology*. Washington, D.C.: National Academy Press, 1995, and most recently, United States Commission on National Security/21st Century [Rudman-Hart Commission], *Road Map for National Security: Imperative for Change*, March 15, 2001.

[37]See Harold Varmus, "The Impact of Physics on Biology and Medicine." Plenary Talk, Centennial Meeting of the American Physical Society, Atlanta, March 22, 1999. At: www.mskcc.org/medical_professionals/presidents_pages/speches/the_impact_of_physics_on_biology_and_medicine.html.

Recommendations

The recommendations that follow are based on the Committee's commissioned research, the presentation statements and suggestions made at the conference, and the Committee's considerable experience and deliberations in assessing government-industry partnerships.

I. Government and industry should expand support of research partnerships and other collaborative arrangements within and among sectors (government, industry, university, and nonprofit) and take other steps to facilitate multidisciplinary research leading to advances in biotechnology and information technology.

A. Support partnerships to exploit the genome. As a first step, the Steering Committee strongly recommends that NIH and other federal agencies work with industry and university experts to identify what technical steps are needed to ensure that the United States can fully exploit the nation's investment in the development of genomic information. For example, what kind of computing power, what advances in software, what new analytical tools (such as DNA-analyzing computing chips), and what improvements in structural biology will the U.S. need? And what research agendas in science and engineering follow from these findings? What are the "grand challenges," the most important complex problems, that arise in these fields? Several existing analyses provide good models for preparing these new roadmaps, including the semiconductor industry's technology roadmaps and those prepared by several industries in partnership with the Energy Department's Office of Industrial Technology.[38]

B. Increase government and industry support for university research in microelectronics and related disciplines. The semiconductor industry has maintained, and even accelerated Moore's Law.[39] This has led to a proliferation of computing power throughout the economy, resulting in increased productivity and economic growth.[40] To maintain Moore's Law, significant increases in funding for physical sciences and engineering—including material sciences, chemistry, physics, and electrical engineering—are needed to build greater understanding of properties of nanostructures underpinning tomorrow's information industries as well to capitalize on advances in biotechnology.[41]

C. Address unresolved questions about research partnerships. Government-university-industry research partnerships have been very productive in both the computing and biotechnology fields. However, as Congressman Boehlert pointed out at the conference, we now face several important questions regarding partnerships.[42] For example, do current partnerships help companies draw upon federally funded university basic research, or are partnerships today increasingly pulling universities away from basic research and towards applied projects of interest to established companies and to startups? How are the new intellectual property rules altering

---

[38]See www.semichips.org for the semiconductor industry roadmap. See also www.oit.doe.gov/industries.html for information on the Department of Energy efforts.

[39]See Gordon E. Moore, "Cramming more components onto integrated circuits," *Electronics*: 38(8) April 19, 1965. Here, Dr. Moore notes that "[t]he complexity for minimum component costs has increased at a rate of roughly a factor of two per year. Certainly over the short term, this rate can be expected to continue, if not to increase. Over the longer term, the rate of increase is a bit more uncertain, although there is no reason to believe it will not remain nearly constant for at least 10 years. That means by 1975, the number of components per integrated circuit for minimum cost will be 65,000." See also, Gordon E. Moore, "The Continuing Silicon Technology Evolution Inside the PC Platform," Intel Developer Update, Issue 2, October 15, 1997, where he notes that he "first observed the "doubling of transistor density on a manufactured die every year" in 1965, just four years after the first planar integrated circuit was discovered. The press called this "Moore's Law" and the name has stuck. To be honest, I did not expect this law to still be true some 30 years later, but I am now confident that it will be true for another 20 years."

[40]See Dale Jorgenson and Kevin Stiroh, "Raising the Speed Limit: U.S. Economic Growth in the Information Age," *Brookings Papers on Economic Activity* 0(1), 2000. Also see National Research Council, *Measuring and Sustaining the New Economy*, Washington, D.C.: National Academy Press, forthcoming.

[41]See Wesley M. Cohen and John Walsh, "Public Research, Patents, and Implications for Industrial R&D in the Drug, Biotechnology, Semiconductor, and Computer Industries," in this volume. Also in this volume, see Kenneth Flamm, "The Federal Partnership with U.S. Industry in U.S. Computer Research: History and Current Concern."

[42]See Congressman Sherwood Boehlert, "Opening Remarks" this volume.

relations between universities and companies and within universities? What is the impact of these new partnerships on education within universities? These are important questions, and both Congress and the Executive Branch might encourage further research on these topics.

D. Support mechanisms for conducting multidisciplinary research. Government agencies and the research community also urgently need better information on what types of research arrangements can best support and perform the new type of multidisciplinary research essential to progress in biotechnology, computing, and other "complex fields."[43] Federal support through NSF and other concerned agencies, e.g., DARPA, should encourage and support innovative university-industry-government research partnerships to address emerging problems and reduce gaps in current U.S. educational capabilities. There are promising models that can and should be evaluated. Federal laboratories, for example, have long experience with multidisciplinary teams addressing complex research problems.[44] Within academia, new models, such as those being pursued by the University of California at Berkeley, seem promising.[45] For example, several universities are building new centers that will physically co-locate researchers from many disciplines to work together on common problems in areas such as genomics. Important questions remain, however, such as:

- How best to train people in interdisciplinary fields such as bioinformatics;
- How generally to get researchers in science and engineering to work together effectively;
- How best to involve industry researchers in these university projects;
- How federal agencies can work together to support research in fields such as bioinformatics that cut across their traditional boundaries.

---

[43]See Marvin Cassman, "Exploiting the Biotechnology Revolution: Training and Tools" in this volume.

[44]In this volume, see Jane Alexander "Biofutures for Multiple Missions" and Paul Horn "Meeting Needs: Realizing the Opportunities."

[45]The Berkeley School of Public Health undertook an extensive reorganization in the fall of 1999 in order to promote these kinds of activities, including the launch a $500 million research initiative linking biological and physical sciences and engineering to spur biomedical advances. UC Berkeley plans to involve as many as 400 researchers—in fields as diverse as public health, psychology, physics, chemistry, engineering, mathematics, and computer science—in multidisciplinary research. Also involved in this effort are scientists at the Lawrence Berkeley National Laboratory. [UC Berkeley News Release 10/6/99—Public Affairs (510) 642-3734]. In this volume, see Ed Penhoet "The View from the Biotechnology Industry," for a description.

Federal agencies, scientific and technical societies, and industry associations all have a stake in developing good answers to such questions.

E. Develop public-private technology roadmaps. Federal agencies should be encouraged to work with industry associations and the scientific and engineering communities to develop technology roadmaps in important interdisciplinary fields such as genomics and bioinformatics, nanotechnology, and advanced information technology, including optoelectronics. The country needs to know what major technical barriers exist and what types of research are thus needed in order to ensure continued progress, and what investments in scientific and engineering research will be required.[46]

II. The scientific community, U.S. industry, and the federal government should explicitly examine the implications of recent shifts in the allocation of federal investment among fields, especially the decline in federal funding for non-defense fundamental research in the physical sciences and engineering, and address possible solutions.

A. Conduct a comprehensive review of the federal research portfolio. Given that future progress in complex fields (such as genomics and next-generation electronics) will depend increasingly on whether the United States is strong in a wide range of scientific and engineering disciplines, the Administration should commission a comprehensive review of how post-Cold War budget trends have affected U.S. research capabilities and personnel in key areas of science and engineering. In the post-Cold War and budget-constrained environment of the early and mid-nineties, many agencies cut or increased their research budgets for mission-related reasons. Those decisions, made independently by each R&D agency in line with its particular priorities, have had major impacts on the nation's overall research capabilities in science and engineering. The Administration could conduct this review internally, either through the appropriate White House coordinating councils, or by contracting with an outside group. In either case, the review should actively solicit analyses and documentation from experts in science and engineering, from U.S. industry, and from government experts. The review must include a review of needs in engineering research as well as scientific research, given the importance of new engineering work in providing the instruments, analytic tools, software, and other new technologies

---

[46]In some cases, existing programs, such as ATP, may prove to be a means of leveraging public-private resources to bring new technologies forward to the market. See the statement by Jeffrey Schloss, "Mission Synergies" in National Research Council, *The Advanced Technology Program: Challenges and Opportunities*, Washington, D.C.: National Academy Press, 1999, pages 56-59.

essential to progress in fields such as genomics, nanotechnology, and computing.

B. Establish an "alert system." The nation would benefit from an "alert system" that would notify government when important disciplines are getting below critical mass.[47] "Benchmarking" projects now underway at, for example, the National Research Council is a possible start in this direction.[48]

C. Make the case for increased support. If scientific and engineering societies and industry believe the federal government should increase funding for the physical sciences and engineering, they will need to make a strong case to the public and Congress for why such increases are in the nation's interest.

III. Federal policymakers should support an infrastructure and create an environment conducive to research partnerships and other collaborative arrangements.[49]

A. Build interdisciplinary competence for multidisciplinary research. Federal agencies can readily undertake some steps that would help build the interdisciplinary competence of researchers in fields such as bioinformatics.[50]

---

[47]Bonvillian, *op. cit,* describes a possible "alert system."

[48]National Research Council, *Experiments in International Benchmarking of U.S. Research Fields*, Committee on Science, Engineering, and Public Policy, Washington, D.C.: National Academy Press, 2000. See also Stephen A. Merrill and Michael McGeary, "Who's Balancing the Federal Research Portfolio and How?" *Science*, vol. 285, September 10, 1999, pp. 1679-1680.

[49]Additional considerations relevant to facilitating multidisciplinary research, such as that required for bioinformatics would include the need for universities to be more flexible with their curriculum since it is unlikely that large numbers of students will seek to gain separate advanced degrees in two or more disciplines. Programs that allow students to develop sound interdisciplinary backgrounds will provide the *knowledge base* for multidisciplinary research. The need to consider the risks firms face in developing software for bioinformatics is also important. Improving data availability, quality, and interoperability can help reduce risks inherent in *funding* multidisciplinary research and can hence help increase the level of private sector investments. In addition, rules on intellectual property can affect the free flow and availability of scientific results. Multidisciplinary research might benefit from *improved patenting rules and policies* as to what constitutes an allowable patent.

[50]The FY 2002 President's Budget request provides funds to the NIH to make high-end instrumentation available to a broad community of basic and clinical scientists, and requests $40.2 million for the National Institute of Biomedical Imaging and Bioengineering (NIBIB). This new institute is to develop new knowledge, create new technologies, and train researchers to integrate quantitative science with biomedical research. In addition, the NIH proposes to hold State-of-the-Science meetings to develop new multidisciplinary research initiatives and to recruit new investigators from bioengineering, biocomputing, and related disciplines to focus on technology development and application. See NIH Press Release for the FY 2002 President's Budget, April 9, 2001.

For example, "glue grants," which encourage researchers from different disciplines to work together on complex problems, can be a low-cost but effective way to encourage multidisciplinary research.[51] Short summer courses that bring biologists and computer scientists together also would be valuable. Any resulting initiatives in bioinformatics should explicitly consider the training of bioinformatics experts as well as R&D needs and opportunities.

B. Increase support for interdisciplinary training of graduate students. A key element of national success in new, complex areas such as bioinformatics is training adequate numbers of high-quality graduate students in these fields.[52] One problem is that while financial support for graduate students is generally available in the life sciences, such support is becoming more rare in important engineering disciplines. One result is that major U.S. companies are now establishing engineering research centers overseas largely because not enough Ph.D. engineers are available now in the United States. In the future, another result could be a lack of good engineering students and faculty to work on problems such as bioinformatics and nanotechnology. NSF and industry associations should work together to gather detailed information on these issues and consider appropriate steps.

C. Review the impact of patent decisions on technological progress. Congress and the Executive Branch also may wish to participate actively in the NRC review of the impact of recent patent decisions (PTO decisions and court decisions) on innovation in both biotechnology and information technology.[53] Uncertainty concerning the scope of new patents in areas such as biotechnology research tools, as well as risks perceived as arising from the proprietary nature of such patents could become deterrents to future innovation in these industries.[54]

---

[51]Cassman, *op. cit.* describes "glue grants" at NIH.

[52]See Paula Stephan, op. cit., in this volume. See also, National Research Council, Bioinformatics: Converting Data to Knowledge. Washington, D.C.: National Academy Press, 2000, p. 9, for a discussion on the need for trained personnel in bioinformatics.

[53]See National Research Council, *Securing America's Industrial Strength*, Washington, D.C.: National Academy Press, 1999. The STEP Board has an active intellectual property rights (IPR) agenda. Its report, cited here, notes that strengthening and extending IPRs are appropriate policies for advanced industrial economies where intellectual assets are a principal source of growth. It also recognizes the growing friction over the assertion and exercise of IPRs, noting claims that in some circumstances, such practices discourage research.

[54]In cases where research and scientific advance is necessarily multidisciplinary, policies adopted to encourage advance in one discipline area can have broader unintended effects. If, for example, proprietary ownership of fundamental biological data poses technology risks for those creating software for bioinformatics, incentives to develop needed research tools may not appear.

CONCLUSION

The Committee believes that if the government, in partnership with industry, universities, and nonprofit research institutes, adopts these recommendations, the nation will experience greater progress in the research that underlies innovation. This progress will occur not only in the fields of biotechnology and information technology, but in other fields as well, because of the interconnectedness of scientific knowledge. The Committee believes that, in addition to changes in the organization of research, the nation must provide greater support for research across a broad portfolio of fields and disciplines in order to capitalize effectively on existing research investments and to ensure continued benefits for future generations.

IV

PROCEEDINGS

Welcome

Dale Jorgenson
Harvard University

On behalf of the Board on Science, Technology, and Economic Policy, it is my pleasure to welcome you to today's conference on "Government-Industry Partnerships in Biotechnology and Computing." This conference encompasses two of our most innovative industries—industries that are key to future economic growth and social advancement. We are delighted to have an especially distinguished group of speakers today. Indeed, the quality of the speakers participating underscores the importance of the topics we will discuss.

I chair the STEP Board, which was established in 1991 to improve policymakers' understanding of the interconnections among science, technology, and economic policies and their importance to the U.S. economy. A distinctive feature of the Board is that it brings together economists, industrialists, and policymakers to explore policy issues of growing importance to economic growth.

Examples of recent STEP work include ***Securing America's Industrial Strength*** and ***U.S. Industry in 2000: Studies in Competitive Performance***. In these volumes, the STEP Board assesses the competitive performance of 11 sectors of the U.S. economy. We carried out the assessment recognizing that a decade ago, dire forecasts for the U.S. economy were quite pervasive. The picture has changed dramatically. We wanted to know how and why. And we wanted to determine what must we do to maintain our current era of strong growth. Our study also revealed imbalances in public-sector investment in R&D. Bill Bonvillian of Senator Lieberman's staff will address this point later this morning.

Today, STEP is releasing two reports. The first is a summary of a workshop held this past March on the Advanced Technology Program (ATP). The volume is titled *The Advanced Technology Program: Challenges and Opportunities,* and we are pleased to have it out just 6 months after the event. It represents one of the few efforts to examine closely the origins, objectives, and current operations of the ATP. As many of you may know, ATP attracts its share of controversy. Partly because of that, the program has commissioned a rigorous assessment effort in cooperation with the National Bureau of Economic Research. STEP plans to draw on this rich assessment program in its review of the program's operations. We plan to produce a report next year with findings and recommendations concerning this innovative program.

The second report is titled *The Small Business Innovation Research Program: Challenges and Opportunities*. Like the ATP volume, this report summarizes a workshop held to review the program's origins and the current research on its operations. The workshop also explored operational challenges faced by SBIR, identified some SBIR successes, and discussed possible improvements in the program. The volume also includes papers by Roland Tibbetts, here today, and Josh Lerner of the Harvard Business School.

One outcome of that workshop was a second symposium on SBIR, in which we looked at a specific effort within the Defense Department to improve the commercialization of SBIR-funded technologies. That effort is known as the Fast Track initiative. The STEP Board commissioned research assessing Fast Track's role in encouraging commercialization of SBIR technologies funded by the Defense Department. That report should be available by the end of the year.

The ATP and SBIR volumes are part of a larger project within STEP called "Government-Industry Partnerships for the Development of New Technologies." Gordon Moore of Intel leads this project, and we are very fortunate to have him with us today and tomorrow. It is under the auspices of this project that our meeting today is being held, and I want to describe for you some of the "Government-Industry Partnership" project's work.

The basic good goal of the project is to identify "best practices" from the many government-industry partnerships underway in the United States and abroad. As a starting point, it is important to recognize the role that government has frequently played in supporting the development of new technologies in this country. Outlined in Alexander Hamilton's 1791 *Report on Manufacturers*, federal initiatives provided Eli Whitney a contract in 1798 for interchangeable musket parts and Samuel Morse's grant from Congress in 1842 for a demonstration project to run a telegraph line between Washington and Baltimore. Following World War I, the government-initiated Radio Corporation of America facilitated the development of radio. This initiative was privatized and became the RCA Corporation.

In recent years, government-industry partnerships have continued to play a prominent role in the growth of the U.S. economy. The development of the Internet stands out as an example. Despite the success of many of these programs—and the failure of others—there has been relatively little effort to take a step back and see why some partnerships have been successful, why some have not, and objectively analyze the reasons underlying success and failure.

To address this, the STEP Board has been holding a series of "fact-finding" workshops on a variety of partnership programs. As I have noted, the reports on ATP and SBIR are products of these "fact-finding" workshops.

We also held a conference last year on international science and technology cooperation focused on the U.S.-E.U. Science and Technology Agreement. The conference resulted in specific agreements to enhance cooperation with our European friends. A summary of the conference with the agreement and an overview of the EU Fifth Framework Program are included in our volume, *New Vistas in Transatlantic Science and Technology Cooperation.*

On September 17, 1999, we issued a workshop report on cooperation between industry and the national laboratories. Sandia National Laboratories asked STEP to gather economists and policymakers to discuss the labs' plans to develop a science and technology park adjacent to Sandia. The S&T park is seen as a potentially effective instrument to enhance Sandia's ability to fulfill its mission through partnerships with the private sector. The report summarizing the workshop is titled *Industry-Laboratory Partnerships: A Review of the Sandia Science and Technology Park Initiative.*

All the publications I have mentioned have been published since June. All are available today at the conference table.

Over the next day-and-a-half, we will talk a lot about biotechnology and computing, with an aim toward identifying synergies between the two sectors, future challenges, and perhaps most importantly, future challenges for policymakers. Based on the deliberations from this meeting, Dr. Moore's Steering Committee will develop findings and recommendations about the issues discussed in the course of the conference. Your participation in this process is therefore especially welcome.

In developing our report, the STEP Board is keeping with its tradition of providing a forum through which pragmatic policy recommendations are channeled to key decision makers in Congress and the Executive Branch. To be effective in doing this, we need the input of the many distinguished individuals on our program and present here in the audience.

I am pleased to turn the podium over to Bill Spencer, Chairman of SEMATECH, who will describe in greater detail what we hope to accomplish today and tomorrow. Bill is the Vice Chair of the STEP Board and also serves as Vice Chair of the Steering Committee for the "Government-Industry Partnerships" project. As many of you know, Bill was CEO of SEMATECH from 1990 to 1996, and before that directed the R&D operations of Xerox. Bill has been instrumental in putting this conference together. It is with pleasure that I present him to you to describe more fully our goals for the next day-and-a-half.

Introduction to the Symposium

William J. Spencer
SEMATECH

Dr. Spencer welcomed participants by observing that the day's symposium was originally conceived as an examination of the broad area of government-industry partnerships in the health sciences and technology as well as information sciences and technology. In an effort to narrow the symposium's scope, the STEP Board held a workshop in the summer of 1999 at which it was decided to focus on biotechnology and computing.

The two areas were chosen because of their rapid growth, their large contribution to job creation, and because approximately two-thirds of the nation's research and development (R&D) budget of $225 billion is spent in these two sectors. The R&D investment in biotechnology and computing is growing more rapidly than R&D investment in the country as a whole. The role of the government, industry, and universities is different in the biotechnology and computing industries. There are, however, areas for potential interdependencies and areas for synergy.

The deliberations of the symposium form the basis of the report herein by the STEP Board containing formal findings and recommendations. Echoing Dr. Jorgenson's statements, Dr. Spencer encouraged the audience to offer questions and suggestions as the day progressed, as this input would aid the STEP Board in developing its recommendations.

Dr. Spencer then introduced the day's keynote speaker, Congressman Sherwood Boehlert (R-NY). As a senior Republican Member on the House Committee on Science, Rep. Boehlert is well positioned to comment on and pose questions about the topics of the symposium. First elected in 1982, Rep. Boehlert has concentrated on what he calls "the three E's"—education, the environment, and the economy. Among his many distinctions, Rep. Boehlert has become a respected voice for environmental protection in his state as well as the nation at large. By virtue of his thoughtfulness and seniority, he has also become a positive force for the nation's scientific enterprise.

Opening Remarks

Congressman Sherwood Boehlert

It's a pleasure to be here this morning to open this timely conference on partnerships in biotechnology and computing.

It would be hard to think of two fields more deserving of the National Academy's attention—or, for that matter, of Congress's attention. The health and information technologies sectors—however defined—are two of the most dynamic in our economy. Some economists estimate that information technology alone has been responsible for fully one-third of our economic expansion since 1992.

But noting the economic significance of these two fields only begins to hint at their impact. Advances in biotechnology and computing have fundamentally changed, and continue to alter the very structure of our lives. To take just one obvious example, our workplaces look, sound, and feel different than they did just a few years ago—and perhaps are in different locations and require different skills—because of the advent of the personal computer.

(As an aside, I should admit that this is not so true of my own personal workplace, in which the color television is the most conspicuous technological advance, but it is true for the rest of my office. And my wife loves the Internet.)

More profoundly, biotechnology and computing are changing the ways we understand the world—altering our sense of what it means for something to be a text, or a piece of information, or an organism.

Like the automobile or the atomic bomb—to take two very different but equally powerful examples—biotechnology and computing will change—are changing—everything. They are literally altering our physical and mental landscapes.

But I'm not here today to discuss the social implications of biotech and computing—although they must be kept in mind—but rather to focus on the federal government's role in advancing these two fields.

As everyone here knows, the federal government has had an enormous role in spawning and supporting these two fields. And the good news is that this support can be expected to continue.

The trickiest questions do not concern whether the federal government should support research in these areas. Rather, the open issues are how much money the federal government should devote to research, how the government should balance its research portfolio, and how the government should structure its research arrangements. Let me give you a sort of "status report" on where the policy debate stands on each of those questions, and then, hopefully, you'll spend the next day-and-a-half moving that debate forward.

On the state of the overall research budget, I can bring you good news and bad news. The good news is that research programs, at least non-defense research programs, have fared pretty well over the past decade. Despite the vagaries of the budget process and the shifting power structure of Washington, non-defense R&D spending (which is mostly R) accounted for a slightly higher share of the federal budget in fiscal 1998 than it did in fiscal 1990—and of course, in 1998, that was a share of a larger budget.

A major reason for that relatively strong showing is that support for research spending in Congress is broad, if not necessarily deep. There is no faction in Congress out to make its reputation by eviscerating research programs, which are widely viewed as "the goose that laid the golden egg"—or to use another popular metaphor, as our "seed corn," which I guess is what is fed to the goose.

That generous assumption about research is holding true this year even as Congress and the Administration have hunkered down for a protracted war of words over the budget. Last week, the President signed the VA-HUD-Independent Agencies appropriations bill, for example, which included a 7 percent increase for the research programs at the National Science Foundation (NSF). That increase, by the way, included major new funding initiatives in biotechnology and computing.

So then what's the bad news? I think it is twofold. One is that a few programs have been notably excluded from the pattern of growth, particularly basic research programs at the Department of Defense, which have been so important historically in bolstering mathematics, engineering, and computing research.

The second piece of bad news is only relatively bad news, which is that I think it's unrealistic to expect a significant acceleration in the growth of research dollars. When the research community starts banking on a doubling of science spending, I think it may be guilty of what Alan Greenspan infamously referred to as "irrational exuberance."

Yes, surpluses are predicted, but spending caps remain in place for the next couple of years, tax cuts of some size remain on the agenda, and there are plenty of competing programs for whatever money is set aside for new spending. Now I know that the science and engineering community tends to feel—with some reason—that every field and sub-discipline is underfunded—that all the children are below average, to paraphrase Garrison Keillor—but money is not likely to become available to "fully fund" every field—even if we could agree on what that would mean.

So that means that we—the research community, the Administration, the Congress—are going to have to continue to make tough choices, to set priorities, to figure out how to balance the federal research portfolio.

I know that one reason the Academy has arranged this conference is because of a growing sense among some that biomedical research is consuming too much of the federal research budget. And indeed health research's share of the federal non-defense R&D budget has grown from less than one-quarter in 1960 to close to one-half this year.

I think it's extremely timely for the Academy to look at this question because I must say the disproportionate growth of health research spending is not exactly causing consternation on Capitol Hill. As we prepare to debate the Labor, Human Services, and, Education appropriation for fiscal 2000, there is a bidding war underway to see who can take credit for adding the most money to the National Institutes of Health (NIH). And the last time Congress asked the Academy to do a report on setting research priorities, it was because the bipartisan team of Senators Hatfield and Harkin thought that biomedical research was underfunded.

It's not hard to explain how we arrived at this state of affairs. First, almost no one in Congress looks at the overall research budget—that's just not how we're set up. Second, everyone in Congress has a body—you may have your own view about whether each of us has a mind, but unarguably each Member of Congress has a body. So the goals of health research are innately understandable and appealing. Third, the political appeal of competing goals—such as fighting the Cold War—has evaporated; only the war against the common cold has been left standing. And finally, and this is of great relevance to your work today, the scientific community—whether academic or industrial—hasn't clearly told us to do anything different.

I don't want to suggest that we're simply awaiting your instructions—we only wait for opinion polls of the uninformed these days. But it is significant that despite the grousing one hears about, no one has clearly made an issue of the balance of the research portfolio. The fairly common understanding in the research community that further advances in biotech may depend on movement in other fields, especially information technology, would come as news to most

Members of Congress. Nor are most Members aware of the multidisciplinary research funded by some of our biomedical programs.

I should point out, though, that there probably is not an era of "ideal balance" on which to look back. Did we strike the right balance in the mid-1960s, when the lion's share of non-defense R&D money went to space research? I don't know. The category of "space" covered a multitude of sins, including a lot of computing research. But the point is one sector was leading the pack by far in that "golden age" of federal funding. And, of course, the balance between defense and non-defense research has been the subject of continuing debate.

Let me also note that computing research has also been a winner in the federal sweepstakes relative to many other areas. In the NSF's proposed budget for fiscal 2000, for example, the computer sciences directorate was slated to get a 41.5 percent increase—far outstripping the runner-up, biological sciences, literally by an order of magnitude.

And one last note on spending: biotech and computing funding raise different policy issues in the minds of many Members of Congress. Biotech—whether in the pharmaceutical or agricultural fields—exists in a web of public regulation—a sign of a public, civic interest in those areas that for better and worse, does not extend to computing. It's no accident that we have difficult ethical debates in Congress about aspects of biotechnology research but not about computing—even though its effect on our daily lives may be equally profound.

Furthermore, biotech research is much more likely to be seen as truly "basic"—whatever that means nowadays—and therefore more clearly deserving of government support. It's more difficult to understand what is "basic" about computing research and why it shouldn't be left to corporations or procurement programs.

The blurring line between basic and applied research, and between R&D, in computing and biotech is one reason for the growing number of partnerships between government, industry, and academia in these areas—another timely focus of your conference. Partnerships have become a "buzz word" on Capitol Hill, as they have in the "outside world." More and more agency programs encourage partnerships, both at federal labs and universities.

But frankly I don't think we've spent enough time considering the policy implications of this newfound interest in collaboration. There is no incentive on either the corporate or academic side of the relationship to assess fully the impact this may be having on our research endeavors, particularly at universities.

Universities have never even admitted how fundamentally the postwar advent of massive federal research funding has changed the institution, often contending that they display the same collegiality and attention to students that they did when they arose in medieval Italy. Life inside the Beltway notwithstanding, a firm attachment to convenient illusions is a poor substitute for analysis.

I think that now is a great time to examine the idea of partnerships—when the idea has been developed and implemented enough to provide some real data, but not so far along as to be frozen in one particular form or to have become irreversible. So here are some questions I have about partnerships, in addition to the obvious one about whether they produce the intended results. Let me emphasize that they are questions; I am genuinely unsure of the answers.

First, are partnerships more likely to be a means for companies to take advantage of basic research or are they more likely to skew government and academic research toward more near-term questions? Businesses are coming to universities, in part, because they recognize the importance of fundamental research but are less likely than ever to fund it. It would be ironic if the end result was to pull universities away from that very research.

Second, in what ways are intellectual property rights concerns altering the nature of the business-university relationship and relationships within universities? In the new world of partnerships have laws like Bayh-Dole become counterproductive by making universities competitors in the intellectual property arena, or more important than ever, by granting property rights to a public or quasi-public entity?

Third, and perhaps most important, how have partnerships altered the relationship between professor and student? Have partnerships tended to broaden a student's experience by exposing him or her to corporate issues, or have they interfered by limiting communications with faculty or publishing rights?

I realize it can be easy to romanticize this last question. A young professor told someone I know that he is always careful in setting up partnerships not to allow his students to become "slave labor for the company." I wondered if that was because in the postwar era, graduate students were supposed to be slave labor for the university. Still, the impact on education is the critical question, especially since companies often claim that what they want most out of partnerships is better prepared students. Are partnerships set up to actually improve graduate and undergraduate education?

I think I'll leave you with those questions. Too often Members of Congress offer up ill-informed answers that leave audiences with only one question: "Did the Founding Fathers make a mistake?"

This conference will offer up the views of some of the leading lights in industry, academia, and government, who will be able to explore these issues in ways that I can only begin to imagine. And I am eager to read the report of what results.

Kurt Vonnegut once defined the "information revolution" as the "idea that human beings could actually know what they're talking about if they really want to." That's an "information revolution" this conference should advance regardless of its conclusions about any particular technologies. Thank you.

QUESTIONS FROM THE AUDIENCE

Dr. Stephen Dahms of San Diego State University recalled a meeting recently in San Diego hosted by the Chairman of Qualcomm on behalf of the Chairman of the House Committee on Science, James Sensenbrenner. Rep. Sensenbrenner discussed new legislation designed to increase funding for information technology R&D, particularly research funded at the National Science Foundation. At the meeting, Rep. Sensenbrenner was asked about funding for biotechnology research. In response, according to Dr. Dahms, Rep. Sensenbrenner noted the importance of biotechnology, but the congressman viewed it as medical research, not as part of science R&D as traditionally understood. Dr. Dahms asked Congressman Boehlert what the scientific community needed to do to break down such interdisciplinary barriers.

In response, Rep. Boehlert recalled advice he had given to the president of Cornell University, which is in his district. The university's leadership and scientists would make the case for funding various projects, and Rep. Boehlert would point out that they had his support, but they had to talk with other Members of Congress. This example, Rep. Boehlert said, pointed to the fact that the scientific community "is not particularly adept at lobbying for its own interests." Scientists spend far too much time talking with Members who are already sympathetic to their interests, but not nearly enough time educating new Members or Members whose committee assignments do not relate to science. Scientists, Rep. Boehlert continued, must explain to rank-and-file Members of Congress just how important their work is to the nation.

Panel I: Biotechnology Information Technologies: The Need for a Diversified Federal Research Portfolio

INTRODUCTION

Clark McFadden,
Dewey Ballantine

Mr. McFadden opened the panel by observing that biotechnology and computing are two of the most dynamic and penetrating areas of technology. The purpose of Panel I is to compare and contrast technology development in these two areas as it has been affected by government-industry interaction. In computing and semiconductor technology, the government was an early instigator, broadly sponsoring technology development and purchasing much of the early products of the computing and semiconductor industries. Over the past several decades, the role of government in the development of information technology has diminished relative to commercial activity. Although more narrowly focused, government-industry collaboration in information technology still accounts for a substantial portion of infrastructure development and, in a variety of ways, has contributed to significant technology developments.

The government's contribution in the life sciences has been even more substantial, and has increased steadily in the last decade. The government's financing and shaping of R&D in pharmaceuticals and biotechnology has had a central impact on these technologies, an impact that has been further accentuated by the government's role in regulating emerging products in these sectors. An understanding of how government-industry collaboration has contributed to technology in both areas is essential to a rigorous assessment of government-industry partnerships. As part of this assessment, it is also necessary to examine whether the government is striking the right balance across these two sectors in its R&D

portfolio. This is an important issue for Congress, Mr. McFadden emphasized, and one that Rep. Boehlert raised in his remarks.

To kick off the discussion on the funding R&D in the biotechnology and computing industries, Mr. McFadden said that the symposium was fortunate to have two men who have made landmark contributions in their respective industries. Both have been innovators and leaders of technology enterprises. Gordon Moore is co-founder of Intel Corporation and serves as its Chairman Emeritus, in addition to serving as Chairman of the Board of Trustees at the California Institute of Technology. His knowledge of semiconductor technology is so intimate that he has characterized the technology's path in a few words, which have become known famously as Moore's Law.[1] Articulating simple but profound insights about technology has been the hallmark of Gordon Moore's career.

Edward Penhoet is Dean of the School of Public Health at the University of California at Berkeley. He is also the founder and was, for many years, the chief executive officer of Chiron Corporation, a leading biotechnology company. Dr. Penhoet combines insight into technology with achievement in technology development. He has a commitment to promoting multidisciplinary research in the biotechnology and computing industries, and, under his leadership, Berkeley's Health Sciences Initiative represents a major initiative to promote such research.

THE VIEW FROM THE SEMICONDUCTOR INDUSTRY

Gordon Moore
Intel Corporation

Dr. Moore said his remarks would focus on his perceptions of how the semiconductor industry—a key input to computers—has developed over his 40-year career watching the industry evolve and contributing to its growth. The birth of the computing industry was driven by the Army's need for ballistics tables during World War II. Calculating the trajectory of a projectile was extremely time consuming; it took a skilled person approximately 20 hours to calculate the one-minute path of a projectile.

Origins of the Computer and the Transistor

At the University of Pennsylvania, the Moore School of Engineering took on the task of trying to reduce the time needed for this calculation. The differential analyzer, invented by Vannevar Bush, had been invented by World War II and could reduce that to 15 minutes. At the Moore School, J. Presper Eckert and

---

[1]See Gordon E. Moore, "Cramming More Components Onto Integrated Circuits," *Electronics:* 38(8) April 19, 1965.

John Mauchly built an electronic computer using vacuum tubes that could calculate the one-minute trajectory in 30 seconds. This was a major breakthrough in an important area for the military. As a result, the military financed the computer in its early days and helped spur the industry.

The transistor had a different development path. It was invented at Bell Laboratories in late 1947 and early 1948 and the transistor's invention was the outcome of a search for solutions to problems in the telephone system. In particular, the transistor addressed the lack of reliability in amplifiers and undersea cables. A solid state amplifier, as the transistor enabled, was thought to offer important technological advantages. Bell Labs had a preeminent position in U.S. technology at this time; as a regulated monopoly, the Bell System was able to maintain and build the laboratories through the rate structure. Even with the advantages of rate regulation, approximately 48 percent of Bell Labs' budget was government supported from 1949 to 1959. Even so, the transistor was developed using Bell Labs' own funds.

One of the most important developments for the commercial semiconductor industry, continued Dr. Moore, was the antitrust suit filed against Western Electric in 1949 by the Department of Justice. It resulted in a consent decree in 1956 that required Western Electric to license all of its patents to any domestic company royalty-free. This included the early transistor patents. For future patents in semiconductors, Western Electric was required to license them to domestic firms at "reasonable royalty rates." Under these conditions, Bell Labs essentially became a national industrial research facility.

This allowed the merchant semiconductor industry "to really get started" in the United States, said Dr. Moore. If one looks at the path of silicon semiconductors, there is a direct connection between the liberal licensing policies of Bell Labs and people such as Gordon Teall leaving Bell Labs to start Texas Instruments and William Shockley doing the same thing to start, with the support of Beckman Instruments, Shockley Semiconductor in Palo Alto. This, noted Dr. Moore, started the growth of Silicon Valley.

The Role of the Government

Up to this point of the story, most of the financing of the transistor's development and the semiconductor was done privately by commercial interests. However, the military "really liked the transistor," Dr. Moore added, and each of the branches provided some support for the early R&D. One result of the military's interest was that semiconductor technology was driven to silicon, not germanium; silicon was functional in a temperature range that suited the military's weaponry needs. The military services actually competed with one another in supporting various research efforts. The Army Signal Corps, for example, put out specifications for about a dozen different types of transistors, and thereby encouraged the industry to compete vigorously "to jump through technological hoops to realize the specifications." As it turns out, none of the specifications ever became important to the commercial industry.

The government R&D funds did, however, redirect the priority of universities and this created long-lasting benefits to the semiconductor industry. Semiconductor technology is about getting electrical signals to function properly using chemical properties. At the time, in most universities, chemists knew relatively little about electricity, and electrical engineers knew relatively little about chemistry. Government R&D resources funded a number of programs in universities that helped build the necessary bridges across disciplines.

As Michael Borrus points out in *Competing for Control*, very few of the innovations in the semiconductor industry came about because of funding from the Department of Defense. Semiconductor technology was advancing so rapidly that people in the commercial sector were in the best position to push the frontiers of the technology.[2] Further, Mr. Borrus made the point that companies pursue research and product development where anticipated returns are highest, and that government support may or may not be helpful in advancing these goals.

From Dr. Moore's perspective, flexibility and timing are critical elements to successful technological development. A government contract takes a long time to negotiate, and often the goal is obsolete by the time the contract is completed. A government contract commits resources with very little flexibility; such contracts tie up key personnel for a year or two without the ability to redirect them if ongoing developments suggest that other research paths are more promising. Moreover, government contracts impose a significant administrative burden. For these reasons, most players in the mainstream semiconductor business took very few government contracts during the 1950s, and when the industry did accept such contracts, it did so reluctantly.

However, the government made one very important contribution to the industry: it served as a market for high-priced semiconductor devices. The cost of the early silicon devices was so high that the consumer demand was insufficient to support volume production. But the military, with its unique requirements, was willing to pay significant premiums for devices. This provided the industry significant resources for R&D that eventually enabled low-cost devices suitable for the commercial market. The existence of the military market in the United States, Dr. Moore said, was far more important than military contracts. For example, at Fairchild Semiconductor, a company Dr. Moore co-founded, the first 100 transistors were sold at $150 apiece to a military program at IBM.

The Minuteman I and Minuteman II missile programs were also very important. Minuteman I had a large demand for silicon semiconductors and the missile required high device reliability; meeting those reliability specifications significantly enhanced the industry's technological capabilities. Among other technol-

---

[2]See Michael Borrus, *Competing for Control,* Cambridge, MA: Ballinger Publishing, 1988.

ogy contributions, Minuteman I helped establish planar technology, a technique for manufacturing integrated circuits that the industry has been building on ever since. In Minuteman II, the military made a strong commitment to integrated circuit technology, and the demand for integrated circuits in large volume greatly contributed to a wide variety of improvements in integrated circuit technology.

The Growth of Commercial Markets

By the mid-1960s, the importance of government programs on the mainstream industry declined. The commercial market forces were by then sufficient to drive technology development. The "insertion time" of new products into military systems was so long that by the time a device was ready for the military, it was obsolete. As a result, the military's needs began to be met by a few specialized companies, or by the in-house capacity of the major defense contractors. This meant that the commercial semiconductor industry developed relatively independently.

Also during the 1960s, government support of R&D in the universities—done mainly by DoD and NSF—increasingly migrated to "non-mainstream projects" such as non-silicon semiconductors. This research investigated interesting physical phenomena, but it had little bearing on the mainstream commercial industry, which was silicon-based. The semiconductor industry benefited from the students trained under the contracts, although companies often had to retrain them a bit to acclimate them to silicon technologies. However, the utility of the research was of relatively little interest to the industry.

As a result, the industry created the Semiconductor Research Corporation (SRC) in the late 1970s, a collective effort by semiconductor firms to address unmet R&D needs. The industry decided to pool the funds it was currently putting into universities, augment it a bit, and support research relevant to the growing commercial semiconductor industry. The general rationale was to provide incentives for university professors to conduct market-relevant research. It was thought, Dr. Moore said, that given a choice between two equally interesting research projects, most professors would choose to pursue the one with practical application. By trying to capture "mind share" within the university environment, the industry hoped to channel more R&D effort to areas of use to the commercial semiconductor sector.

These efforts have been successful, Dr. Moore said. Today, through the SRC, industry funds $30 million per year in university R&D, and over time the SRC has helped ensure that a substantial portion of semiconductor research conducted in universities is relevant to industrial needs. In sum, Dr. Moore considered the SRC to be a very successful program.

The Challenges of the 1980s

The decade of 1980s was easily the most difficult one that the U.S. semiconductor industry has faced, said Dr. Moore. By every measure, Japanese firms were doing a better job of manufacturing semiconductors than U.S. firms. Using the same equipment, Japanese firms would maintain production 98 percent of the time, while U.S. firms would maintain production 85 percent of the time. U.S. firms would produce 20 wafers per hour, while Japanese firms would produce 40 wafers per hour on the same equipment. The United States trailed the Japanese in manufacturing by virtually every measure. A disadvantage in manufacturing technology is extremely alarming in an industry such as semiconductors, in which volume and reliability are crucial determinants of competitiveness.

The U.S. semiconductor industry chose cooperation as a means to address common problems, and the industry sought government support to help address these common problems in manufacturing technology. An important catalyst to cooperation was the National Cooperative Research Act (NCRA), which was passed in 1984 and allowed firms to talk to one another about research problems. This was very important, Dr. Moore said, because prior to that, the potential antitrust ramifications were simply too great to allow firms to discuss common research challenges. As a result, semiconductor firms' R&D efforts went in separate directions, with a tremendous amount of duplication.

The Role of SEMATECH

Aided by the NCRA, and as a result of studies by the Defense Science Board and the Semiconductor Industry Association (SIA), the industry established SEMATECH in 1987 to address competitiveness issues that the semiconductor industry faced. Initially, the industry planned for SEMATECH to work on process technology and build a fabrication facility for the industry. That turned out not to be the right approach, Dr. Moore recalled, because it was expensive, duplicated a variety of internal efforts in the industry, and required the sharing of proprietary information that individual firms were not willing to release to competitors. After several iterations of mission statements, SEMATECH settled upon a mission that involved working closely with semiconductor equipment manufacturers, a sector on which the semiconductor industry was very dependent and in which quality had been suffering.[3]

---

[3]SEMATECH's mission set targets for developing equipment that would manufacture chips at line-widths comparable to the worldwide state of the art. By the end of 1992—the promised date—SEMATECH did meet the target of 0.25 microns. The fact that SEMATECH was able to change its mission in response to changing industry needs has been viewed as one of the consortium's strengths. For a comprehensive assessment of SEMATECH, see John B. Horrigan, "Cooperating Competitors: A Comparison of MCC and SEMATECH," monograph, Washington, D.C., National Research Council, 1997.

The initial government commitment to SEMATECH was $100 million per year for five years to be matched equally by industry; that commitment was renewed after the initial five-year period. After an additional three years, the industry decided in 1996 to forgo government funding. In total, the government contributed $850 million to SEMATECH from 1988 to 1996 with industry matching that figure. Since the termination of government funding, industry has continued to support SEMATECH at approximately $130 million per year. Obviously, Dr. Moore observed, the industry is quite satisfied with the SEMATECH program, otherwise it would not tax itself at that level to support it. The consortium, Dr. Moore said, "certainly has helped us [the U.S. semiconductor industry] regain our leadership in the semiconductor industry." Now, when one looks at various measures of manufacturing capability, the United States semiconductor industry is equal to any in the world.

SEMATECH was also a good investment for the U.S. government. Currently, each quarter, Intel alone returns more in taxes to the U.S. government than the entire amount of federal funding that went to SEMATECH. Neither Intel's current corporate health, nor that of other SEMATECH members, can be attributed to the consortium, but it is a fact that the U.S. semiconductor industry was losing significant market share at a rapid rate when SEMATECH was founded. SEMATECH was an important part of the industry's turnaround.

The Industry's Technology Roadmap

Shortly after SEMATECH's founding, a presidential commission, the National Advisory Committee on Semiconductors (NACS) was established to make additional recommendations on how to address competitiveness challenges facing the U.S. semiconductor industry. The chairman of NACS was Ian Ross, head of Bell Labs. Among NACS recommendations was for the development of a "technology roadmap" by the semiconductor industry. The SIA took up the task of developing the roadmap, which asked what the industry needed to do over the next 10 years in terms of research to maintain the industry's historical rate of progress. The roadmap broke up the various components of future challenges into pieces, assembled the necessary expertise in the industry, and laid out a research path.

The roadmap has been an excellent vehicle for coordinating research within the industry—laying the track, if you will, ahead of the locomotive to ensure steady progress in the semiconductor industry. The roadmap has been updated four times since 1988 and it is now essentially continuously updated. One reason that the roadmap has been useful to the semiconductor industry is that the technology path is fairly predictable; the industry seeks to make devices smaller and smaller, and ever more complex. The roadmap, Dr. Moore added, is fairly broad, cuts across a wide swath of the industry, and has aided in directing government R&D investment in semiconductor technology.

Future Research Paths

Even with the industry's resurgence today, cooperation remains an important theme for the industry. Current research in extreme ultraviolet (EUV) lithography technology is an example. EUV lithography is the industry's preferred technology for moving beyond optical lithography to make small structures. EUV technology originated in the Star Wars program of the 1980s, but it is now showing promise to be a commercially viable alternative to optical lithography in the foreseeable future. The EUV research program is moving at a rapid pace, but much work remains to be done in developing appropriate x-ray light sources, specialized lasers, and mirrors whose specifications exceed those of the Hubble Space Telescope. EUV lithography is a very complex system to develop, and it is a prime example of how future technology development in the semiconductor industry will draw on a wide range of science and technologies.

Conclusion

Directly targeted government programs, Dr. Moore said, did not nearly have the impact on the semiconductor industry's development that government purchases of advanced devices did. Key advances, such as the original invention of the transistor, came about from broad government support of science and technology. Narrowing the scope of R&D support has not proven to be successful. As the EUV example suggests, the industry remains dependent on a wide variety of advances in physics, chemistry, materials, metrology, and other areas. Improvements in computing power underpin advances in most of the areas mentioned above, and government can play a very positive role in pushing the frontiers of computing power.

For example, with respect to high-performance computing, Dr. Moore observed that much to his surprise, the market is largely the government. The government needs high-performance computing for purposes as diverse as weather forecasting, weapons development, and the space program. There is a limited commercial market for very high-end computers. University researchers, for their part, would love to have access to high-performance computers. It is therefore important that government continue to play its role as a sophisticated demander and buyer of high-performance computers.

With its origins in the United States, semiconductors remain the archetypal high-technology industry, and it has evolved into an important industry domestically and worldwide. Much of the industry's success can be traced to broad support for science and technology by the U.S. government, whether from R&D funds, support for semiconductor R&D in universities, or from government purchases of high-end devices. As in the past, cooperation among industry and between industry and government will play a role in maintaining the industry's rapid pace of technological development. The relationship between government and industry in semiconductors may offer some lessons to the biotechnology industry, or may offer interesting contrasts. The remainder of the conference presents a great opportunity for the two industries to learn from one another, and discover possible new synergies.

THE VIEW FROM THE BIOTECHNOLOGY INDUSTRY

Edward Penhoet
University of California at Berkeley and Chiron Corporation

Dr. Penhoet opened his remarks by saying that, unlike the computer and semiconductor industries, the biotechnology industry has been almost entirely dependent on the government for basic research support. In a sense, the history of the biotechnology industry to date has involved the commercialization of technologies that were funded almost entirely by the National Institutes of Health (NIH). Partnership, therefore, has been a major theme of the biotechnology story—partnership between government, industry, and universities. Much of the technology commercialized by industry was developed in the university setting using NIH grants.

That situation is changing dramatically today and that change, Dr. Penhoet said, is what he would concentrate his remarks upon. In particular, the technology base has been broadened in the biotechnology sector, and today a number of different technologies affect the biotechnology industry.

Origins of the Biotechnology Industry

Historically, the first 20 years of the biotechnology industry have been driven substantially, if not exclusively, by advances in molecular biology. The field was started in the 1970s by the invention of recombinant DNA, which has fueled much of what we think of today as biotechnology, and also the invention of techniques that allowed the production of monoclonal antibodies. If you examine the first wave of biotechnology products, they can be traced fairly directly to the application of these two inventions to a wide variety of uses in health care.

As we move into the late 1990s and the next millennium, the dominance of these two technologies is being challenged, or perhaps aided, by a wide variety of new technologies that are at least as important as recombinant DNA and monoclonal antibodies. The breadth of the new technologies, which are important both for biotechnology and pharmaceuticals, points to the need for a diversified federal research portfolio. To convey the need for adjustments in the federal R&D portfolio, Dr. Penhoet said he would concentrate on the University of California at Berkeley's Health Sciences Initiative.

Berkeley Health Sciences Initiative

The Berkeley Health Sciences Initiative combines a wide variety of scientific disciplines with an aim toward addressing health problems. The initiative is a working demonstration of the proposition that advances in the health sciences today require a multidisciplinary approach.

At Berkeley, the health sciences initiative involves the construction of a new building that will house scientists from a wide variety of disciplines, such as chemistry, biology, physics, engineering, and others. The objective is to use physical proximity as a catalyst for collaboration. The molecular engineering group, for example, will bring together molecular biologists, chemists, engineers, and physicists. Collaboration across these disciplines, Dr. Penhoet said, will address fundamental problems in the health sciences.

Collaboration in the Berkeley initiative is also designed to address practical problems in the health sciences field. Recalling Dr. Moore's observation that academicians turned increasingly to practical problems in semiconductors in the 1970s (driven by the creation of the SRC), Dr. Penhoet said that faculty at Berkeley today have a growing interest in pursuing research with practical applications. One reason for the focus on practical applications is that the links between university laboratories and the private sector have been well developed over the past 20 years.

With respect to specific technologies, magnetic resonance imaging (MRI) is among the most important tools in the biosciences, as it allows scientists to see structures within the human body. The current technologies are both expensive and invasive. At Berkeley, a group of scientists is working together to develop new imaging technology that may enable an MRI without the magnet, which makes the technology expensive and the experience of having an MRI rather uncomfortable.

Robotic surgery and microelectromechanical systems (MEMS) are technologies that Berkeley's engineers are pursuing aggressively in cooperation with biologists. Many drugs produced using biotechnology require injection at very specific time intervals, placing the burden on patients to adhere to schedules. Technologies such as MEMS can measure metabolism on a real time basis, and then regulate the release of appropriate doses of drugs at appropriate times. The development of sensing and delivery devices demonstrates the benefits of collaboration between engineers and the pharmaceutical industry.

With the latest generation of drugs now comes the field of structural biology, and molecular biologists are now able to provide the raw material for structural analysis. Molecular biology can now produce pure proteins for structural analysis. Structural analysis, in turn, has been aided by tools such as powerful x-ray sources that enable the x-ray crystallography of a protein to be determined very quickly. Improved computational power now allows diffraction patterns generated by the x-ray beam to be reduced to structures in a very short period of time.

The first protein to be analyzed by x-ray crystallography was myoglobin, with hemoglobin following shortly thereafter. Those projects each took 20 years to complete. Today, due to improved computing power, we are able to determine a protein's structure within days of having the crystal. All of this greatly quickens the pace of drug development and indeed the knowledge base in the field of structural biology is growing at a breathtaking rate. In these examples, tremendous strides in biology and health are being driven by contributions from physics (x-ray light sources) and computer science. The fields of multimedia and informatics are also driving innovation in biotechnology, simply because they improve the quality and quantity of information exchanges among researchers.

Berkeley has also built a second facility to house a number of different disciplines as a way to spur collaboration. The second facility focuses on neuroscience, genomics, molecular biology, public health, and cancer research. Cancer research now encompasses many fields. Whereas cancer treatment once was primarily chemotherapy, it now draws on advances in immunology and diagnostic techniques.

Neuroscience is making rapid strides in understanding how the brain functions and develops, and these strides are aided by many disciplines. Geneticists, psychologists, vision specialists, computer scientists, and other disciplines are working together to understand the complex system that the brain is. At Berkeley, the neuroscience group defines its mission as understanding the brain from molecular level to behavior.

The Demand for Computing Power

Within the biotechnology world, genomics progress depends mostly on advances in computing power. The human genome has 13 billion nucleotides and 100,000 genes. Simply collecting and sorting the information in the human genome is an enormous computational undertaking. As we begin to understand what genes mean to any living organism, the next challenge is to determine how genes are regulated and how they interact with each other to affect the metabolism of cells or the functioning of the brain. Diagnostic tools are beginning to emerge that depend on understanding the human genome. To make these tools operational is, however, a massive computational undertaking.

The most widely used genetic analysis technique today is the so-called gene chip, and this perhaps best illustrates the convergence of biology and computing. Gene chips are made by spotting individual genes on small chips, and then "interrogating those chips with laser beams." With the readout from the chip, it is possible to understand the expression of every gene on the chip. The technology was invented by Layton Reid, and his idea originated after he attended a semiconductor lithography workshop. It occurred to Reid that one could use lithography techniques to make specific gene sequences directly on silicon chips.

Today, as many as 20,000 different gene tags are on a single chip. Gene chip technology is, therefore, a hybrid using biology, chemistry, and semiconductor manufacturing processes. Gene chips also require huge computational resources, as analyzing data from gene chips requires taking 20,000 gene tags in 20,000 combinations.

Conclusion

In the future, Dr. Penhoet said that progress in the health sciences will require that researchers address challenges that cut across standard disciplinary boundaries, as opposed to the more traditional approach of advancing narrow academic disciplines. That said, each of the disciplines, if they are to contribute to progress in health sciences, will have to develop in parallel in order to have maximum impact. In agreeing with the remarks of Rep. Boehlert, Dr. Penhoet said there is no rational way *a priori* to know what the optimum balance in research spending is at any given point in time. It is difficult to predict the importance of any given field 10 years into the future. Nonetheless, it is clear that a convergence in the fields of physics, chemistry, biology, and computing will continue to occur. Broad support of science is, therefore, very important to sustaining progress.

While acknowledging the importance of NIH funding to his field and industry, Dr. Penhoet expressed his belief that future advances in biotechnology will require a multidisciplinary research approach. A wide range of fields, from mathematics and statistics, to biology, physics, and chemistry work in concert today to develop new solutions to health problems. Cooperation across disciplines, concluded Dr. Penhoet, as well as cooperation between government, industry, and universities, hold the key to biotechnology's future.

DISCUSSION

A questioner asked Dr. Moore why university R&D on semiconductors was disconnected from industry needs in the early years and whether the research that universities pursued during this period resulted in any long-term impacts on either the semiconductor or other industries.

Dr. Moore responded that early university research on semiconductors created a number of important developments. Gallium-arsenide semiconductors, for instance, were an outgrowth of some of this research, and such semiconductors are used widely today in cell phones and other applications. Other specialized devices were developed as well. It is not that the university research of this era was not important, Dr. Moore said, but it was not directly relevant to an industry that today has almost $150 billion in revenue worldwide and that is almost exclusively silicon based.

A questioner asked Dr. Penhoet whether there are any risks associated with encouraging interdisciplinary training at the graduate level. Dr. Penhoet responded by acknowledging the risk of training a generation of generalists if the trend toward multidisciplinary research went too far. Collaboration across disciplines tends to work best when individuals involved are highly skilled in their particular field. At Berkeley, the co-housing is designed not to train generalists, but to make students aware of the opportunities to apply work in their disciplines to other fields.

Dr. William Spencer of SEMATECH asked Dr. Penhoet about the relative roles of government and private funding in developing the computational, software, and mathematical tools for Berkeley's multidisciplinary initiatives. Dr. Penhoet responded that there is a hybrid of government and private funding for these tools. The Computational Biology program at Berkeley is perhaps most active in pursuing a variety of funding sources, and this program explores biostatistics, statistics, molecular biology, and mathematics. There is dearth of talent in this field, and this presents a challenge in training undergraduates in computational biology. NIH has provided some financial support to encourage more training in this field and the bulk of funding in computational biology comes from public sources.

A questioner asked Dr. Moore why the semiconductor industry grew in the United States and not Europe, wondering specifically whether demand from the government for high-end devices played a role, or whether a greater entrepreneurial spirit in the United States has been the reason. Dr. Moore noted that the United States was better prepared to exploit a new technology such as semiconductors because we have a culture that encourages the formation of small firms. "New technologies are not easily exploited by large existing companies," Dr. Moore said. In semiconductors, large firms initially pursued semiconductor technology, but they eventually dropped out of the industry. New entrants were the important players in semiconductors. The fact that the semiconductor was invented in the United States, Dr. Moore added, also gave U.S. companies a huge advantage in pursuing business opportunities. The availability of the government market—willing to buy large quantities of devices at high prices—also propelled the industry in the United States.

Dr. William Long of Business Performance Associates said that EUV technology received support from the Advanced Technology Program, specifically in the development of the mirrors necessary to successfully deploy EUV technology. Dr. Long asked Dr. Moore where he thought EUV technology would be without government support for the lasers, mirrors, or other supporting technologies.

Dr. Moore said that it is unlikely that EUV technology would have developed—or at least developed at the pace that it has—without initial support from the Star Wars program. Only because some of the basic research was conducted during the Star Wars program has EUV today become a viable lithography alternative for the industry. Other approaches, such as electronic beam, short wavelength, and shadow x-rays, might have garnered most of the R&D support from industry without the support for EUV from Star Wars. The existence in the national laboratories of basic research on EUV has greatly enhanced its attractiveness to industry.

A questioner observed that the amount of data being generated by the Human Genome Project is growing extremely rapidly. He asked Dr. Moore whether computer storage would keep pace with the growth of genome data.

Dr. Moore observed that the growth of storage capacity has maintained an exponential rate for approximately 40 years, a truly remarkable fact. Dr. Moore said he expected this growth to continue for several more years, as there remains room for more advances using traditional techniques. Dr. Penhoet commented that biology students at Berkeley spend about 25 percent of their time in front of the computer manipulating databases. Comparative biological research can now be conducted using databases alone, underscoring the need for improvements in storage capacity and computer technology.

Dr. Kathy Behrens observed that Dr. Moore had cited numerous examples of government-industry and intra-industry interactions in the development of the semiconductor industry. Dr. Behrens asked Dr. Moore if there were specific examples that might be useful today in sustaining the advances we have witnessed in the past 30 to 40 years.

With respect to EUV, Dr. Moore cited the industry's relationship with Lawrence Livermore National Laboratory as a government-industry collaboration that has been very mutually beneficial. Lawrence Livermore has brought tremendous assets to the development of EUV, and the semiconductor industry has invested $100 million annually to support the program. The EUV partnership, Dr. Moore observed, is probably the largest Cooperative Research and Development Agreement in place. In the computing area, government needs to continue to support high performance parallel computing. Such computers have widespread scientific applications, but only the government can afford them.

A member of the audience remarked how accurate Moore's Law has proven to be, and asked Dr. Moore how much longer we could expect to see the law hold. Dr. Moore responded that the fact that materials are made of atoms would eventually hinder progress; problems at the atomic level would come into play within a decade. But that would not signify the end of progress; larger chips, advances in interconnection technology, and other advances will continue to improve the capacity of semiconductors.

Panel II:
A Historical Perspective:
Federal Partnerships in Computing and Biotechnology

INTRODUCTION

Patrick Windham
Stanford University

Mr. Windham said that the second panel would try to provide more details about the historical evolution and today's relationship between government and industry in the biotechnology and computing sectors. In particular, the panel will try to understand better the forces that are affecting both industries today and think about the forces that are likely to influence them in the future.

The panel today is very distinguished, consisting of Kenneth Flamm of the University of Texas, who has written extensively on the computing and semiconductor industries; Leon Rosenberg of Princeton University, a distinguished pediatrician who has also served as director of research at Bristol-Myers and as Dean of Yale Medical School; and William Bonvillian, legislative director for Senator Joseph Lieberman (D-Connecticut) and one of the most respected senior staffers on Capitol Hill.

PARTNERSHIPS IN THE COMPUTER INDUSTRY

Kenneth Flamm
University of Texas at Austin

Dr. Flamm said that his remarks would update some of the data he compiled in writing two books in the 1980s on the development of the computer industry in the United States.[1] Rather than discuss the history of the computing industry, Dr. Flamm said he would focus on current trends in federal funding for R&D in the computer industry. In light of recent changes in funding levels, as well as the growing importance of information technologies in the economy, updating his data may contribute to this conference's deliberations, as well as the broader policy debate.

Evolution of the Computer

To set the stage for his discussion of data trends, Dr. Flamm touched on a few topics from the computer industry's history. Unlike the semiconductor industry, nearly all computer development in the immediate postwar era enjoyed significant federal support. It was not only that government served as the market for computers, but the government also provided substantial funding for computer development. From 1955 to 1965, the commercial market for computers grew rapidly, although the government continued to fund most high-performance computing projects and government-funded computer development projects served to push the leading edge of technology. In the 1965 to 1975 period, the growth of the commercial market for computers accelerated even more, and the government role centered primarily on funding the very high end of computer technology development. During the next fifteen years, from 1975 to 1990, commercial markets grew far larger than government markets and pushed the preponderance of technology development. However, government played an important niche role in funding what might be called "the exotic leading edge" of computer technology development. The period from 1990 to 1997 is marked by some surprising, and in some ways disturbing, trends in computer R&D.

Before turning to describing these trends, Dr. Flamm made an observation about the role of cryptography in the invention of the computer. The ENIAC computer at the University of Pennsylvania is generally credited as being the first electronic computer ever built. However, Dr. Flamm said that a case can be made that the first electronic stored-program digital computer was built in England in 1943 and used by the cryptographic community to break German codes.

In fact, the National Security Agency (NSA) in the United States and its antecedents were the chief funders of much of the advanced computer R&D in the 1950s and into the mid-1960s. The reason for NSA's interest was cryptography, and indeed the cryptographic community continues to drive innovation in high-end computing as a way to make and break codes for national security purposes.

---

[1]Kenneth Flamm, *Targeting the Computer: Government Support and International Competition*, Washington, D.C.: The Brookings Institution, 1987; and Kenneth Flamm, *Creating the Computer: Government, Industry, and High Technology*, Washington, D.C.: The Brookings Institution, 1988.

In addition to NSA, the Defense Advanced Research Projects Agency (DARPA), from the 1960s onward, played a very important role in driving innovation in computing. Among the well-chronicled contributions to computing funded by DARPA include time-sharing, networking, artificial intelligence, computer graphics, and advanced microelectronics. The Defense Department's record is not unblemished, Dr. Flamm noted. Projects such as the Ada programming language and the Very High Speed Integrated Circuit (VHSIC) initiative did not pay off.

The Atomic Energy Commission and the Department of Energy have also been major players in advanced computing over the years. Their role has not just been in providing R&D funds, but also in serving as a market for high-end computing machines. The National Aeronautics and Space Administration has played a small role in computer development, but an important one in certain niches, such as computer simulation, image processing, and large-scale system software development.

The National Science Foundation came relatively late to the support of computer research because it was bound by traditional academic disciplines in the 1950s and therefore relatively unreceptive to a new field such as computer science. It was not until the 1970s that computer science was seen as a separate discipline and incorporated into NSF's grant structure. In the 1990s, NSF began to play a very prominent role in high-performance computing.

At the National Institutes of Health, the role has traditionally been rather modest, but as the remarks of Ed Penhoet show, the need for advanced computer systems to interpret data and aid in diagnostics is growing rapidly.

Trends in Federal R&D Support for Computing

Adding up the contribution of all federal agencies to computer R&D, a picture emerges of a declining role of federal R&D support in the computer sector, which Dr. Flamm defines as firms in the Office Computer and Automated Machinery (OCAM) sectors. In the 1950s, the federal government funded 60 percent of computer R&D. As the commercial sector grew in the 1960s, the figure fell to about 33 percent, falling further to 22 percent in 1975, 13 percent in 1980, increasing slightly to 15 percent in 1984, and 6 percent in 1990. As businesses invested heavily in computers throughout the 1990s and as the market for personal computers exploded, the federal share in R&D shrunk dramatically; in 1995, 0.6 percent of computer R&D was federally supported, a number that fell to 0.4 percent in 1997 (Table 1).

Taking a more expansive look at the federal role, in which funding for math and computer science in universities and expenditures for federally funded research and development centers (FFRDCs) are included, the decline in the federal role is less precipitous. Dr. Flamm presented data showing that, when OCAM and these additional categories are considered, the federal share fluctuated be-

TABLE 1 Shrinking federal role: Federal share of OCAM R&D dollars

1950s	60%*
mid-1960s	33%
1975	22%
1980	13%
1984	15%
1990	6%
1995	0.6%
1997	0.4%

* estimated

tween 16 percent and 26 percent between 1975 and 1995, before falling to 10 percent in 1997 (Table 2). Much of the decrease is attributable to a decrease in funding for math education, an issue policymakers may want to take into account.

TABLE 2 Shrinking federal role: Federal share of OCAM and mathematics and computer science dollars in universities and university FFRDCs

1975	26%
1980	19%
1984	21%
1990	16%
1995	22%
1997	10%

NOTE: Mathematics funding has been sharply reduced recently; see trend toward "targeted" federal research noted below.

Analyzing Trends in Federal R&D Support for Computers

In assessing these figures, Dr. Flamm said that one might view them as simply a natural progression for a maturing industry. There may be some optimal amount of money the government should spend on such research, and as the industry grows and the commercial market expands, it should not be surprising or alarming that the overall federal share declines. In other words, the pie may be expanding so greatly that the federal slice will naturally shrink.

Dr. Flamm said, however, that this is not the case. In presenting a chart that tracked actual expenditures for computer R&D—in 1992 constant dollars—Dr. Flamm showed that the amount of R&D for hardware (which is essentially OCAM) has remained more or less constant since 1984 (Figure 1). Dr. Flamm conjectured that the large drop in R&D spending on hardware in 1995 occurred because IBM's business classification was changed from hardware to software due to greater sales volume of software. The increase in 1997 may be because

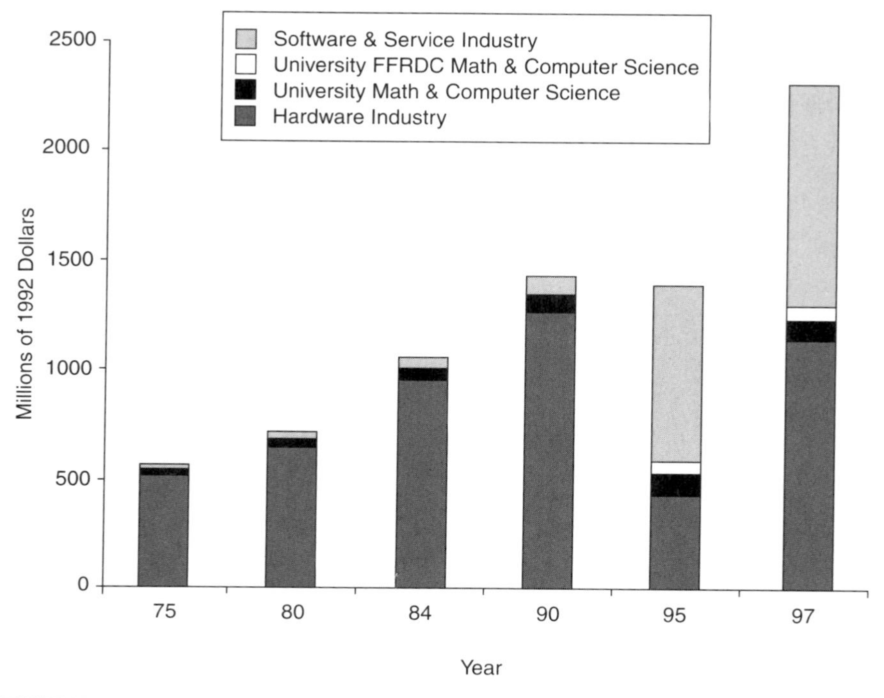

FIGURE 1 Computer R&D.

IBM returned to the hardware category. In sum, the generally flat spending on hardware R&D by industry indicates that the declining federal share in R&D spending in OCAM cannot be explained by growing overall expenditures for computer hardware R&D.

Taking another perspective on the data, Dr. Flamm displayed a chart showing that the federal share in R&D funding for math and computer science in universities and for FFRDCs has remained fairly constant since 1972 (Figure 2). For other categories, such as R&D funding for hardware and software, the federal share has declined markedly. Basically, Dr. Flamm concluded, there is almost no federal R&D being devoted today to computer hardware. This raises the question, in light of major initiatives on high-performance computing, of how these resources are being channeled to industry. Dr. Flamm suggested that procurement is the likely candidate; R&D funds for computer hardware are included in government contracts to develop and purchase high-performance computers for industry.

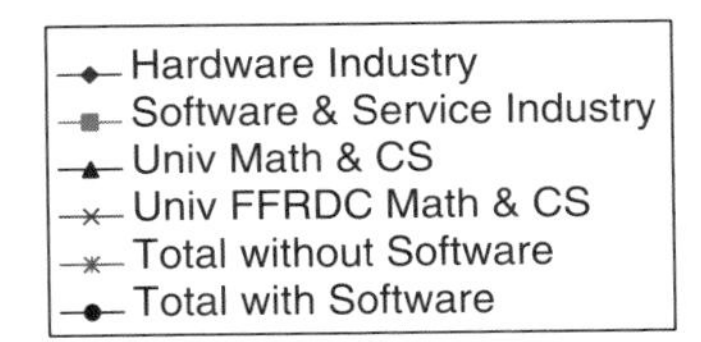

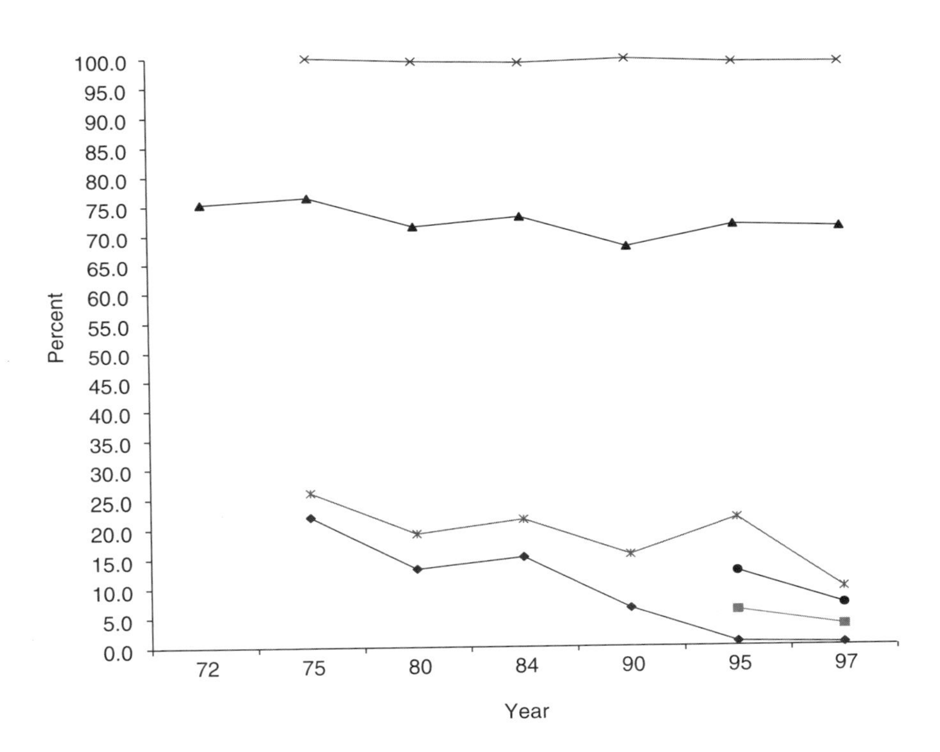

FIGURE 2 Federal share of computer R&D.

The Distribution of R&D Among Computer Firms

Dr. Flamm then turned to a discussion of the distribution of R&D among firms in the computer hardware industry. With respect to hardware sales among the top 20 R&D performing hardware firms, Dr. Flamm presented snapshots of the sales distributions in 1984, 1990, and 1997. As Figure 3 shows, the industry is concentrated, as the top four firms have about as much revenue as the remaining firms in the sample. In a cross-year comparison, it appears that concentration has declined, as firms that are not among the top 20 R&D performers accounted for relatively more sales in 1997 than in 1984. In terms of total R&D in constant

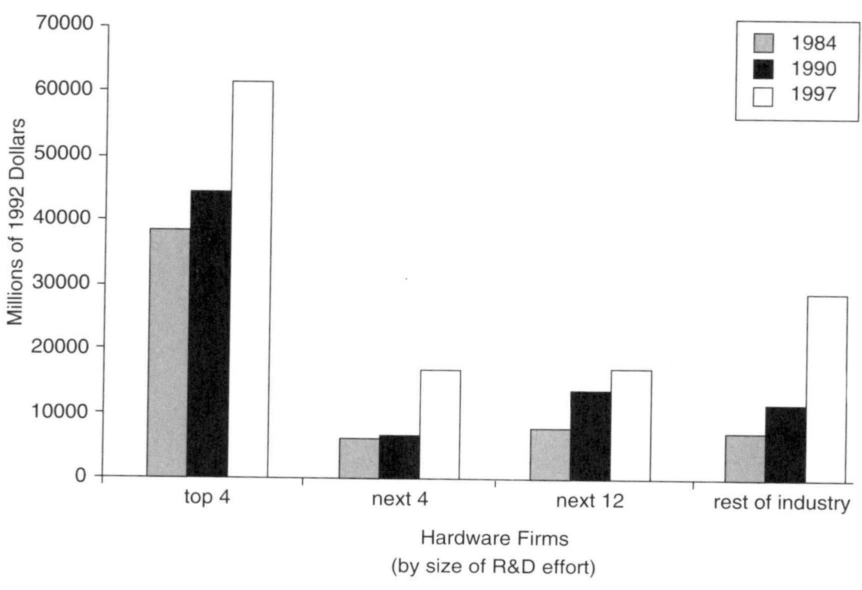

FIGURE 3 Distribution of hardware sales.

1992 dollars, R&D spending among the top four firms was roughly the same in 1997 as in 1984 and down from 1990 levels (Figure 4). The remainder of the industry shows only modest increases, indicating a flat trend since 1984, and an overall decline since 1990. Turning to R&D as a percentage of sales, Figure 5 shows that this has fallen sharply in recent years, especially among the top four R&D performing companies. Finally, Figure 6 demonstrates that in 1984, federal R&D funds for computer hardware firms were a significant source of funding for the top R&D-performing firms in particular. Federal R&D dollars were never a large source of funds for firms outside the top tier, but by 1997, federal R&D dollars for all segments of the computer hardware industry had ceased to be a meaningful amount.

Hardware versus Software

Comparing R&D across hardware and software firms is still another way to assess the data, Dr. Flamm continued. In 1995, the Bureau of the Census and the National Science Foundation began to collect data specifically on software companies. This new data source creates the opportunity for interesting comparisons. Among the top R&D performers in the industry, hardware firms outpaced software firms in terms of sales, $140 billion to $85 billion in 1997.

The most notable contrast from the data is in the distribution of sales, total R&D funds, and total R&D as a percentage of sales across the hardware and

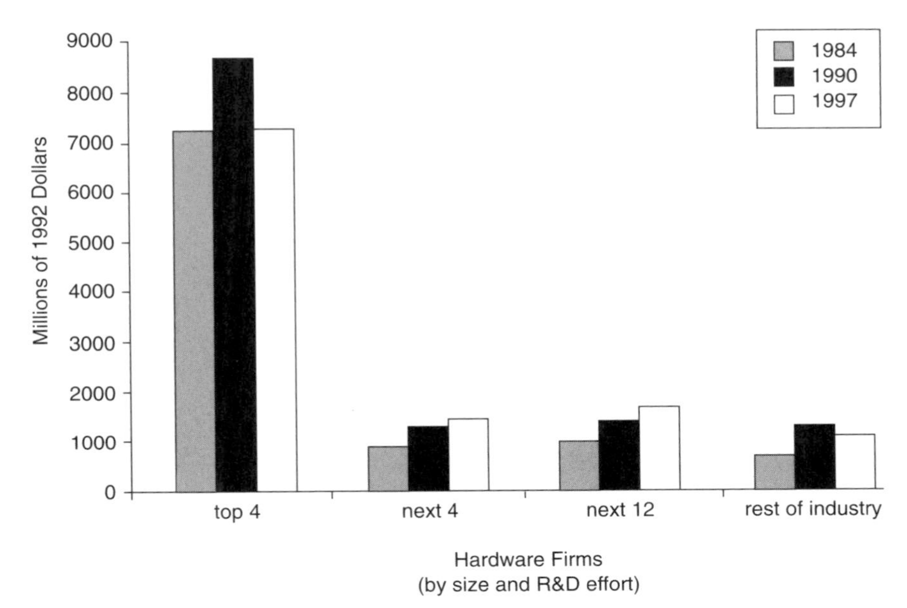

FIGURE 4 Distribution of hardware R&D.

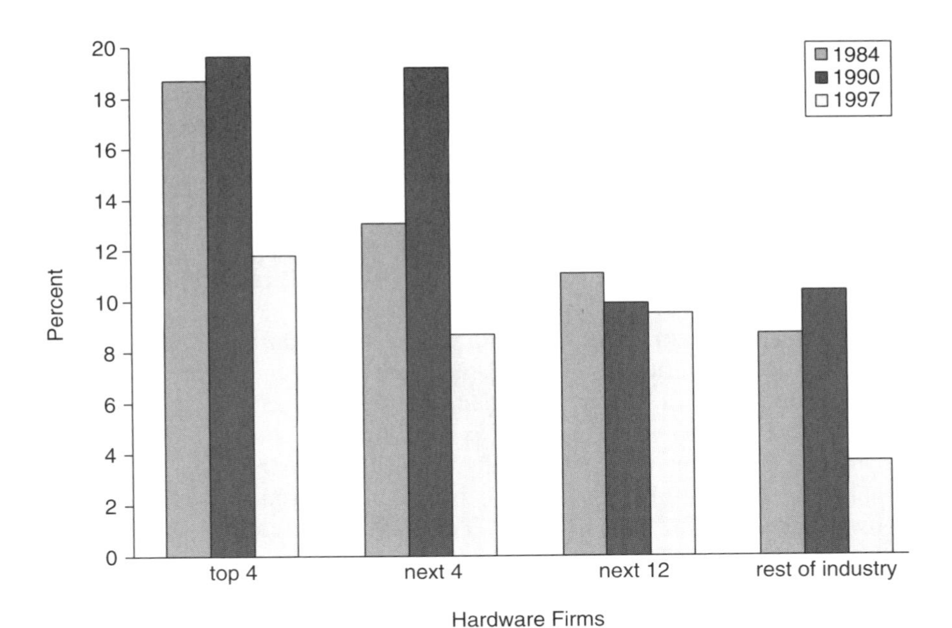

FIGURE 5 Total R&D as a percentage of industry sales.

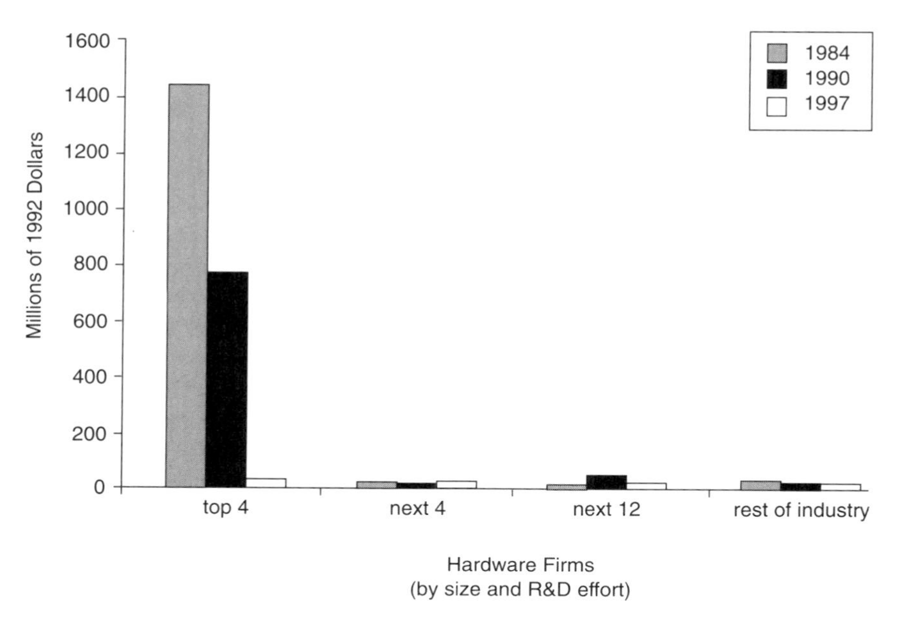

FIGURE 6 Distribution of federal R&D to industry.

software industries. As Figure 7 shows, the top four R&D-performing hardware firms account for a large share of total sales, while for software, the "rest of the industry" (that is, small firms that rank low in terms of total R&D spending) accounts for most of the sales. When the total amount of R&D spending is added across the same categorizations of hardware and software firms, the same pattern emerges (Figure 8). Finally, when examining R&D as a percentage of sales, Figure 9 shows that software companies are more R&D intensive by this measure than hardware firms, and the distribution of R&D intensity for software firms is more even than that for hardware companies.

The distribution of R&D activity in the software industry raises questions, given the size and influence of Microsoft. One conjecture is that some small firms conduct R&D in the hope or anticipation of being acquired by Microsoft, meaning that much of what is counted as small-firm R&D is eventually fed into Microsoft. Dr. Flamm said he knew of at least one example in which "hope-for acquisition" by Microsoft was an explicit part of the business plan for a small software start-up. As for the higher level of R&D as a percentage of sales in the software industry, Dr. Flamm speculated that a wider range of activities could be counted as R&D in the software industry than in the hardware industry. Anyone writing code, for example, could be seen as engaging in development activity.

In terms of federal funding for software R&D, Figure 10 shows that small software firms receive the vast majority of federal R&D dollars; however, this is still a meager amount of overall software R&D. As Figure 11 shows, for small

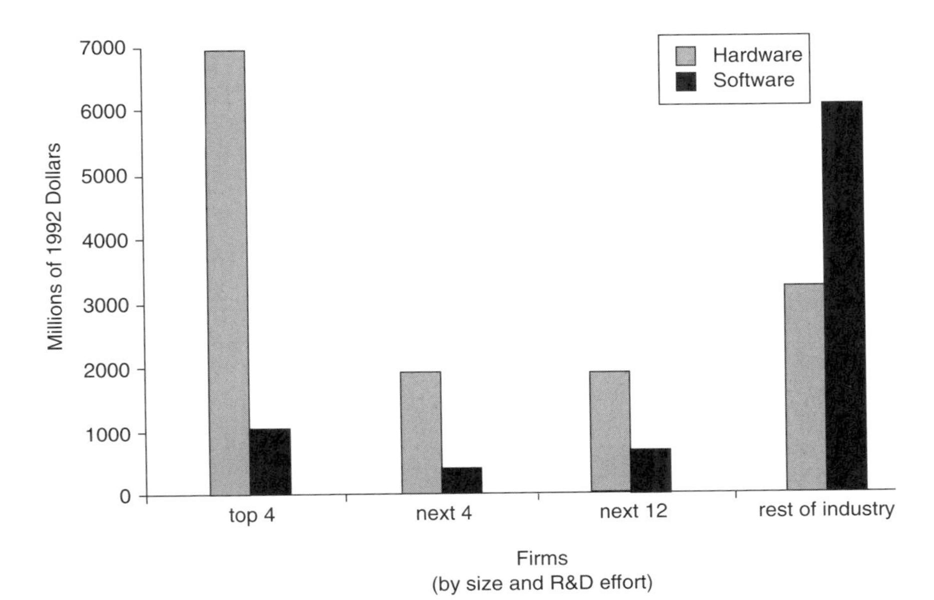

FIGURE 7 Hardware and software sales distribution, 1997.

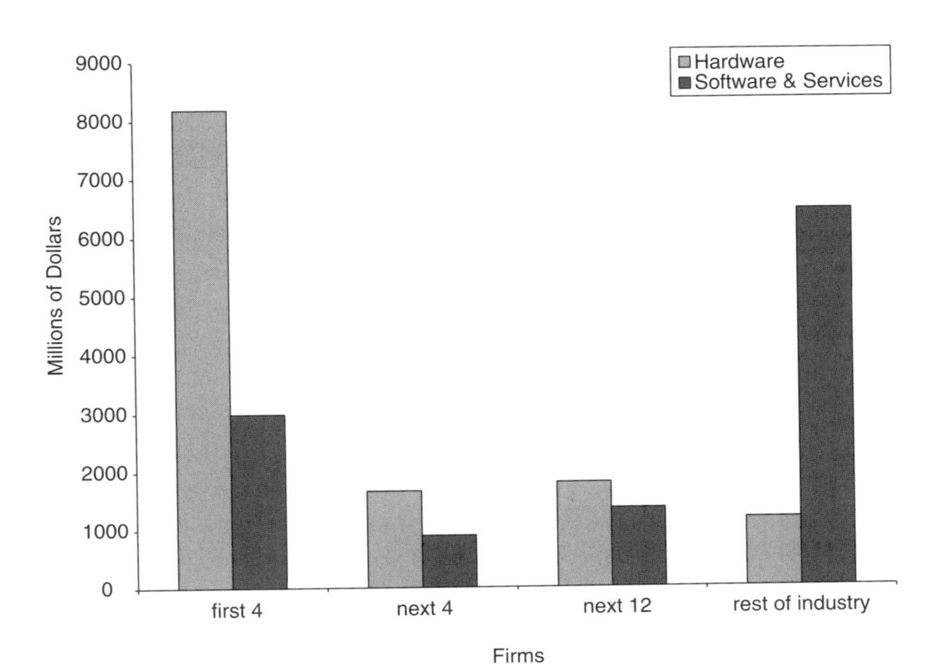

FIGURE 8 Distribution of total R&D funds, 1997.

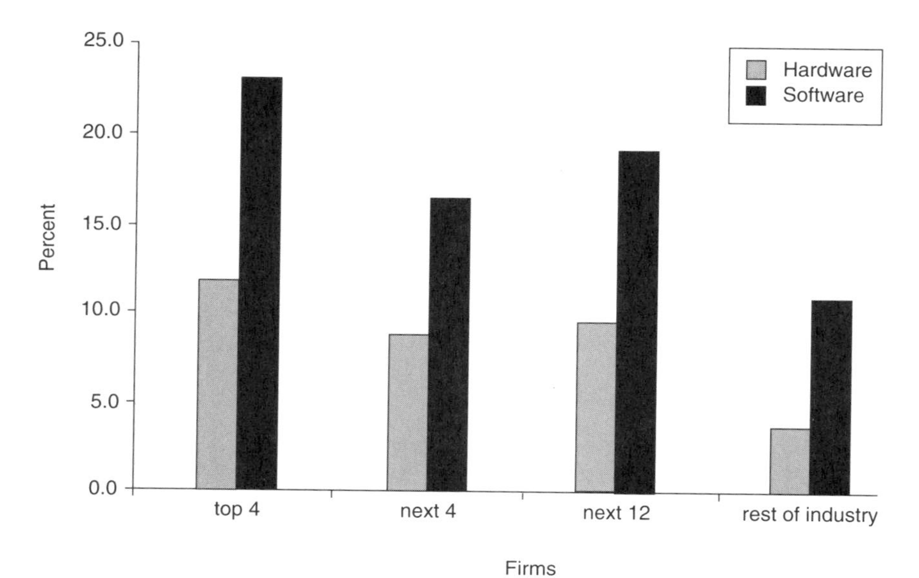

FIGURE 9 Total R&D as a percentage of sales, 1997.

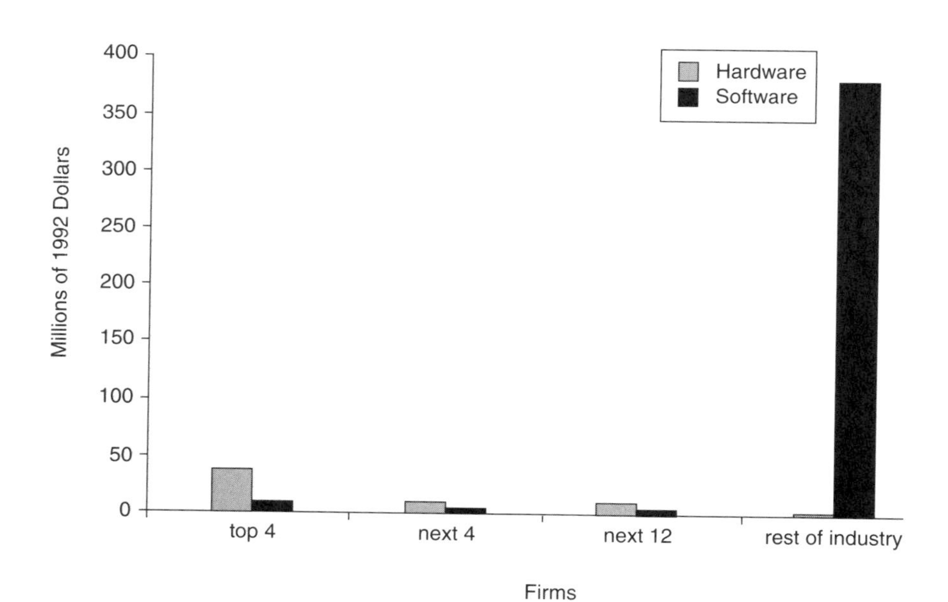

FIGURE 10 Distribution of federal R&D funds, 1997.

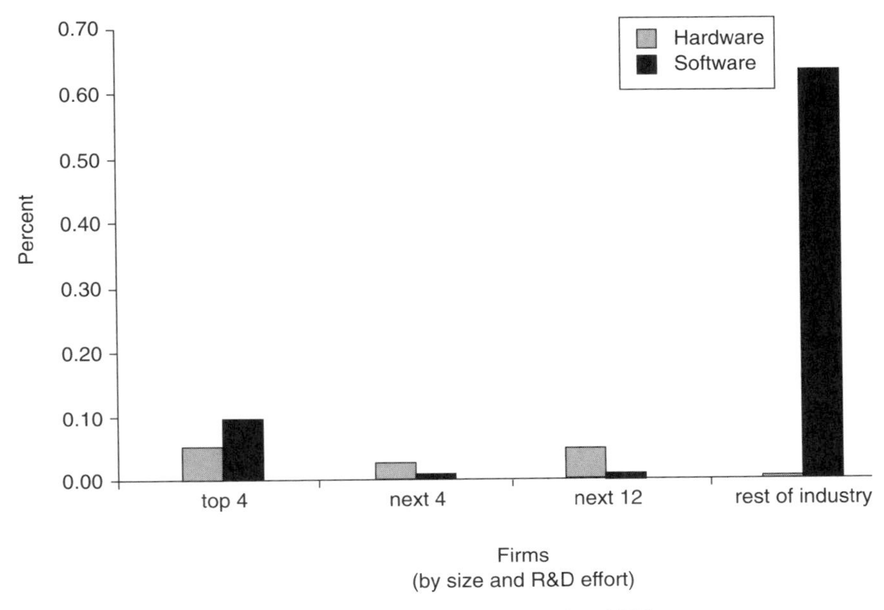

FIGURE 11 Federal R&D funds as a percentage of sales, 1997.

software firms, federal R&D as a percentage of sales is only 0.65 percent, and much smaller for larger software firms. So federal funding is not very important for software firms.

One possible explanation for the decline in R&D in the computer hardware industry is that R&D is migrating from the computer industry into the semiconductor industry. Figure 12 offers support for this notion. Although both sectors saw R&D intensity decline in the early 1990s, the semiconductor industry has witnessed a sharp rebound in R&D intensity since 1995, while the computer industry has experienced a decline in recent years. The distribution of computer versus semiconductor R&D also supports the idea that R&D has shifted to semiconductors from computers. Figure 13 shows that the semiconductor and computer industries now conduct about the same amount of R&D, whereas in the late 1980s computer firms conducted approximately $4 of R&D for every $1 conducted by semiconductor companies.

Summary

The data indicate that industrial R&D in the computer industry is declining, and that federal R&D funding (measured as a percentage of sales) is declining quite sharply. Moreover, fundamental research seems to be bearing the brunt of overall R&D reductions, whereas targeted research—funded both by industry

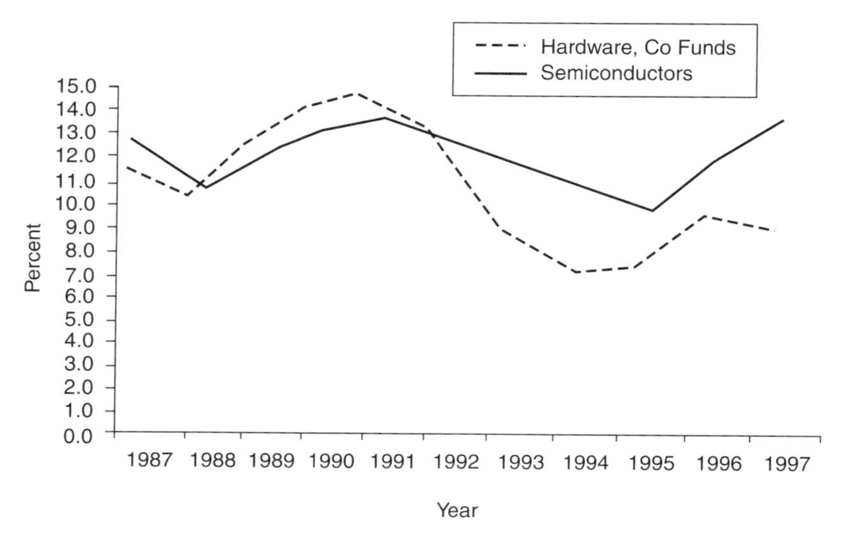

FIGURE 12 R&D as a percentage of sales.

and government—seems to be on the increase. Dr. Flamm pointed out that while it is true that venture capital funding is more plentiful these days, venture capital should not be confused with R&D funding. Finally, a traditional source of fundamental R&D, namely DARPA, is devoting fewer resources to this type of R&D.

A number of reports have highlighted these disturbing trends, most notably one produced by the President's Information Technology Advisory Council (PITAC). PITAC has recommended that federal computer R&D support be doubled by 2004 to a total of $1.24 billion, with approximately half going to hardware and half to software. In particular, PITAC recommends that this funding emphasize fundamental R&D as opposed to targeted R&D.

In conclusion, Dr. Flamm said that a strong consensus has developed that information technologies have a substantial impact on the economy. Both government and industry have concluded that the seed corn for future progress in information technologies—long-term research and development—has been underfunded in recent years. Additionally, government and industry have grown to understand that each party should fund different aspects of information technology R&D. However, there is little consensus at this point over the proper level of funding, the proper division between hardware and software, program structures, and overall objectives. Each of these topics is likely to be very controversial and generate vigorous political discussion as Congress considers how to allocate scarce federal resources.

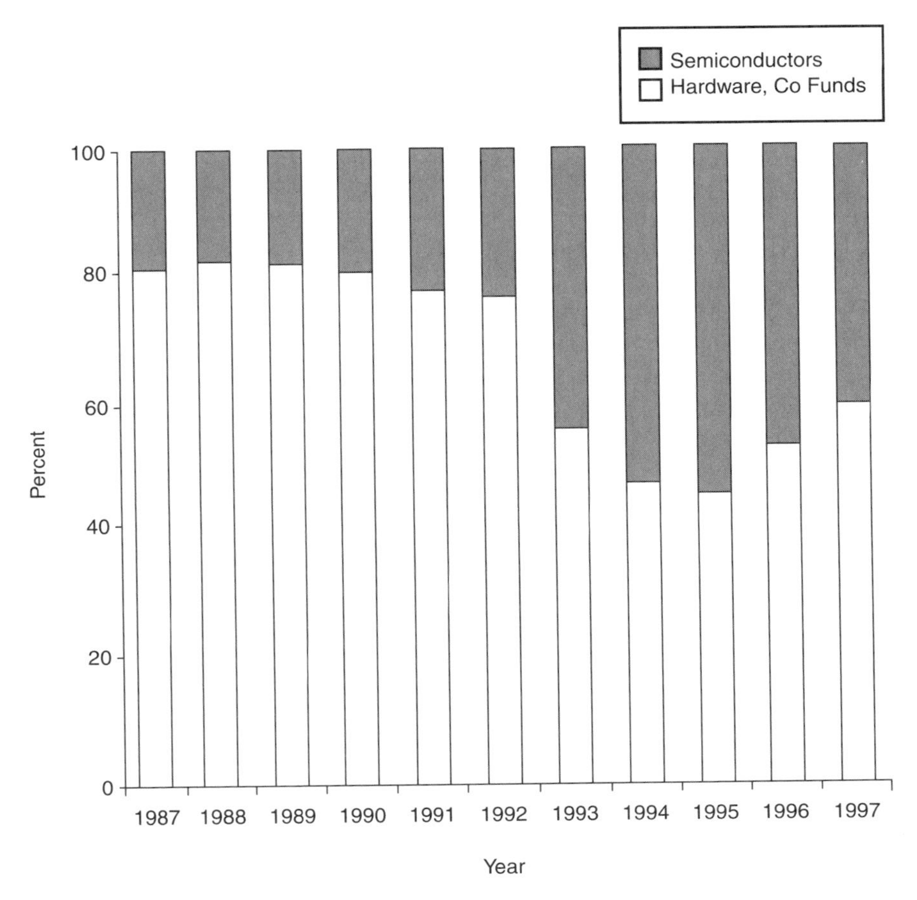

FIGURE 13 Distribution of hardware and semiconductor R&D.

PARTNERSHIPS IN THE BIOTECHNOLOGY INDUSTRY

Leon Rosenberg
Princeton University

Dr. Rosenberg opened his remarks by observing that it was important to have an ongoing conversation on the issues that academics and industrialists must face in the biotechnology and computing industries. Looking at the audience assembled, Dr. Rosenberg said he believes that this does not happen much. Seeing very few familiar faces in the audience—a surprise to Dr. Rosenberg since he spends a good deal of time in Washington and at the National Academy of Sciences—suggests that these gatherings are rare, but also bodes well for this conference's deliberations.

To take full advantage of the opportunity the conference offers, Dr. Rosenberg presented the definitions of several terms used in medical research and biotechnology to give participants a common understanding of key concepts, including the following.

- **Medical Research:** Science-based inquiry, both basic and applied, whose goal is improvement in health and eradication or mitigation of disease and disability. It is synonymous with health research.
- **Biotechnology**: A group of techniques and technologies that apply the principles of genetics, immunology, and molecular, cellular, and structural biology to the discovery and development of novel products.
- **Biotechnology Industry**: This industry is composed of nearly 1,300 private companies that use various biotechnologies to develop products for use in health care, food and agriculture, industrial processes, and environmental cleanup. The subset of companies engaging in health research makes up an important part of the medical research enterprise of this country. It is also important to recognize that most of these 1,300 companies are very small. Two-thirds have fewer than 50 employees. All but 20 of these companies are "unencumbered by revenues." The biotechnology industry is a very entrepreneurial industry, but one with relatively few commercial successes.
- **The Pharmaceutical Industry:** This industry is composed of approximately 100 research-based private companies whose purpose is the design, discovery, development, and marketing of new agents for the prevention, treatment, and cure of disease. The term "bio-pharmaceutical industry" is in some ways a more accurate description of the industry, because pharmaceutical companies are today very reliant on biotechnologies.
- **Partnership**—the state or condition of being associated in some action or endeavor. In the business context, this means the provision of capital by two parties in a common undertaking, or a joint venture with the sharing of risks and profits.

Dr. Rosenberg pointed out that to some people, government-industry partnership is a misnomer, because the two parties have very different missions and roles in society. The term "alliance" might be preferable to many people, or perhaps the more generic term "interaction."

Medical Research in the United States

The federal government and the private sector account for over 90 percent of research investment for medicine in the United States today. The remainder comes from universities, non-profit organizations, foundations, and voluntary health agencies. In the private sector share, bio-pharmaceutical firms account for nearly all of the private sector R&D support. In 1999, $45 billion was invested in medical R&D, with about 60 percent from industry, 35 percent from NIH, and 5 percent from other sources. To keep track of medical research, therefore, one must keep an eye on industry and NIH primarily, even though the other sources are important.

The scientists who produce research from these funding sources, not surprisingly, are located mainly in industry and universities, with a small share from NIH's intramural program and from independent research institutes. Yet the work that is carried out in universities differs from that carried out in industrial research laboratories. Academic scientists still are fundamentally concerned about the frontier, asking questions that may not have utilitarian purposes. Generally, these researchers compete for peer-reviewed funds in the form of grants from NIH. For academic researchers, approximately 90 percent of their grants come from the federal government, with the remainder coming from industry. That ratio is not, at present, changing much. For NIH, 90 percent of its R&D funds are spent in academic laboratories, with the remaining 10 percent spent in the intramural laboratories in Bethesda.

Company revenues, of course, support industry research. Today, in biotechnology, the industry spends between 15 and 20 percent of revenues on R&D, a very high share by the standard of other industrial sectors, and even other high-tech industries. Taking exception to a point mentioned by Dr. Flamm, Dr. Rosenberg said that this percentage is not dropping. The bio-pharmaceutical companies recognize that it is in their interest to maintain a high level of R&D spending.

Between industry, universities, and government over 200,000 people are engaged in research in the bio-pharmaceutical field. That number continues to grow.

Advocacy for R&D in Medical Research

Dr. Rosenberg noted that little had been mentioned today about the role of advocacy in support of federal funding for R&D in the computer and semiconductor industries, but that he believes it has been crucial to the development of the bio-pharmaceutical industries. Effective lobbying from the Association of American Medical Colleges, Research America, Funded-First, disease groups, and others, have been responsible for the growth in support for federally funded medical R&D over the past 30 years, and the past 10 years especially. In 1990, the NIH budget was $8 billion; in 1999 it was $15.6 billion with likely continued strong growth. That is essentially a doubling of NIH funding over 10 years, and there are many calls for another doubling of NIH's budget over the next five years.

From 1990 to 1999, private R&D investment in the bio-pharmaceutical industry has grown from $10 billion to $24 billion. The bio-pharmaceutical industry has more than kept pace with increases in public R&D budgets as the industry has continued to use platforms developed in academic laboratories to develop new drugs. In general, the division of labor among industry and academia is such that industry conducts most of the applied research, while university researchers carry out most of the basic research. The bridge between basic and applied research—the transitional R&D that positions the research for commercial use—is conducted to varying degrees by all parties.

The interaction among participants of the medical research enterprise is important. Informal contacts between industry, government, and university researchers take place all the time at scientific conferences. Money does change hands in these interactions, but these informal collaborations are thought to be among the most fruitful for the advancement of the field.

Partnerships in the Biotechnology Industry

In the bio-pharmaceutical sphere, the discussion of partnerships cannot be limited to government-industry collaboration, simply because university research plays such a critical role. Indeed, universities are virtually always the third party in the bio-pharmaceutical R&D arena. Dr. Rosenberg added that grants and contracts play an important role in partnerships in the bio-pharmaceutical field. These grants and contracts may be from government to academia or industry to academia. If one were to exclude grants and contracts from a discussion of medical R&D, then one would greatly skew one's understanding of partnerships in the field.

Large companies frequently enter into sponsored research agreements with universities, and for them, exclusive access to the research results is very important. Patents are pursued and defended at every level in such arrangements, and licensing agreements are closely overseen for compliance. From the perspective of universities, grants are the most preferable type of collaborative arrangement, because they allow for the maximum amount of academic autonomy in the conduct of R&D and use of its results. For academic scientists, grants are more desirable, but contracts and agreements play a significant role in academic science—a total of $1.5 billion worth of contracts and agreements were entered into between the bio-pharmaceutical industry and universities in 1999.

Dr. Rosenberg added that small companies—especially those founded by academic researchers—need all the sources of support that they can muster. Grants and contracts from government, collaborative relationships with government and industry, as well as taking or giving licenses to intellectual property—these are all useful sources of funding for small, academically spawned start-ups.

The Rationale for Partnerships in the Bio-Pharmaceutical Industry

The major participants in collaborative relationships enter into these arrangements for different reasons. Academics and small companies want more money for R&D, while large companies want access to new technologies and intellectual property. The different needs have led to a very healthy relationship among participants and they have led to greater interaction between the university, government, and private sector. All research-intensive academic institutions have a rich network of interaction with industry and government agencies. Government agencies such as NIH have steadily expanded their contacts with universities, established companies, and small start-ups. For example, in 1998 NIH and private companies entered into 166 cooperative research and development agreements (CRADAs), which allow for the sharing of compounds and other research materials and results, as well as for the exchange of funds. This is the largest number of CRADAs ever between NIH and the private sector. Through the Small Business Innovation Research (SBIR) program, NIH awarded $266 million in grants to small firms for medical and bio-pharmaceutical research. It is expected that the SBIR program at NIH will exceed $300 million in 1999. In 1999, the Commerce Department's Advanced Technology Program (ATP) awarded $29 million to small biotechnology companies. It is hoped that that number will grow.

To demonstrate that the flow of funds is a two-way street, Dr. Rosenberg said that in 1999, NIH received $40 million in royalties from its licenses, and universities received more than $300 million in revenues from licenses for their innovations in health and medical R&D.

Conclusion

Dr. Rosenberg concurred strongly with Dr. Penhoet that continued advances in biotechnology depend on advances in information technology. Examples are numerous, from the human genome, to computational neurobiology, on-line databases, gene profiling, and patient records. For those close to academia and industry, it has become clear that there is a dearth of trained people in bioinformatics. From his experience in government, universities, and industry, Dr. Rosenberg noted the different cultures within these various institutions. Nonetheless, Dr. Rosenberg said incredible advances have been made over the years. Many concerns have arisen over whether academic culture could withstand greater collaboration with industry, and whether academic research would become more applied as a result of interaction with industry.

In fact, Dr. Rosenberg said, academic research has not become more applied, and the culture of academic research in the health sciences has survived nicely. By and large, the worst-case scenarios—technologically, ethically, or sociologically—have simply not played out. Tensions and potential conflicts of interest have emerged, and they must be addressed as government, industry, and academia increase the frequency of collaboration. However, Dr. Rosenberg said that collaboration has benefited all parties in recent decades, resulting in better academic research and a more productive industry, and ultimately a better-off public. This should give us confidence that we can effectively address the challenges facing government-industry-academic collaboration in the next millennium.

TRENDS IN FEDERAL RESEARCH

William Bonvillian
Office of Senator Joseph Lieberman

Mr. Bonvillian began by drawing an analogy of federal funding for science and technology. Imagine, Mr. Bonvillian said, a character called "Dr. Joe Science" who is standing in the middle of a 16-lane federal interstate highway, a highway called "the domestic discretionary funding interstate." Dr. Science is in the middle of it, with cars and trucks speeding by. The interstate has a number of huge 18-wheel trucks on it, and Dr. Science has to dodge one called "Social Security entitlements." Another 18-wheeler approaches called Medicare entitlements, and another truck called the "tax-cut express" careens toward him. The trucks are eating up the road as Dr. Science tries to stay out of harm's way. Joe Science is definitely stuck on a tough road.

The question is: Will Dr. Science be able to construct his own 18-wheel truck in order to survive on the road? As Dr. Rosenberg described, the life sciences community is busy building an 18-wheeler that seems well positioned to navigate the 16-lane highway. For the physical sciences, however, no one has even begun to assemble a truck. In other words, Mr. Bonvillian said, neither political party has etched science on its stone tablet; the ideologies of each party simply do not include science in any prominent way, at least in the ways in which entitlements and tax cuts have been included. The life sciences community is making efforts to have its agenda carved onto the parties' stone tablets, but the physical sciences have not been as successful as the life sciences in this enterprise. The different abilities of the two scientific communities to take their messages to policymakers have profound ramifications for the life sciences and physics, particularly as they become more interdependent.

Trends in Federal R&D Spending

Recent trends in federal R&D spending reveal a steady decline in federal R&D spending as a share of Gross Domestic Product (GDP). From a high of 1.8 percent in the 1960s, Mr. Bonvillian said, federal spending on R&D has fallen to 0.8 percent of GDP today. In the so-called "knowledge economy" of today, financial analysts look increasingly at factors other than earnings to evaluate firms' future profitability. Certainly, R&D spending is one important non-earnings-related measure of firms' potential. However, in the context of the federal budget, R&D seems to be taking a back seat. From 1992 to 1997, federal R&D spending declined by 9 percent in real terms, and there was an additional scare this year when House appropriators passed a bill reducing civilian R&D by 10 percent. Subsequently, the House settled on a figure that approximates last year's spending on civilian R&D. Mr. Bonvillian suggested that this year's action by the House signals future political problems in the R&D arena.

The Distribution of Reductions in R&D Spending

R&D spending has not been cut equally across all sectors, Mr. Bonvillian said. There have been winners, and Dr. Rosenberg earlier showed that the research budget at NIH has been the beneficiary of effective advocacy by medical schools, patient groups, and other groups. The National Science Foundation has seen its R&D spending increase slightly in recent years, but R&D funding for some of the large agencies, such as the Department of Energy and the Department of Defense, has declined sharply.

Because different agencies emphasize different science fields, certain scientific disciplines become winners while others become losers. Work published by the STEP Board has found that 15 science research fields are declining, while 11 are rising.[2] The fault line for winners and losers is life sciences versus physical sciences, with the life sciences, as already noted, doing much better. Defense R&D cuts play a central role in these trends, simply because the Defense R&D budget is so large; in real terms, DoD's R&D budget is down by about 30 percent over the past six years. At the time of this report, the Clinton Administration is considering another 14 percent reduction in Defense R&D for the next fiscal year. Over the past 50 years, the Defense Department has funded the research of 58 percent of Nobel Prize winners in chemistry and 43 percent of the Nobel Prize winners in physics. The decline in DoD R&D therefore has profound implications for the physical and life sciences community.

Understanding Changing Patterns of R&D Spending

To some extent, one could argue that the shifting pattern of R&D expenditures reflects progress in different fields and new opportunities in them. The shifts may then be seen as natural adaptations to changing circumstances.

---

[2] Michael McGeary and Stephen A. Merrill, "Recent Trends in Federal Spending on Scientific and Engineering Research: Impacts of Research Fields and Graduate Training," in National Research Council, *Securing America's Industrial Strength*. Washington, D.C.: National Academy Press, 1999, pp. 53-98.

Mr. Bonvillian argued, however, that the shifts are largely a result of organizational changes in government. In the immediate post World War II era, Vannevar Bush envisioned a single Department of Science that would rationally allocate funding across competing scientific research needs. Of course, the United States never established such a department; only a small piece of it was incorporated into the National Science Foundation. Perhaps the rich mix of mission agencies and general funding of basic science has served this country better than a single Department of Science would have. With the end of the Cold War and the changing relative institutional strengths of various science-funding agencies, there has been a substantial reallocation of R&D funding. And neither Congress nor the Executive Branch has come to grips with these changes.

Perhaps the best argument for better balance in science and R&D funding across disciplines comes from the preceding discussions—namely that many of the most important scientific advances today come from research with a multidisciplinary component. In biotechnology and computing, future advances are likely to be based on inherently multidisciplinary R&D, so there will be major societal consequences if this research is not funded adequately and in proper proportion.

Challenges for Policymakers

A crucial question for policymakers is how to determine which R&D efforts must be funded to ensure future progress, in other words, "to separate the crown jewels from the paste" in funding certain parts of the overall federal research portfolio. This is a challenging question to say the least, because it is very difficult to project what will be important 20 years from now. In 1969, for example, two great technological developments occurred: the United States landed a man on the moon, and in the last day of the fiscal year, the Advanced Research Projects Agency (ARPA) put ARPANet—the forerunner to the Internet—into operation. At the time, the entire world knew about a man walking on the moon, but only a couple dozen knew about ARPANet. In retrospect, we need to ask what was more important. Even though it seems obvious to some today that ARPANet has had more impact, the answer to the question is very much open.

"Random disinvestment" in various scientific fields—which is arguably occurring today—is an issue that greatly affects the talent pool for the nation's overall scientific enterprise. Federal R&D funds not just specific science and technology projects but also graduate education. Such grants to graduate students help them develop not just textbook knowledge but also experiential knowledge on how to become good scientists and engineers. These grants, directly or indirectly, affect industrial innovation.

In confronting these issues, Mr. Bonvillian said that policymakers should try to develop "an early warning system" on the federal R&D portfolio. Policymakers must devise a way to assess when funding in a particular area has fallen below a critical mass level, below which future innovation may be at risk. The issue is not just a matter of funding levels but, as Dr. Rosenberg's and Dr. Flamm's presentations indicated, also the balance of R&D funding. Some "alert system" is necessary, but a host of measurement issues arise. How do we measure success and failure in a scientific discipline? Numbers of patents, citations in scientific journals, R&D investment levels, and scientific personnel are all possible measures, but each are imperfect in some way.

Mr. Bonvillian said that Dr. Dolores Etter, the Deputy Undersecretary of Defense for Acquisition and Technology, is concerned about these issues because they affect our defense industrial base. Dr. Etter has asked her staff to develop a "global technology watch" to track trends in worldwide R&D using some of the measures mentioned above. Dr. Etter's effort will focus in particular on trends in fundamental research. Additionally, a number of private organizations, the Science and Engineering Indicators at the National Science Foundation, the National Critical Technologies panel, and others, track research trends. Mr. Bonvillian suggested that the information from these various efforts could be combined to create an "R&D alert indicator" for policymakers. Developing an alert system does not ensure success in guarding against precipitous drops in R&D spending in some fields; but without at least assembling this information, we are far more likely to be caught off guard.

Mr. Bonvillian also suggested that the science and engineering community give more thought to how the nation can get more mileage from its R&D investments. The innovation process has been changing, and Mr. Bonvillian mentioned a recent article by Don Kash and Robert Rycroft arguing that innovation in complex technologies is now done by a network of innovators, not the sole researcher. They also note that complex technologies made up 43 percent of exported goods in 1970; in 1995, complex technologies made up 82 percent of exported goods. Moreover, 73 percent of patents filed in the United States in 1997 referred to publicly funded R&D; this indicates the growing importance of public R&D and networks of innovation.

The biotechnology and computing industries are prime examples of how constant interaction between government, industry, and academia has contributed greatly to innovation. The Internet, Mr. Bonvillian noted, evolved largely through the rich, continuous interaction between government, university, and industrial researchers. Indeed, the Internet's evolution points to the fact that innovation takes place through the exchange of information among networks of scientists, engineers, and users. However, federal R&D resources seem to be allocated on the basis of an earlier model of innovation—the so-called "linear model"—rather than today's era in which networking and multidisciplinary research is crucial. Policymakers must become more attuned to how innovations are actually developed and how federal R&D can support an innovative environment.

Mr. Bonvillian added that networking does not have to be based only on formal institutional arrangements. Companies are increasingly turning to expert software systems to enable them to ask simple questions such as "who knows what" about a particular area. This growing field of "knowledge management" is improving companies' capacity to maintain awareness of innovation. Federal R&D managers would do well to improve their "knowledge management" to keep pace with developments in the private sector and universities.

Summary

Mr. Bonvillian made the following points in summarizing:

- Science—especially physical sciences—face serious funding challenges. Physical scientists must learn from their counterparts in the life sciences how to build political support for R&D funding.
- An "alert system" must be developed to warn policymakers when funding levels for some scientific disciplines drop below critical levels.
- The federal R&D system must be organized to take advantage of burgeoning networks of innovation in the private sector and universities.
- The federal government must take better advantage of "knowledge management" to enhance the efficiency of the federal R&D enterprise.

THE CORNUCOPIA OF THE FUTURE

Daniel S. Goldin
National Aeronautics and Space Administration

I am really pleased the Academy is following up its June report on industrial competitiveness with this week's events. Science and technology drive the world's economy and help provide a healthier, safer, and better world. I am also glad you are emphasizing the links between biotechnology and computing. I believe those links will become even stronger as we enter the new millennium—just not in the way you may think. For NASA, that link will be crucial to everything we're planning to do in the twenty-first century.

Last week, I was flying to a meeting in Norfolk, Virginia. I was looking at the full moon through the airplane window and thought of the Apollo era and people walking on the moon. I thought about what an amazing achievement that was when you consider that the computer on Apollo was less powerful than the one in your car today. Yet it helped take humans to the moon and brought them back safely. And the Apollo era provided an incredible bounty of technological achievements that we are all still using. But that also reminded me of the great paradox we face today.

Today's electronics are much more advanced than what we used on Apollo missions. Those computers only ran at a few megahertz and cost millions of dollars—about 100 times slower and 10,000 times more costly than today's desktop machines. And yet, for all that, our software is more and more complex and much less error tolerant. We still interface with computers by typing characters into a keyboard, rather than a more natural interaction in a fully immersive, multi-sensory, and fully interactive environment.

Think about how many people in the course of a business day gather around a desktop computer just to figure out where this interface problem is coming from. I'm sure that has never happened to any of you.

And the basic platforms for aircraft and spacecraft in 1999 are no different than they were in 1969. The 707 was the last leap forward in aviation, and the rockets we fly today are the same technology we had 30 or 40 years ago. That's a real paradox.

The shuttle represented a major step forward. It opened the first era of reusability. We no longer throw all the hardware away, but we practically rewrite all the software between each and every mission. Your home computer is more powerful than the ones that first went into the shuttle and you can buy better flight simulators at a toy store than NASA had for the Apollo astronauts.

We have come a long way in some areas, but we still require thousands of people to process the shuttle between each mission and hundreds to conduct launch and mission operations. This operational dilemma I'm talking about is not unique to space or the shuttle program. All you need to do is take a look at today's air traffic control system or large petrochemical facilities or heavy manufacturing, look into the operations and see the price we're paying for present-day hard, deterministic computation.

Why the paradox? *We do too much by brute force.*

Then I began to think forward instead of looking backwards. Some things won't change. Reaching orbital speeds will always be a great accomplishment. However, if "past is prologue" as is written across the cornice of the National Archives, the future is very exciting. The last 350 years have shown the power of science and technology to shape our society. With each new plateau came great accomplishments—not foreseen by even the greatest of visionaries.

Newton's mathematical formulation of gravity and the laws of force and energy ushered in the era of modern technology. Maxwell's mastery of electromagnetism in the 1800s brought the Industrial Revolution to full bloom. The discovery of atomic structure, quantum mechanics and Einstein's theory of relativity in the 1900s gave us an understanding of the universe and the forces that shape it at its largest and smallest scales.

The twenty-first century will bring the age of bioinformatics and biotechnology, and the physical scientists cannot sit back and argue about why not to spend money on biology. Most people don't understand biology's potential to dramatically change electronics, computational devices (both hardware and software), sensors, instruments, control systems, and materials—or the new platform concepts and systems architectures that will bring them all together. The terms of the future are biomimetics, bioinformatics and genomics, and they will be as common as transistors and microchips are today.

So far, our ability to emulate biological functions has been limited to what we can do with silicon microchips, space age materials and chemical reactions. This is simply not good enough. In the future we want our systems to be biologically inspired. This means we can mimic biology or embed elements of biology to create hybrid systems, or they can be fully biological and life-like.

The greatest attribute that biological systems have over solid-state systems is the ability to change on their own. These systems will adapt to different operating environments, to accomplish different tasks or to renew and repair themselves. A robot on a distant planetary surface shouldn't walk off a cliff simply because someone in mission control on Earth pre-sent a code to move it forward 10 paces. Now this is not fiction, because we just had a robot out in the desert to simulate an operation on Mars. We pre-programmed the robot, and it walked right over a dinosaur footprint.

NASA has set its sights on understanding our universe in greater depth. Eventually, we want to send humans to the edges of our solar system and beyond. This requires a dedication to pursuing the unknown. We have to learn as much as possible about our home planet. We need to explore other planets and return the data—or the human explorers—to Earth safely, relatively quickly, and inexpensively. However, we cannot use brute force to achieve these goals. It is too clumsy, too costly, and too slow. What we need are systems that work with people, to enable a few people to do what many people do today. We need to develop new computing systems to take over routine and mundane tasks, to monitor, to analyze, and to advise us. This means they must be capable of more than just following a set of hard, deterministic pre-programmed instructions.

Today's computers are a lot like moving a ball on a flat table. People decide direction and velocity, and they have to keep maneuvering the ball on the table. That's how today's computer coding works. The codes describe an action, computation, or comparison, and the computer simply executes what it is told to in a linear, deterministic order—one step after another.

Today our complex systems contain thousands of microchips controlled by millions of lines of code all written by hand and structured around a systems architecture that has very little fault tolerance. Since we can only write, verify, validate, and qualify a few lines of code per hour, the cost and time for software development is huge. And we need so many software coders—there aren't enough in America—that it is holding back our progress because we keep sticking with old tools. The software concepts we are using today were basically developed in the 1970s.

Yet this software is so complex that we can never be sure of finding all possible failure modes. We keep patching and patching. In addition, when everything has checked out and been proven flightworthy, a subtle manufacturing change in any one component can introduce a new failure mode that may not be uncovered during check-out and quality control procedures. We are starting to see this problem become more and more serious each year, and it's costing American industry dearly. It's costing the space program dearly.

This interdependency between software and hardware implementation has, in effect, built design obsolescence into our products. It is becoming almost impossible to easily and safely upgrade systems as technology advances. This paralyzes the ability of our designers to utilize platforms that are upward-compatible with technological advances.

In our space program, this has resulted in robotic operations that require direct human oversight, whether it is a rover on a planetary surface or an astronaut controlling the shuttle's robotic arm to move objects in and out of the payload bay. It requires many months of expensive simulation and training to plan and conduct these space operations.

We need to move toward the next era of space robotic operations that places the human into the role of a systems manager, not a real-time controller as they are today. This is a crucial difference.

Levels of intelligence need to be incorporated into our systems so that they can conduct routine tasks autonomously. Ultimately we want "herds" of machines to function as a cohesive, productive team to explore large areas of planets, build structures in space, and perform continuous inspections of our most critical systems.

We also want those robots to be able to express emotion and be able to overcome some of the tremendous challenges we are facing. They can also perform the most dangerous tasks and keep our astronauts out of harm's way. Within the foreseeable future, intelligent machines will not totally replace people. They simply let people do what they do best—engage in creative thought—in the safest possible environment.

NASA believes one of the most fruitful approaches for getting us there is to look to biology for inspiration. Mother Nature holds all the best patents that already exist. Nothing approaches the inherent intelligence, power efficiency or packing density of a "brain." "Eyes" can almost respond to a single photon and biological electronics sensors are extraordinarily sensitive. Fireflies convert chemical energy to light with near-perfect efficiency, and living membranes, such as skin, will self-heal when injured. No other technology we know of has a comparable ability to self-organize and reconfigure.

We need to look to nature for solutions.

We cannot have a power cord going all the way to Mars—the space program cannot work that way. So our computers must become thousands of times faster, denser, and less power consuming. We will talk about teraflops per watt, not per megawatt, and sizes in terms of cubic centimeters, not cubic meters. The fastest computer today has a teraflop of speed, but it takes a megawatt of electricity to operate it. The brain is a million times faster and takes a fraction of a watt.

Think about it.

We will have robots and spacecraft that are self-diagnosing and self-repairing. They will be able to learn and act, respond, adapt, and evolve. They will also be able to reduce vast amounts of raw data to useful information products to assist humans, not replace them.

We will have a new era of human-machine partnerships. But this time, HAL of "2001: A Space Odyssey" won't get depressed. Nature is clearly superior, so we can use our understanding of cells and systems as guides for our next-generation devices.

The computational systems of the future will be very different. They will not be designed or built anything like today's computers. They will be more like the brain, with millions to billions of relatively simple but highly networked nodes. These computers will solve problems by absorbing data and being inherently driven to assimilate solutions to our most difficult problems. Our goal is to keep these computational systems focused on what we want them to do. They will work like water flowing downhill, changing directions, flowing around large obstacles, and over or through smaller ones—rapidly finding their own way to solutions. The computer will capture all relevant physics, biology, and real phenomena, including complex transient behavior.

Even more exciting will be the development of hybrid systems that combine the best features of biological processes, including DNA and protein-based processes, with optoelectronics devices and quantum devices. These hybrid systems will be extraordinary in performance and functionality, and not achievable by any one technology. We get the best of both worlds, but only if physicists learn to cross the biological divide.

Just imagine the day when computers behave more like we do. We will communicate with them using all our senses through totally immersive environments including natural language—not simply through characters typed on a keyboard. They will understand our intentions and even sense our physical state.

Sound like science fiction? Researchers at NASA Ames have successfully tested a revolutionary bio-computer interface using electromyographic signals (tiny electrical impulses from forearm muscles and nerves) to fully control the takeoff, flight and landing of a commercial airline in a high-fidelity simulation. They don't have keyboards for the computer, just the sensors on their arms and hands. The eventual goal of this research is to develop direct bio-computer interfaces using total human sensory, and two-way interaction and communication, including even haptic feel. As strange as it sounds, we will be moving back to the analog computer, except now the analog is the human brain. Biologically inspired computing tools will also enable us to develop intelligent robotic systems.

Before we send humans to explore beyond our planet, we will first send robotic colonies to set up livable systems. They will need to behave just like humans. They will have to adapt automatically to the environment. They will be autonomous, and they will be able to do jobs in non-traditional ways. Like humans, these robots will employ biologically inspired sensors and motor control, anticipate future events, cooperate with other systems, select and depose leaders, and be motivated to explore, repair, and adjust to meet changing needs or to respond to an emergency.

Giving robots and computers the capacity to learn and act requires a move from conventional deterministic software to "soft computing," which accounts for uncertainty and imprecision. Neural networks—one element of soft computing—could be the basis for the futuristic computers I described earlier. Neural networks assimilate vast amounts of data and extract information—trends, patterns, solutions—the kind of thing we do when we learn and think as humans.

Today we can build systems with hundreds to thousands of neural connections. In the future, they will have millions of connections in a package the size of a sugar cube.

For many applications, neural networks reduce supercomputer time by more than an order of magnitude while providing more accurate analysis than conventional approaches.

We are simulating turbine engines on computers driven by neural nets to help engine manufacturers regain the critical edge they need. But they need biology to do it. NASA also has applied neural networks to the flight controls of an F-15 aircraft. The traditional flight control software system—based on conventional, deterministic software methods—required one million lines of code. We reduced it to about ten thousand. We demonstrated improved performance. We introduced software faults, and the system identified the problem and self-corrected in seconds. We even simulated loss of aircraft control surfaces, and the system adapted and returned aircraft control authority within seconds. Now that is an intelligent system. And we are just getting started.

We will move from data . . . to information . . . to knowledge . . . to intelligence.

When we are ready to send humans deep into space, astronaut health and safety will be our top priority. Biologically inspired technologies will enable a human-machine partnership that is advanced enough to ensure the well-being of astronauts, perhaps on a 2- to 4-year trip to Mars.

If you get appendicitis 50 million miles from Earth and travelling at 25,000 miles an hour, you will not be able to come home. And if you are the only doctor on board, you had better have some of these biologically inspired tools. You cannot take a hospital with you. It is too heavy.

We could develop nano-scale sensors with sensitivities and detection capabilities at the molecular level. They would be like little monitor cells, and they would transmit the information they acquire to other systems outside the body or inside our spacecraft. While we will not sense that such devices are in our bodies, we will certainly know they are there because of the continual stream of information they provide about what they are doing and finding. Such on-board devices will be crucial to health care in space or on the ground. An astronaut's medical emergency could become a tragedy if we were forced to rely on responses from Earth, which might involve round-trip transmission times of 20 to 40 minutes.

Health and safety are much better served with real-time assessments and decisions, which could easily be done with on-board machines. A smart robot, acting as a health-monitoring "buddy" to a human, will free an astronaut from ongoing self-monitoring, leaving much more time for humans to think, create, and experiment.

Humans will still be the ultimate decision makers, using information from the robots and sensors. However, we will be much more informed decision makers than we are today. The idea of a "one size fits all" cure like "take two tablets every four hours" just won't do when we send astronauts to Mars. We will have diagnostic and treatment procedures that are adaptive in unknown environments. Our machines will respond to our needs and our moods. By measuring nerve activity on the surface of the skin, machines can determine if a person is calm or agitated. Or we might measure hormone or neurotransmitter levels as an indicator of our emotional state or stress level.

They will also make us more productive and improve safety by alerting us to mistakes before we make them, letting us know when we are showing signs of fatigue—an enormous problem in corporate America today. This is just a sampling of NASA's vision of the amazing benefits biotechnology will bring us in the future.

And we can use these technologies to improve life on Earth. For instance, the tiny health monitoring sensors I mentioned could give constant health updates and instant diagnoses. We are already working with the National Cancer Institute (NCI) to attempt to develop such sensors. We are bringing our physics-based technologies to researchers in the NIH who need these tools. Science and technology are intertwining. We cannot separate them.

NASA sees great potential for enhancing astronaut health during lengthy missions, and the NCI is obviously interested in using the sensors to detect indications of cancer. The National Cancer Institute would like to detect the first mutations of a cell. We don't know if we'll get there, but micro-electric sensing may be the way to do it. This system could also be used to control the delivery of drugs. The sensors could analyze blood chemistry—or other material *in vivo*. Equally important, the sensors could be placed at specific locations where measurement is optimal or where measurement is critical. We would know that a drug may be accumulating at too high a rate in a sensitive part of the body before there is any adverse response.

We would not wait for the body to tell us there is a problem at the macroscopic level; we would know it beforehand. This sensor suite could monitor all critical bodily functions, account for gender differences, and use this data to guide therapies. Biologically inspired technologies also have incredible potential beyond health-care applications.

Eye-like devices of the future would not just work in the visual. They would work across the entire spectrum. Just imagine how much intelligence that would give our factories, especially "eyes" with single-photon sensitivity. These "eyes" could open up the entire bandwidth of vision and greatly enhance processing and control of products by detecting even the tiniest material defects both on the surface and far below the surface.

We could use "noses" to protect workers in hazardous situations by "sniffing out" potential dangers. This would revolutionize plant safety by providing layered levels of protection and warning, not achievable with today's mass spectrometry. As I said, the twenty-first century will be the age of biotechnology. It promises a bounty of life-enhancing technologies that will eclipse America's incredible achievements during the last five decades—the decades of physics.

Let me take you on a futuristic journey. Our entire notion of designing and building systems could change. We would like to have the power to pre-build entire systems in "cyberspace" totally within the computer with geographically distributed teams using heterogeneous systems that use physics and biologically-based calculations to enhance their work.

We could plan and develop every step of every mission from concept to disposal before we ever place a single order or bend a single piece of metal. But in the end, we will still order parts and bend metal unless we have a revolution. Instead of committing 90 percent of our resources early in the process when we only have 10-percent design knowledge (this is why programs overrun in industry and government), we could simulate everything first and focus our resources where they are needed most. When ready, we will have bounded 90 percent of uncertainty and have full confidence in the total system life cycle, cost, and performance.

That is what NASA wants to do a decade from now. However, the ultimate power of biology will be to design the molecules that contain the coding, or blueprint, for complex space systems—that is, spacecraft DNA—that will initially build critical parts, and ultimately, the entire spacecraft. We will simulate the entire process to be sure we have it right, but once we "plant the seed," spacecraft genetics will take over in our factory of the future. Think of it as hydroponic farming for robots.

I am talking about truly multi-functional capabilities. Machines would fully integrate mechanical, thermal management, power, and electronic systems. They'd be more like humans than simply black boxes and wiring harnesses placed on structures. These biologically based systems may not look exactly like the systems we have today, but they will perform the same basic functions. They will just be better, faster, and cheaper. When I talk of planting the seeds for a new revolution in technology, I mean it in literal terms.

Let's go another step further.

With the ability to grow spacecraft systems, our craft could grow new parts from raw materials, or better yet, consume the parts it no longer needs to make the parts it does need.

This metamorphosis of single living entities or local ecosystems is taken for granted—it happens in nature every day, every minute, every second, everywhere. To some, it is science fiction to think of spacecraft in this way. But think of the implications upon modern manufacturing and the radical new products we can make on Earth to enhance our society.

As I said, we need to develop machines that will work with our people in the future. But today, we have to work together with a common purpose to make this vision a reality and to make sure America achieves its destiny in the new millennium. Every member of the partnership—government, industry, and academia—needs to contribute to researching and developing revolutionary technologies.

And I want to leave you with this thought. This is not going to happen with wishful thinking. This is not going to happen by scientific cannibalism. This is going to happen because this country must understand that unless it looks to the future, unless it does not take a vacation from long-term R&D, we may have problems beyond the twenty-first century. But I believe without a doubt that the American public will get it. The potential is there. All we have to do is reach out.

Panel III: Biotechnology: Needs and Opportunities

INTRODUCTION

Edward Penhoet
University of California at Berkeley and Chiron Corporation

Dr. Penhoet convened the panel by observing that much of the day's previous discussion had underscored the importance of the biotechnology and computing sectors to the economy, and how each field will fuel advances in the other. This panel would focus on what the computing and biotechnology sectors need to sustain the current rate of advancement, and how the government and private sector could best work together to usher the fields into the next century.

EXPLOITING THE BIOTECHNOLOGY REVOLUTION: TRAINING AND TOOLS

Marvin Cassman
National Institutes of Health

The marriage of biology and technology, said Dr. Cassman, is creating an incredible pace of change in the biotechnology field. It is also creating a glut of raw information that is threatening to engulf everyone in the field. Three technologies are driving the fast rate of change:

- Molecular genetics
- Structural biology
- Genomics

The first two are disciplines that have turned into tools, and the third is a tool that has turned into a discipline. All three have brought us to the post-genomic era, a phrase that connotes the plethora of information that is pouring in upon researchers. Much of the information comes from genomics, but not all of it.

The primary driver for modern biology, Dr. Cassman said, is the "great engine of molecular genetics." That quote came from a report in synchrotron radiation, which is something that until about 15 years ago was the exclusive province of high-energy physicists. Now, however, synchrotron radiation is an important tool in biology for conducting research on macro-molecular structures at high resolution. This has been a revolutionary development in biology and it permits sophisticated research on DNA and proteins. Before the adaptation of synchrotron radiation for biology, it would take years to understand the structure of a protein. Such an effort might comprise an entire dissertation for a doctoral student. Now, however, understanding one protein would be one part of a larger research program.

On the whole, the rate of advancement in structural biology has been extraordinary recently. One of the most important papers in structural biology in the last year, which appeared in *Nature*, concerned the potassium channel. None of the authors was a "card-carrying" crystallographer; traditionally, such research has required the specialized expertise of a crystallographer and biologists have been relatively unconcerned about structure. This example shows biologists' newly found interest in structure, and advances in information technology have enabled this.

Connected to biologists' research into cell structure are advances in genomics. Genomics provides a baseline of understanding of the total complement of information in a cell. Of course, there are other things happening in the cell, but understanding the baseline is the starting point for research into the cell.

With the tools of molecular genetics, structural biology, and genomics, the question arises about the discipline's future path. A reasonable progression in biology would be to first understand the components of cells and their function, then how they link together, and finally to learn how cells interact as complex systems. This is an idealized progression—advances in biology do not take place in such a neat sequence—but it provides useful guidance nonetheless.

Metaphors to Motivate Progress

A Parts List

The ultimate goal, Dr. Cassman continued, is to understand biology as a complex system. To lay out a research program to attain that goal, several metaphors are helpful. The first is the periodic table or a "parts list" of how the parts of living systems operate and the intrinsic function of, say, a particular protein at the molecular level. The next step is to sort the parts into the proper bins, which is not at all simple and on which only limited progress has been made. Presently, only between 30 and 50 percent of the "coding regions" of most organisms' genome are identified. Even once a coding region is identified, it may be possible to identify its biochemical properties, but its cellular function is still not obvious.

A Wiring Diagram

Advancing from biochemical functions to cellular functions will take more than a "parts list." It will take an understanding of cells' connectivity, and this is why a "wiring diagram" is an appropriate metaphor to advance biology further. Knowing the genes that contribute to a phenotype is necessary but not sufficient for understanding a cell's function. Understanding connectivity is required, and this is why a "wiring diagram" is needed. The pace of advancement in understanding the "wiring diagram" is good, though not as rapid as for the "parts list," but much remains to be done.

For example, at one time understanding the signaling pathway mediated by a hormone was the staple of the practice of biochemistry. In the past, this was conceived of as a linear pathway, but in fact the path is not linear. It is a network and networks of connections at the molecular level govern most regulatory processes in cells. Dealing with networks is a far more complex task than dealing with a linear path, and we need more sophisticated tools today to understand properly how these networks function. Similarly, dealing with human diseases will require understanding how multiple components interact, not just identifying a single gene defect. In fact, identifying single gene defects has been the basis of much of modern biotechnology.

Networks

The third stage in advancing the frontiers of biology involves understanding complex systems. This stage derives from the fact that living organisms and cells are not static; they exist in space and time. The network shown in a "wiring diagram," for example, conveys nothing about a living system's spatial or temporal resolution. Living systems are dynamic, and biology is still a considerable distance away from fully understanding living systems in a dynamic environment.

Appreciating the cell as a system requires first an understanding of cellular components and their interaction. To make progress in understanding biological systems, we will need a quantitative understanding of biological materials, and this means that truly interdisciplinary tools will be called for from chemists, mathematicians, physicists, and engineers, as well as biologists. Rather than looking at biology as loosely linked molecular devices, researchers will have to think of living organisms as systems.

Challenges to Progress

One important challenge in biotechnology is the need to increase the use of quantification and mathematical modeling. However, these tools are not commonly used by biologists. In the field of molecular genetics, in which dramatic strides have been made, the extent of mathematical sophistication is limited to determining whether there is a spot on a cell or not. Second, multidisciplinary research is more often honored in the breach; more people talk about it than engage in it. Collaboration exists in biology, but it is based more on a mutual need than a common goal. A cell biologist will readily seek out a structural biologist to address a particular problem, but this is far from bridging the chasm between mathematicians and biologists. To bridge this gap, an accommodation to different professional practices and cultures is needed, a difficult undertaking. Finally, biologists have adopted the paradigm of the single gene defect for understanding phenomena. A certain degree of retraining will be necessary to think of biological phenomena as part of complex systems.

Nonetheless, Dr. Cassman said that moving to an understanding of cells and organisms as complex systems in space and time is difficult, but not impossible. Such research is already underway, and the effort at Berkeley described by Dr. Penhoet is one example, among others around the country, of that. The goal is to arrive at an understanding of the cell's complete function in such a way so that the phenotype can be modulated. By that, Dr. Cassman meant phenotype in its most general sense, that is, the expressed characteristic of a system or an organism. For example, a biologist wants to know what happens when a certain hormone is added to a complex system, such as a cell. Understanding how the hormone affects the function of a single cell is presently difficult, but it is crucial to understanding how to develop therapies that will modulate a cell's function.

Biology and Complex Systems

The challenge is how to accelerate biologists' understanding of organisms as complex systems. Dr. Cassman mentioned some discussions under way at the National Institute of General Medical Sciences over the past 3 to 4 years to address this issue. The Institute of General Medical Sciences has brought over 100 researchers together in these discussions, along with a set of broad operational categories that seek to address challenges associated with understanding complex systems. With respect to expanding the "parts list," more funding is the key. Reams of data are pouring out from genomics research and the challenge is to develop new database tools to classify and store the information. Aside from additional funding, database development does not require a new targeted initiative. With respect to the "wiring diagram," Dr. Cassman said that steady progress is being made in understanding cell functions as networks.

The largest challenge lies in understanding cellular functions as complex systems that exist in time and space. Dr. Cassman says that we are far away from meeting this challenge, and an important question is how to properly define a complex system in biology. At the Institute of General Medical Sciences, a complex system is defined as "one in which the behavior or expressed characteristics of a biological system is determined by the multiple interactions of components whose quantitative expression may vary in time and space." Dr. Cassman noted that, though useful, this definition was not fully satisfactory; indeed, Dr. Cassman observed that *Science* magazine devoted an entire issue to complex systems, but it was unable to develop a definition of a complex system.

In addressing complex systems, Dr. Cassman said that scientists as yet really do not have a method to adequately do so. There are no obvious ways to model complex biological systems, and Dr. Cassman was unaware of anyone who has tried to do so. It is important to start with simple systems and build from there. Recently, bacterial chemotaxis was successfully modeled, a system with only four or five components, but complex nonetheless. Dr. Cassman noted that the investigator who worked on bacterial chemotaxis is a physicist who has turned to biology. This is an example of someone with the mathematical and modeling tools turning to biology. More of that must occur.

NIH Initiatives to Promote Collaboration

The National Institutes of Health have made efforts to increase collaboration among biologists, physicists, and mathematicians through funding collaborative research. While a useful beginning, Dr. Cassman said that presently the initiative suffers from "too many good intentions and not enough knowledge." There is a substantial gulf separating the two disciplines; biologists are unaware of mathematical tools available from physicists, while physicists are unaware which biological systems can usefully be investigated. It will take long and hard work, said Dr. Cassman, to overcome these barriers.

A second part of NIH's efforts to increase collaboration is something called "glue grants." Dr. Cassman said that for many difficult research problems in biology, the material and intellectual resources do not exist in a single laboratory. Even if a laboratory could employ 20 top-flight post-docs on a single project, this would not be sufficient to address some problems. NIH's "glue grants" do not support underlying research, as support for that is assumed to be in place, but are intended to facilitate interaction among diverse sets of researchers. It was surprising to Dr. Cassman that a large number of applications were submitted for "glue grants"; this suggests that the research community has significant interest in expanding collaborative activities.

The third area is the development of and access to research tools. This may seem obvious, but as in the example of x-ray crystallography, today's specialized research tools are becoming useful for a wide variety of purposes. Specialists are no longer the sole users of advanced research tools such as nuclear magnetic resonance and mass spectrometry; in recent years, for example, there has been a rapid growth in the use of mass spectrometry among cell biologists. Today, cell biologists not only need mass spectrometry tools, but also people trained in how to use them.

The issue of tools also extends to data sets and material resources. In structural genomics, the research agenda is to arrive at the three-dimensional structure of every protein known. To accomplish this extremely ambitious task, researchers plan to parse the human genome into various families and sub-families to pick representatives of other proteins. Homology modeling will then be used to determine the structure of the proteins. This research program will be international in scope. In fact, a meeting will be held in Europe in Spring 2000 to ensure that collaborators in England, Germany, France, and Japan are working in concert as progress is made in modeling 10,000 structures in 5 years. This will require not only funds for laboratories and researchers' salaries, but also the ability to analyze and share large data sets.

Finally, NIH is trying to develop new training programs in computational biology and bioinformatics. Dr. Cassman cautioned against programs—at NIH or elsewhere—that train computational biologists and specialists in bioinformatics, but do not develop ways to promote true cross-fertilization.

In conclusion, Dr. Cassman said that the biology and the biotech industries are today confronted with more information than they can assimilate. Biologists really have no alternative but to draw on tools from chemical engineering, physics, and computer science in order to construct a quantitative dynamic structure for biological systems. The barriers to doing this are cultural as well as scientific, but the effort to bridge the gap separating disciplines must urgently be undertaken.

Discussants

Dr. Dahms noted that there is a continuum of training needs among computer scientists and biologists. That is, there may be different relative payoffs from training biologists in computer science versus training computer scientists in the life sciences. Dr. Dahms asked Dr. Cassman how, from the perspective of universities, training programs should be developed in light of potentially different payoffs to training in the different disciplines.

Dr. Cassman responded that whatever the relative payoffs in interdisciplinary training, the important thing is for "bodies to be exchanged" across disciplines. Computer scientists must spend time in biologists' labs, and vice versa. Dr. Cassman has not witnessed much of that, although he said that there have been exchanges among physicists and biologists. There will always be only a few people sufficiently trained in physics and biology to perform experiments in both fields at a high level. It is more important, Dr. Cassman said, to have an adequate number of people conversant in the language of the two disciplines so that meaningful collaboration can occur. Training such people is feasible, although it will take time to produce enough trained individuals. In the meantime, NIH has developed one- to two-week intensive courses designed to improve understanding among biologists, physicists, and computer scientists about each other's disciplines.

Dr. Penhoet asked why, in the age of the Internet, it was necessary to promote face-to-face interaction among scientists, as has been the thrust of many comments during the conference. He noted the paradox of vastly improved communications technology existing alongside apparently growing geographical concentration of innovation in certain regions. Dr. Cassman responded that even before the Internet, it was possible, in principle, for researchers to read all the relevant journal articles and then go replicate or expand upon a certain experiment. Almost no one works that way. Dr. Cassman added that he told students to first read a journal article, and then go work in the lab of the researcher if the student wanted to learn how to do a particular experiment. There are too many informal modes of communication that a journal article or the Internet cannot capture and convey.

A questioner commented that biology departments often segregate internally, meaning that molecular biologists may rarely communicate with ecologists, and that this undermines the interdisciplinary goals Dr. Cassman has discussed. The questioner asked whether NIH has considered this problem. Dr. Cassman agreed that despite all the talk of multidisciplinary research, collaboration remains difficult among sub-disciplines in a field. To facilitate collaboration, a culture of collaboration must take hold. Developing such a culture is hard to do, and takes time. Dr. Penhoet added that our entire educational system penalizes teamwork; it is often called cheating in universities and results in academic penalties. In industry, the payoffs to teamwork are well known. Dr. Penhoet noted that the educational system appears to be doing a better job today of promoting teamwork.

THE NEW FRONTIER: BIOINFORMATICS AND THE UNIVERSITY

Rita Colwell
National Science Foundation

Dr. Colwell began her remarks by observing that two decades ago, computers and biology were rarely used in the same sentence, let alone research. In her days as a graduate student, Dr. Colwell was considered a pioneer because she wrote a computer program—in machine language on an IBM 360—to classify bacteria. A model of the computer she used, housed at the time in the attic of the chemistry building at the University of Washington, is now on display at the Smithsonian. Today, doctors and hospitals automatically classify and identify bacteria by computer.

In conveying the distance that biology has traveled since Watson and Crick discovered the double helix in DNA in 1953, Dr. Colwell said that we now know that the Arabidopsis genome is 1.8 inches long and contains 135 megabases of information. The human genome, estimated to be 3.3 feet long, is being elucidated today. Sequencing the human genome, as is well known, will require significant improvements in biologists' ability to manage information.

Dr. Colwell said that bioinformatics is not just about genomics or managing information, but it has led to new ways of communicating information among biologists. In taking a broad look at information technology, advances in that field will result in faster, more secure, and more reliable software for biologists. In the National Science Foundation's (NSF) FY 2000 budget, Congress granted the agency's request for significant funding for information systems that will provide improved tools for biotechnology research. In order to sustain the rate of innovation predicted by Moore's Law, Dr. Colwell pointed out, it will be necessary to develop new research tools for biotechnology.

Biocomplexity

The promise of information technologies in biotech research has to do with allowing scientists to study entire biological systems. "Biocomplexity" is the next challenge for biologists—studying the complex interdependencies among various systems in the environment. As biologists turn to these areas of inquiry, the challenge is not in collecting data, but in managing it. As we improve our ability to manage biological data, scientists will be able to make significant strides in biocomplexity.

A host of innovations have created a flood of data about the Earth's complex biological systems and the processes that sustain them. From DNA chips, to geographic information systems, to bio-sensors, to ecological monitoring devices and satellite imaging systems, we have more data than ever before about the Earth and its inhabitants. Because of these new demands, NSF will invest $50 million in FY 2000 in a new focused research initiative on biocomplexity.

As an example of NSF's efforts to use information technology to better understand complex biological systems, Dr. Colwell described NSF-funded projects that explore global and regional distribution of temperature, precipitation, sea level, water resources, and biological productivity. Dr. Colwell displayed a map that, using sophisticated satellite and sensing technology, showed the distribution of the population of a particular bird in Mexico. The map was created using information collected by 24 museums throughout North America. In the past, these data were kept in separate databases, making research on the bird and the ecological system that supports it nearly impossible. Extending this research will mean understanding the interplay among species and how different species coexist and co-evolve over time. This will place new computational demands on biology.

Biocomplexity has evolved from the integration of disciplines—finding places where biology and physics explain each other—to include chemistry and geology in understanding the environment. The interconnection of disciplines is best captured by a quote from John Muir: "When we try to pick out anything by itself, we find it hitched to everything else in the universe." Biocomplexity will present many research challenges, and universities will play a prominent role in this research. Universities will also play a crucial role in training the workforce that will conduct multidisciplinary research.

Twenty-first Century Workforce

Dr. Colwell raised the larger question of U.S. leadership in innovation and information technology, and how the nation must prepare for a future in which more workers will need to be literate in science, engineering, and mathematics. This is why NSF's Twenty-first Century Workforce Initiative is important to the agency's overall mission and the nation's economic future. This program establishes partnerships with the private sector and universities to expand training programs in math, science, and engineering.

One focus of NSF in this initiative is the issue of information "haves" and "have-nots"—the so-called Digital Divide. Dr. Colwell said that social scientists have identified demographic groups in which telephone, computer, and Internet access lag well behind national averages. In general, according to NSF's Science and Engineering Indicators, there is a strong correlation between income level and computer use. Information gaps exist among nations as well, and Dr. Colwell observed that most people in the Third World have never used telephones. Less than 2 percent of the world is on the World Wide Web, and if the United States and Canada are subtracted, the share is less than 1 percent. The inequality in access to cyberspace should be of concern not only for humanitarian reasons, but for economic and political ones as well.

An important element of the Twenty-first Century Workforce Initiative is to better understand the nature of learning. The initiative will take a quantitative approach to behavioral science research to improve our fundamental knowledge of how individuals teach and learn. In the arena of research using supercomputers, NSF has funded brain-imaging research at the University of California at Los Angeles that has developed a "brain template." This research, though at a very early stage, is seeking to determine which parts of our brain are most active when we are learning. People learn in a variety of different ways, and those who appear to have difficulty learning may simply learn in a different way. If the different ways in which people learn can be isolated, then teaching techniques can be adjusted to individual needs. Research into the structure of the brain, usually conducted by NIH, can therefore be usefully connected to research at NSF on the nature of learning.

Bioinformatics and NSF

To fuel the growth of bioinformatics, more must be done at the level of graduate education. As has been mentioned earlier, many of the research challenges in biology—from gene analysis to drug discovery—are in fact computational ones. As we all know, there is a desperate shortage of trained specialists able to meet these challenges. To address the problem, NSF has funded a "career awards initiative" to prepare graduate students in science and engineering not only for careers in academia, but in the private sector as well. Dr. Colwell said that with the NSF initiative, she expected to see many more graduate students doing internships in private industry in the future.

In closing, Dr. Colwell noted that NSF celebrated its fiftieth anniversary in early October. The upcoming challenges in bioinformatics mean that NSF's mission is ever more urgent, in terms of training graduate students to develop the new computational tools that will be necessary to exploit fully the promise of the biotechnology revolution. The scientific community must embrace the new methods and approaches in biotechnology research, and NSF stands ready to play a support role in encouraging this.

Discussants

Dr. Dahms pointed out that there have been some projections that the bioinformatics industry might be $2.5 billion by 2005, a 12-fold increase from today's level. He also observed that the president of Compaq has predicted that by 2005 40 percent of all biotech companies will be in bioinformatics; essentially, two-fifths of all firms in biotechnology will sell information. This raises questions about workforce preparation, namely whether there will be enough people trained in bioinformatics to fill outstanding jobs. Dr. Dahms said that the industry may need as many as 20,000 workers in bioinformatics by 2005, a substantial increase. Dr. Dahms asked what NSF is doing to address these challenges.

Dr. Colwell said that the NSF's Twenty-first Century Initiative would be critical to meeting this challenge and mathematics education will be the important underpinning to this effort. In FY 2001 and 2002, mathematics will be the subject of a major NSF initiative—pure mathematics, applied mathematics, and statistics. In biology, Dr. Colwell expected to see a reemergence of mathematical biology, in addition to collaboration among physicists, biologists, and chemists. One initiative at NSF involves encouraging graduate students in math, science, and engineering to become involved in kindergarten to twelfth grade (K-12) education. Under that initiative, an NSF stipend will allow graduate students to spend up to 20 hours a week in K-12 classrooms under the supervision of a teacher to promote science education. Dr. Colwell says that she believes that there is a "valley of death" between grades 4 and 12 during which we lose the interest of many students in math and science. The NSF is attempting to address this problem in the United States and elsewhere. The United Kingdom, she said, also faces this problem and is attempting to address it. Finally, Dr. Colwell said NSF would launch regional science and technology centers, modeled after engineering research centers, to be funded at between $3 million to $5 million per year for several years, and then be potentially renewable.

Another initiative that Dr. Colwell would like to implement is to raise the salaries of outstanding instructors in introductory science courses in universities. It is especially important to encourage quality teaching of science to students whose major may not be in the sciences. Another area to explore is how to permanently increase the salaries of teachers recognized for excellence. Dr. Colwell said it was oxymoronic to give outstanding teachers a plaque and a semester off from teaching. It would be better to reward them with a permanently higher salary.

Greg Reyes of Schering-Plough observed that his industry was already overwhelmed by data. He asked Dr. Colwell specifically how NSF planned to facilitate allowing chemists and biologists to better manage data. Dr. Colwell said that NSF's information technology initiative will help in this regard, investing $105 million primarily in new software and research into how to link disparate computer systems together to share biological data. NSF also received $36 million in funding for high-speed computing. Dr. Colwell also said that advances in computing power should enable dramatic improvements in social and behavioral sciences. With respect to translating some of this research into useful information for industry, Dr. Colwell said that programs such as the Small Business Innovation Research program is an excellent vehicle for outreach to industry by NSF. Partnerships between industry, government, and universities will be necessary to get the most out of the new NSF initiatives in mathematics and information technology.

EMERGING OPPORTUNITIES AND EMERGING GAPS

Paula Stephan
Georgia State University

Dr. Stephan said her remarks would focus on four dimensions of emerging gaps in the supply of workers trained to handle the vast quantities of biological data being produced. These are:

- An indication of strong demand;
- A summary of what is in the pipeline at present;

- Explanations for the sluggish response to growing demand; and
- Possible solutions.

Strong Demand

Not only is mapping of the human genome generating a great deal of data in biology, other types of data are being created as well. There is widespread agreement that we are experiencing just the "tip of the data iceberg" today. Because of the huge amounts of data, a number of companies have begun to recognize the possibilities of computational science and in particular the potential to develop drugs from models based on biological data.

As indications of strong demand, the scientific press frequently reports on students being "grabbed" from graduate programs before they have completed their degrees. Faculty members have been lured from universities as well, creating worries that bioinformatics is "eating its own seed." Moreover, large salaries—on the order of $65,000 for master's graduates and $90,000 for new Ph.D.'s—indicate surging demand for individuals adept at manipulating biological data.

To get a better handle on the nature of the problem, Dr. Stephan and her co-author, Grant Black, examined ads for positions in bioinformatics in *Science* magazine and surveyed institutions of higher learning to examine their responses to the changing demand conditions. From analysis of ads for bioinformatics and computational biology positions in *Science*, Dr. Stephan said that the number of ads was generally higher in 1997 versus 1996. Looking at the data more closely, the number of distinct position announcements grew by 68.6 percent from 1996 to 1997, with position announcements from firms growing by 70 percent, universities growing by 29 percent, and other non-profit institutions growing by 133 percent.

Another way to assess demand is to explore the placement of students in formal or informal bioinformatics programs at universities. Dr. Stephan and her colleagues surveyed 21 programs in bioinformatics, of which 16 responded, and found that placement rates were very high; over 50 students from undergraduates to post-doctorates found employment. Only one student was "grabbed" by industry before completing the degree program, undercutting the notion that industry is raiding university bioinformatics programs for talent. In general, salary levels are quite impressive, with several undergraduates earning more than $50,000 and several masters, doctorates, and post-doctorates earning more than $100,000.

In summarizing demand, Dr. Stephan said that demand for bioinformatics is strong and growing, although still small relative to other areas in biology. A large amount of demand comes from industry, with salaries high in industry relative to other areas in life sciences. Graduates of formal or informal programs do not fill the majority of jobs in bioinformatics; indeed, graduates from those programs filled at best 15 percent of positions advertised in 1997. Individuals not formally trained in the field fill many jobs in bioinformatics and a number of jobs remain unfilled.

The Pipeline

According to Dr. Stephan's survey, as of March 1999 formal training programs had 23 undergraduates enrolled, 35 master's students, 86 doctoral students, and approximately 25 post-doctorates. The strong demand for bioinformatics and growing enrollment occurs at a time of a "crisis of expectations" for young life scientists. The career outlook for young life scientists is not bright; concern is sufficiently high so that the National Research Council has established the Committee on Dimensions, Causes, and Implications of Recent Trends in the Careers of Life Scientists. As a member of that Committee, Dr. Stephan said that the Committee was very concerned about young life scientists' career prospects. One recommendation was to restrain the rate of growth in the number of graduate students in life sciences.

An important issue to address is whether it is contradictory to have a "crisis of expectations" along with a strong demand for specialists in bioinformatics. Why, for example, are there only 9 doctoral programs in bioinformatics and computational biology, while there are 194 programs in biochemistry and molecular biology and over 100 in molecular and general genetics? Dr. Stephan said that there were four possible explanations for the imbalance:

- Low incentives for individual faculty to recruit students in the area;
- The educational system responds differently when demand is driven by industry;
- The interdisciplinary nature of the field creates disincentives; and
- A possible quick fix—turning life scientists into computational biologists.

Dr. Stephan said she would discuss in detail each possible explanation.

Lack of Incentives

Dr. Stephan said that it is a fact of academic life that the need for external grants to support research makes faculty very responsive to research funding opportunities. One way to encourage students to enter the bioinformatics field is to target research funds to that area; this gives faculty members incentives to populate their labs with graduate students doing research in bioinformatics. From the evidence collected by Dr. Stephan and her colleagues, research funding agencies are only beginning to direct grants to bioinformatics. In effect, agencies have placed all their computational eggs in one basket, namely training.

This may be the best solution in the long term, but it does not address the short-term needs to shift the distribution of graduate students in biology programs toward computation. Training grants signal "collective bodies," Dr. Stephan said, while research grants signal individual investigators. A collective response from universities is needed to expand programs in bioinformatics/computational biology, but the individuals within universities are driven by where their next grant may come from.

Industry-Driven Demand

The fact that the demand for computational biologists is driven by industry, according to some, is not much of an incentive for universities to respond by expanding programs in this field. The educational system is generally poised to respond quickly to changes in research funding portfolios, but moves inherently more slowly in response to industrial demands. In the life sciences, the tradition has not been to place people in industry, but rather to train people to do research in academic or non-profit laboratory settings. Moreover, recently industry has been hiring away promising faculty to conduct research, thereby "eating the seed corn" for training graduate students in bioinformatics.

Interdisciplinary Challenges

Dr. Stephan said that the interdisciplinary nature of computational biology creates disincentives to establish new training programs. Coordination across fields is intrinsically difficult, and it is exacerbated by geographical separation of departments within universities. Such things as allocating costs, credit for teaching hours, and finding classroom space are among the many details that makes coordination difficult. Moreover, the fields involved often have very different career goals. In biology, a master's degree has traditionally been seen as a "consolation prize" for some students, whereas in engineering and information technology, master's degrees are prestigious terminal degrees that successfully launch people's careers.

No Quick Fix

Dr. Stephan concluded by suggesting that a "quick fix" to address the supply shortfall in computational biology was unlikely. First, salaries in computer sciences are uniformly higher than those in biology, life sciences, and health sciences. Therefore incentives do not exist for one potential quick fix—encouraging people in computer sciences to do post-docs in biology in order to become computational biologists. Computer scientists do well financially as it is, and have little reason to become computational biologists.

The other possible quick fix—biologists becoming adept in computer science and math—is also unlikely. Dr. Stephan's research showed that biologists generally did not have the training or aptitude for transforming themselves into computational biologists. In their survey of top life sciences programs, Dr. Stephan and Grant Black found that most had no math prerequisite for entry and that few had any math courses as part of the degree program. Moreover, data indicates that students intending to enter biological and medical sciences have lower mathematical aptitudes than their counterparts planning to enter computer sciences and mathematics programs. Using Graduate Record Examination scores from 1993 to 1996, Dr. Stephan pointed out that students planning to enter math and computer science programs usually score substantially higher in mathematics than biology or medical science students (Table 3). A recent *Science* article discussed a UCLA survey that found that faculty in engineering programs used information technology most heavily, followed closely by physical scientists, and that biologists were in the middle in terms of use of information technology, slightly ahead of scholars in the humanities.

Possible Solutions

Dr. Stephan said that one way to increase the supply of computational biologists would be to find ways for faculty at urban campuses to interact across disciplines and institutional boundaries. Proximity of space really matters in promoting collaboration. It is also necessary to provide incentives for new interdisciplinary programs, and additional training and research awards would facilitate the formation of new programs. Additional research awards in particular, said Dr. Stephan, would give faculty the right incentive to undertake multidisciplinary research and, therefore, increase the supply of biological researchers adept at using information technology.

Dr. Stephan also recommended that faculty do more to provide students

TABLE 3 GRE Scores by Intended Field of Graduate Study, for Seniors and Nonenrolled College Graduates, 1993-1996

Intended graduate field of study	Test sections	Mean score	Percent of test takers with score above 700	Percent of test takers with score of 800
Biological sciences	Verbal	501	3.6	0.1
	Quantitative	595	20.7	1.1
Health and medical sciences	Verbal	449	0.7	0.0
	Quantitative	515	5.8	0.1
Computer and information sciences	Verbal	483	5.4	0.2
	Quantitative	672	52.2	5.6
Mathematical sciences	Verbal	502	6.5	0.2
	Quantitative	698	60.6	8.8

SOURCE: 1997-1998 *Guide to the Use of Scores*, Educational Testing Service, 1997.

with information on career outcomes. This requires faculty to do a better job tracking students' progress after graduation. Dr. Stephan was astounded that so few faculty members knew where their students were working and what salary levels were. Finally, it is important to recruit early in the educational process the right kind of student to work in computational biology. The field is new and presents different intellectual challenges than either biology or computer science alone do. Identifying students with promise to bridge the gap between the fields will not only increase supply, but also make the field more attractive to peers.

Discussant

Greg Reyes of Schering-Plough made a distinction between how drugs were discovered before 1990 and since. Prior to 1990, medicinal chemistry was the primary method, and small molecules (or natural product as a source of molecules) were placed in a biological screen and a biological effect was observed. That was the starting point for drug optimization. There are limits to that method of drug discovery; for example, it does not identify the mechanism by which the drug works, so one cannot know *a priori* what potential side effects may be.

Since 1990, however, a number of new technologies have greatly aided drug discovery, and they are all information-based, such as sequence databases. Micro-array technology, gene chips, combinatorial chemistry, and other technologies all generate new drugs and tremendous amounts of data. To maximize the value from these technologies, bioinformatics is crucial.

Bioinformatics also enables a new set of questions to be asked in research and drug development. For example, it will be easier to identify targets for drugs and validate the targets. A question one might pose is what genes are uniquely expressed in cancer cells, and to answer that researchers need sequence data from a normal source and a cancerous source. Researchers can then compare the differences and explore whether the normal cells have, for example, different regulator genes. Posing such questions enables researchers to develop priority gene targets for drug development. All of this involves the generation of large amounts of data and therefore requires computational capability.

With the host of new drug discovery technologies, a major challenge is how to integrate them to capitalize fully on their ability to discover new drugs. Each of these technologies may be the expertise of separate biotechnology companies, and the challenge for a large pharmaceutical company is either to partner with firms to exploit the technologies or develop the in-house expertise in the technologies. A typical pharmaceutical company will generally use a combination of both strategies.

At present, a limitation in drug development is in the confirmatory biology to study a treatment once the target has been identified and the development is fairly far along. This means, Dr. Reyes said, that informatics provides enough capability to generate compounds, libraries, and targets. A drug's impact is probably better explored by investigating the molecule itself, so trained biologists remain a key part of the equation for the industry, not just skilled computational biologists.

Dr. Reyes said the future is best considered by asking the question: What if we could design a molecule to uniquely fit the active side of a membrane and not cross-react with related enzymes? This requires greater multidisciplinary collaboration between chemists and biologists. Using tools in existence, it should be possible in the future to predict a drug's pharmacology in humans.

Beyond 2000, *in silico* drug discoveries will be key for the pharmaceutical industry. Using silicon-based information technologies, researchers will be able to do things such as study the structure of proteins faster than ever before. Hypotheses about proteins will be testable and verifiable using informatics. It will be important, however, for researchers to be able to produce the protein in adequate quantities to obtain the crystal and conduct studies. Highly trained biologists, Dr. Reyes reiterated, will be very important. That said, Dr. Reyes added that at Schering-Plough, the bioinformatics group has generated 10 years worth of data for biologists. In other words, Schering-Plough could shut down it bioinformatics operations today and keep its biologists busy for the next 10 years.

DISCUSSION

A questioner said that a plausible explanation for the shortfall in supply of trained individuals in bioinformatics is that the field is very new. The questioner asked Dr. Stephan how the newness of the field factored into her analysis. Dr. Stephan responded that, even though the field was new, many people contacted in the course of her research said that they had found it difficult to start computational biology programs within their academic institutions. Thus, even with very strong demand in industry, structural difficulties in universities have inhibited responses from academia. In fact, some people she interviewed left universities out of frustration when starting a bioinformatics program was impossible.

A questioner recalled a meeting that he attended at which large companies such as Merck and IBM said that for bioinformatics jobs, the companies do not need people that have graduated from formal degree programs, but rather people with certain skills. Dr. Stephan agreed that developing people with proper skills was the right goal, more so than necessarily graduating more people from formal programs. She added that students must be made aware of job opportunities in the field; this is not done adequately now, so students who may be interested and ideally suited for bioinformatics remain unaware of the field. That is why, Dr. Stephan said, more outreach must be done.

Dr. Kathy Behrens commented on Dr. Stephan's observation that faculty members in the life sciences do not know where students are employed upon graduation; she suggested that granting agencies require universities to establish a tracking system as a condition for the grant. Dr. Stephan agreed that this would be useful, and she added that institutions should be required to track students after post-doctorate employment.

Panel IV: Information Technology: New Opportunities–New Needs

INTRODUCTION

David Goldston
Office of Congressman Sherwood Boehlert

In opening the discussion, Mr. Goldston said that having heard about some of the possibilities of the biotech revolution, it was time to discuss public and private initiatives to push the frontiers of information technology in ways that support the biotech industry. The panel would offer insights into current efforts to address some of the information technology challenges, especially when it comes to biotechnology, and how public policy can best provide a supportive environment to respond to these challenges.

BIOFUTURES FOR MULTIPLE MISSIONS

Jane Alexander
Defense Advanced Research Projects Agency

Dr. Alexander said that the Defense Advanced Research Projects Agency (DARPA) has a great deal of interest in biotechnology, computer science, information technology, and microsystems (i.e., microelectronics, photonics, optoelectronics, sensors and acuators, and the microsystems made of these things). DARPA's interest is driven by its identification of these systems as important to the nation's future security. This means moving a whole range of technologies forward, and the relatively recent recognition that biology can play a role offers great potential for the Department of Defense. DARPA has been involved in the emergence of new fields (e.g., computing in the 1960s and materials in the 1970s), and the agency believes it can play a similar supporting role in new developments in biotechnology and microsystems. DARPA plans to move broadly into the area of biotechnology and microsystems, as opposed to focused technology initiatives that are often how DARPA pursues technology development.

The combined thrust of biology, chemistry, physics, and information technology is therefore very exciting for DARPA. But it requires true collaboration among disciplines on new problems, not simply the juxtaposition of physicists and biologists in the same lab. The objective is to find the places where the fields are strongest and the likelihood for meaningful impact on important problems is the greatest. The hope is to foster synergy that creates a whole greater than the sum of the parts. Moreover, there are already "interesting things brewing" at the intersection of biology and information technologies—a number of foundations, for example, are funding research in this area. DARPA wants to help the field attain a critical mass so that research programs can really take off. DARPA also understands that many young people are enthusiastic about the field, and by investing early, the agency can help develop a pipeline of talent in the field.

Current DARPA Initiatives

DARPA already has activities under way at the intersection of biotechnology and computing. The "electronic dog's nose" initiative—the ability for canines to sniff out explosives—tries to understand that process at a very high level of sophistication. The objective is to develop electronic devices that can perform this task as well as dogs.

DARPA is also working in the area of controlled biological and biomedical systems. This involves interfacing directly with living creatures, such as insects, and altering them at the larval stage. For a certain type of wasp, for instance, exposure to specific vapors in the larval stage will enable it to detect explosives when it develops into an adult. DARPA is also considering having insects carry electronic chips, so that their hunting patterns become search algorithms for DoD sensors.

DARPA is also funding R&D in tissue-based biosensors. The agency is exploring whether cells and tissues can be used to detect toxins in the environment. This can enable certain cells to determine whether something is dangerous and alert people to the danger before it is too severe. Other initiatives involve DNA computing, microfluidics (i.e., how to move organic materials around on a chip), and diagnostics for biological warfare agents.

These DARPA programs are focused technology initiatives, but the agency also believes that it has to look at some fundamental scientific questions as well. DARPA has labeled this area of inquiry "Biofutures." Exploring how biology interacts with microsystems and information technology is the cornerstone of the effort.

With respect to biology and microsystems, microsystems technology offers biology the ability to create interesting interfaces at the micron and nanometer level. For example, microsystems allow biologists to measure a number of things that have been difficult or impossible to measure until now. Single sensing systems will be able to measure optical properties, cell adhesion, temperature, and other cell properties. The challenges for biologists will be to integrate into silicon and machine-based microsystems the biological materials and materials that are suitable to asking biological questions. For microsystems technologists, biology offers tremendous opportunities; for example, biology offers very reliable functions from fundamentally unreliable parts. A problem will be how to do engineering in biology. The field of bioengineering already exists, but to take it to the next level of abstraction will be no easy task.

In terms of biology's interaction with information technology, many of the challenges have to do with the complexity of systems that integrate the two technologies. Information sciences allow biologists to scan databases with lots of information—genomic databases are good examples—but the complexity of the biological systems on which data is being collected is beyond the capabilities of information sciences. Dr. Alexander, a physicist by training, said she managed a biology group at DARPA several years ago and she was struck by the cultural gap between the two disciplines. Physicists use mathematical tools for research, but these tools are presently not capable of addressing the inherently complex systems in the biological world. At the same time, each field could benefit the other greatly. Information security is a huge concern today, and biological systems may offer new ways to address such problems. Immunology, for example, is about identifying external threats to a biological system and responding; these are exactly the characteristics we want in security for information systems.

Future DARPA Initiatives

In preparing to address fundamental questions in biology and information sciences, DARPA has talked with people in industry, universities, and the policy community to determine emerging needs. The result has been an expansion of efforts to develop underlying tools to enable fruitful collaboration between the two disciplines. Such efforts will be less focused than a typical DARPA initiative, because of the uncertainties involved in an upstream undertaking. DARPA plans to release a "research announcement" in the near future with the goal of beginning the first round of research grants by the Spring of 2000.

DARPA will also hold a large conference to further refine research ideas with the hope of developing a focused research agenda at the intersection of biology and information technology. That conference is scheduled for June

2000. DARPA plans to partner with the National Science Foundation and the National Institutes of Health in this process, and the agency will also reach out to the technology community in developing these workshops.

In conclusion, Dr. Alexander asked for the engagement of participants at the conference. DARPA has no labs of its own, and must therefore work with the scientific and technological communities to be effective. She challenged those in attendance to develop excellent research proposals as DARPA expands its program in biology and information technologies.

MEETING THE NEEDS, REALIZING THE OPPORTUNITIES

Paul Horn
IBM Corporation

Dr. Horn posed the following scenario in opening his remarks: a patient walks into a doctor's office, describes his symptoms, and then responds to several questions asked by the doctor. The doctor scrapes some DNA off the patient's cheek, places it in an analyzer, and quickly gives the patient a precise diagnosis of his problem. Next, the doctor prescribes drugs that are tailored to the patient's DNA to treat the illness. Dr. Horn said that this scenario was not science fiction, but rather likely to be upon us with surprising speed.

In a few years, Dr. Horn said, we will have completed sequencing the human genome. However, we are still a long way from taking a sequence of amino acids and turning it into the fundamental structure of the proteins that make up the human body. The movement from sequencing genes to having biologically useful information with which to treat humans is an information technology challenge. In the not-too-distant future, Dr. Horn predicted, we will be able to provide the kind of diagnosis and treatment options for patients that he described in his scenario.

Developments in information technology hold great promise for turning advances in biotechnology into practical and operational ways to improve patients' lives. Using the analogy of disk drives in the computer industry, Dr. Horn pointed out that in 1954 IBM built the first disk drive, a sizeable piece of equipment that stored one gigabyte of data at the cost of $10 million. Today, it costs $10 to store a gigabyte of information in a much smaller device. By 2020, one terabyte of storage will cost just a few cents. A terabyte is the amount of information the brain can store, but the processing speed of the brain is far faster than a single personal computer, or even a number of them linked together. In 10 years, high-end supercomputer processors will have the operational power of the human brain—10 pedaflops. The amount of change in computational capabilities will be truly breathtaking over the next decade, Dr. Horn said. Changes in computing will have profound impacts on every facet of our lives.

The Economic Impacts of Computers

Dr. Horn noted that the Chairman of the Federal Reserve, Alan Greenspan, testified before Congress in May that computers—by enhancing productivity—were contributing greatly to U.S. economic growth. As we enter a fundamentally information-based economy, digital information will become increasingly important to everyone. By 2003, Forester Research estimates that worldwide $1.3 trillion worth of goods and services will be purchased over the Internet. That is 5 percent of the world economy; for the United States, electronic commerce will amount to 10 percent of the economy.

Companies will also realize tremendous efficiencies through the use of digital technology. Taking an example from IBM, Dr. Horn said that in 1999 his company purchased about $12 billion worth of goods and services over the Internet. By using the Internet, IBM estimates that it will save $260 million—a substantial contribution to IBM's bottom line.

The Bioinformatics Challenge

The bioinformatics challenge is to make sense out of the vast amount of information available from the Human Genome Project. As of mid-1999, IBM had 1.4 million pages of information on the World Wide Web. Dr. Horn said that he probably really needed to read at least three of them, but was hard pressed to know which ones. In the world of bioinformatics, the challenge is to manage huge amounts of data; indeed, today the Human Genome Project involves a great deal of data, but a meager amount of information. Turning this data into usable information will be the paramount challenge for biology in the next century.

Information technology, said Dr. Horn, will be the language of biology in the twenty-first century, much like mathematics became the language of physics beginning in the late nineteenth century. Without information technology, it will be impossible to engage in meaningful biological research. To think that we will train future researchers for the bio-pharmaceutical industry through a few bioinformatics institutes, argued Dr. Horn, dramatically underestimates the importance of information technology to the field.

The Role of the Government

Given the growing importance of information technology to biology, the magnitude and nature of government support must be carefully thought out. Dr. Horn listed three elements that would be important in the future:

- Support for *basic research* will continue to be important for biology and the federal government has traditionally provided the bulk of this support.

This research will be most valuable if it is multidisciplinary, that is, at the boundaries of information technology, biology, and physics.

- *New bridges* must be built between basic research and industry. In an information-based economy, technical information of commercial relevance flows rapidly and often freely around the globe. Corporations with strong ties to universities are well positioned to absorb such information and thereby capture economic benefits. Indeed, this is how economic value is created and captured by a region or country. Government programs, such as the ones funded by the National Science Foundation, the Advanced Technology Program, and the Defense Advanced Research Projects Agency, can help speed up the flow and absorption of these ideas.
- Government can assist in building the *information infrastructure*. Today, the government is the primary funder of computers with incredible computational power, such as the teraflop computer at Sandia National Laboratories. As we advance toward the pedaflop computer, government can aid in the research and development of such new computing power. With a pedaflop computer, scientists will be able to do fundamental protein folding to determine a protein's structure and function nearly on a real-time basis.

Using what is called "deep computing" at IBM—computers with enormous memory and processing speeds—Dr. Horn predicted that it would be a matter of roughly two decades until computers can process the hundreds of thousands of proteins in the human genome to determine their structure. For medical purposes, the human genome will then be fully understood. Thus over the next 10 to 100 years, the combination of information technology and biology will revolutionize society.

Discussion

Commenting on the challenges of improving quality and quantity of cross-disciplinary contacts between biologists and computer scientists, Dale Jorgenson asked whether trying to train computer scientists in biology would be a wise strategy. Dr. Horn responded that the conference heard about possible difficulties in teaching biologists the quantitative skills necessary for computer science. Without passing judgment on whether overcoming these difficulties would be insurmountable, Dr. Horn said he believed that biologists in the future will have to become skilled "information workers" able to handle databases and computer software. Dr. Horn said that even today a biology graduate student has to be facile in finding, manipulating, and synthesizing information. It is not so much a matter of training biologists in hard mathematics, but rather in ensuring that biologists will be able to use information technology in a reasonably sophisticated way.

Dr. Alexander observed that 15 to 20 years ago, materials science faced these same issues in its nascent stages. As the field developed, physicists, chemists, and engineers each played a role, but it was unclear which discipline would take the lead in training. Eventually, materials science emerged as a discipline, and it attracted a different brand of person, distinct from physics or chemistry. If the intersection of computer science and biology pays off as anticipated, Dr. Alexander believed that a new hybrid field would attract a different brand of person than has been traditionally been drawn to biology or computer science.

NEW INFORMATION TECHNOLOGY INITIATIVES

Tom Kalil
National Economic Council

Mr. Kalil said he would offer a case study of one of President Clinton's science and technology initiatives, and do so in the context of the imbalance in support for biomedical research versus funding for research in the physical sciences and engineering. NIH's budget increased by $2 billion in FY 1999 and should increase by approximately $2.3 billion in FY 2000. In contrast, one does not see Members of Congress "falling on their swords" in defending research funding for physics, chemistry, computer science, and engineering. The Clinton Administration, therefore, has been looking for ways to increase support for research in physical sciences, computer science, and engineering.

To do this, the administration proposed a 28 percent increase in information technology (IT) research for FY 2000. This was to support research in three areas:

- *Fundamental research* for information technology to improve increasing software reliability and network technology (i.e., developing networks able to support billions of information devices). Other areas include exploring the ability to visualize large quantities of data, real-time foreign language translation, and DNA computing.
- *Tera-scale computing* for civilian science and engineering. The government spends significant sums for high-end computing for defense-oriented stockpile stewardship. But these machines are not available on the civilian side, and tera-scale computing will support a variety of modeling and simulation projects.
- *The ethical, legal, and social aspects* of the information revolution. A program such as this exists for the Human Genome Project, but nothing analogous exists for information technology. Although this area had the smallest funding request, the topic is nonetheless extremely important.

The Rationale for Increased IT Spending

One reason for the Clinton administration's requested increase in research funding for IT has its origins in the President's Information Technology Advisory Council (PITAC). That panel, headed by Irving Wildavsky-Berger of IBM, consisted of a number of computer scientists. While it may not come as a great surprise that a group of computer scientists recommended an increase in computer science research funding, PITAC's distinguished membership made its findings well worth taking seriously.

A second reason is that the administration believes that historically payoffs to research funding for information technology have been very high. This year is the thirtieth birthday of ARPANet, and many innovations in IT—and not just in networks—have come about because of DARPA, NSF, and other agencies. Computer time-sharing, reduced instruction set computing (RISC) technology, the first graphical Web browser, and many search engine companies (which came out of the government's digital library initiative), have origins in government funding. Looking across investments in these activities, it was easy to conclude that the returns had been very high.

The third reason for the administration's interest is that leadership in IT is critical for economic growth. Over the last several years, information technology has accounted for over one-third of U.S. economic growth. Jobs in this sector pay roughly 80 percent more on average than other jobs. The United States has been the center of innovation in IT, and the administration wants to help maintain U.S. leadership.

The fourth reason for increasing investment in information technology is national security. U.S. national security relies on technological superiority; a cursory inspection of U.S. doctrine shows phrases such as "dominant battlefield space awareness," indicating a strong reliance on IT.

The fifth reason is that the administration believes that advances in information technology will aid discovery in all disciplines—from chemistry to engineering. More and more disciplines employ modeling and simulation, which rely greatly on IT. Taking an idea of the National Academy of Engineering president, Dr. Wm. Wulf, Mr. Kalil noted that "collaboratories," in which scientists, large databases, and remote scientific instruments are all linked by high-speed computers, depend on IT to be successful, and may provide a model for research with teams that are geographically separated.

Preparing a technically trained workforce was the sixth reason that the administration requested an increase in funding for information technology R&D. Industry has been vocal in expressing its concern that the lack of a trained workforce is a major constraint on economic growth. Because university professors usually hire a team of graduate students to conduct grant-funded research, increasing the supply of research dollars will have payoffs in terms of training in addition to research outcomes.

Finally, information technology will lead to a number of economic and societal transformations that are important. From electronic commerce, to telemedicine, to electronic government, IT is an important enabler for a variety of developments that can improve people's lives.

Budgetary Outcomes

The Clinton administration experienced "a bit of a mixed bag" when it came to having its IT funding requests fulfilled. On the one hand, the administration had bipartisan support for much of its proposal; House Science Committee Chairman Sensenbrenner introduced legislation authorizing the administration's request for the agencies under his committee's jurisdiction. There was also widespread industry support for the initiative. PITAC strongly supported it and TechNet, one of the leading high-technology political organizations, sent a letter signed by 37 CEOs from the technology industry advocating the administration's initiative. The American Association of Universities and the Council of Scientific Society Presidents, in addition to several trade organizations, also endorsed the proposal. The scientific community had some concerns, however, about the size of the increase for IT relative to other disciplines.

In the end, President Clinton's initiative received approximately $126 million of the $146 million requested for the NSF. Of the $100 million requested for the Defense Department, $60 million will be funded; $40 million for DARPA will not be appropriated. The Department of Energy received none of what the administration requested, primarily because the department receives substantial funding for high-end computing in modeling and simulation for stockpile stewardship. Congress chose not to fund civilian and defense high-end computing in the Energy Department.

In conclusion, Mr. Kalil said that the administration regarded the budgetary outcomes as a useful beginning, and he added that the administration hopes to do better next year. Information technology research is critically important for the United States, and the Clinton administration is committed to building on efforts to increase research funding in this area.

Discussion

A member of the audience asked Mr. Kalil how the administration planned "to do better" in procuring more funding for its information technology initiative next year. Mr. Kalil responded that one likely approach would be to collapse all IT research funding into one package. This year, confusion arose over the base level of spending versus the increment in the new initiative; this did not help the administration's cause. Most important, Mr. Kalil said, will be making the case individually to Members—early and often—about the importance of research funding for IT.

Mr. Kalil was asked if he could further explain why the Energy Department received no funding from this initiative. Mr. Kalil responded that the administration has to do a better job of explaining why high-speed computing is important for civilian purposes, not only stockpile stewardship. He noted that the Energy Department has a number of research projects under way—from climate change, to materials by design, to developing more efficient combustion engines—that require high-end computing capabilities.

DISCUSSANTS

Charles Trimble
Trimble Navigation

Dr. Trimble provided an overview of the Global Positioning System (GPS), something he regarded as a very successful government-industry partnership. The government did not fund the commercial end of GPS, but it did fund important elements in basic research and infrastructure, which facilitated the commercial development of GPS. There were also important multidisciplinary aspects to GPS. It may not seem like it today, but 30 years ago hardware, software, and electronics were very separate areas of inquiry. Multidisciplinary research at universities brought those fields close together.

Most importantly, the government provided an early market for the GPS, as the U.S. Geological Survey required that government surveys be done using this technology. In his industry and others, observed Dr. Trimble, government's role in providing an early market has tapped into the country's entrepreneurial spirit.

These three components—basic research, university support, and providing an early market—have been important in other fields, such as the Internet and increasingly the Human Genome Project. In biotechnology and computing today, it is important to push universities in the direction of multidisciplinary research and then "let industry mine" this for commercial purposes. With respect to computing, Dr. Trimble emphasized that the research community remains "hungry" for high-speed computing, and the NSF should continue to support and expand development of high-speed computers for the university research community.

Richard Rosenbloom
Harvard University

As a long-time observer of the technology industry from the university setting, Dr. Rosenbloom said he would focus on turbulence in the information technology industries in the United States. If this conference had been held 10 years ago, the discussion would have focused on a very different set of topics and actors than is the case today. IBM is the only company represented today that would have been on a similar agenda 10 years ago.

The extraordinary rate of change in the computing industry is the hallmark of the computer and IT industries. This gives rise to what Dr. Rosenbloom called the "paradox of Moore's Law." Moore's Law, of course, predicts steady and remarkable improvement over time in computing devices in terms of processing power and cost decreases. However, leaders in the computing and information industries consistently fail to accurately predict or prepare for the changes that are brought about by Moore's Law. It is difficult for leaders in any industry to cope effectively with revolutionary technological change.

Dr. Rosenbloom said that the quantitative improvements in computing power have been accompanied by qualitative changes in industries. Twenty-five years ago, the information industry could easily have been described; it consisted of IBM and "The Bunch"—Control Data, Burroughs, Univac, NCR, and Honeywell—as well as AT&T, semiconductor firms, and a few firms making minicomputers. "The Bunch" has been wiped out, IBM nearly was, companies such as Digital Equipment have been bought out, and AT&T exists only as a brand name. The ranking of semiconductor companies has changed dramatically, even though many of the same semiconductor firms survive in some form today.

Public policy must recognize that the information technology industries are extremely dynamic. As we develop policies for the future, they should be robust to withstand the tremendous turbulence that characterizes the information technology industry. Ken Flamm commented on this turbulence earlier, Dr. Rosenbloom said, and one lesson he took away from Dr. Flamm's discussion is that government statistics often cannot keep pace with the change brought about by turbulence. Only recently have government statisticians come to recognize that a software firm such as Microsoft was conducting R&D and that government statistics must account for it. And companies such as Dell and Cisco were on no one's radar screen in 1989, and they are certainly important companies today. Dr. Rosenbloom argued for caution in relying on data to identify trends in the face of such turbulence.

If so much effort is going to be devoted to encouraging a fusion of biotechnology and computing, Dr. Rosenbloom said, we should expect changes in both industries' structures comparable to what has occurred in the computing industry in the last decade. Dr. Rosenbloom said that he hoped new IT initiatives at DARPA and other agencies would be fully funded. If history is any guide, handsome payoffs will occur, and the companies exploiting them are likely to be ones whose names we have not yet heard. Dr. Rosenbloom urged policy makers in attendance to take into account turbulence in the IT arena when formulating policy. The pharmaceutical companies, for instance, have been remarkably stable even as their methods for drug discovery have changed. Dr. Rosenbloom conjectured that the future development of computational biology might have a profound impact on the industrial structure of pharmaceuticals. Within a decade, the pharmaceutical sector may be as turbulent as the information technology sector is today.

DISCUSSION

A questioner asked whether any of the panelists were aware of important developments elsewhere in the world in biotechnology and computing, and their growing interdependency. Dr. Horn responded that prowess in information technologies is certainly not the sole province of the United States. He said that the United States is probably the best country at adopting new technologies, whatever their origin, and the United States spends the largest share of Gross National Product in IT. Led by Silicon Valley, the United States is best at capitalizing on new technologies, but Dr. Horn cautioned that the rest of the world is not far behind.

Mr. Kalil said that, in terms of foreign competition in IT, Europe is particularly strong in mobile and wireless telephony. The unique advantage of the United States lies in its strong university research base, the venture capital industry, and a culture that encourages entrepreneurship and tolerates risk. Here involvement in a failed start-up is seen almost as a "badge of honor," Mr. Kalil said, whereas in other countries that sort of failure is a black mark on someone's career. The turbulence that Dr. Rosenbloom spoke about is in fact an asset in the United States, argued Mr. Kalil, and perhaps not one that we appreciated in the late 1980s and early 1990s.

Dr. Alexander said that one of the strengths of the U.S. academic system is its flexibility relative to other countries. The U.S. system encourages junior faculty to start new lines of inquiry, and although things may not always move as quickly as some would like, the situation in the United States is better than elsewhere.

Dr. Jorgenson asked panelists about two policy areas that had not been discussed, standard setting and antitrust. Mr. Kalil responded that the entire issue of standards is "very tricky." One can point to the European example of the GSM standard for wireless communications as a stunning success; by gathering European companies into a room to agree on a standard, wireless penetration rates have taken off. It is easy to ask why the U.S. government did not do the same thing. In the specific U.S. context, the wireless industry was split on the appropriate standard. Qualcomm said it should be CDMA and a number of other carriers said it should be TDMA. The Federal Communications Commission asked: On what basis could the agency make the decision? Why should the FCC believe its judgment is better than the market? The FCC chose not to choose in this case, because there was no policy principle to which the FCC could appeal to make a choice.

That said, Mr. Kalil continued, there are examples in which that approach failed. The Japanese government's promotion of an analog high-definition television standard stimulated a lot of investment, but it has gone nowhere. In the United States, one part of the government mandated that a standard called OSI be used for data interfaces. Another part of the government ignored OSI and adopted TCP/IP, which has become the standard for the Internet. Because of conflicting examples, it is difficult to conclude that a "top down" government-imposed approach to standards is the correct approach. When industry wants to work on a standard, government can be supportive in important ways, such as providing funding for test beds.

In the area of competition policy, Mr. Kalil said that promoting competition in telecommunications has been one area in which the government has made sound choices. Because of the Telecommunications Act of 1996, competitive local exchange carriers—so-called CLECs—are now capitalized at $30 billion. CLECs, competitors to the Baby Bells, are rolling out new services and challenging incumbents to respond in kind in terms of service price and quality.

Dr. Trimble commented that with the European standard for wireless communications, it is true that this appears to be a success, but added that competition in wireless telephony is far from finished. In general, Dr. Trimble thought it was unwise for government to "pick winners and losers" in standards situations, because government is likely to pick the low risk option and established companies. The entrepreneurial push in the United States has in fact fostered much innovation and wealth creation in this country.

Dr. Kathy Behrens observed that by the end of 1999, over $30 billion—perhaps close to $37 billion—in institutional venture capital will have been invested in the United States. More than 80 percent will go to information technology and Internet firms, while only $1.1 billion will be invested in biotechnology. This, she said, will certainly contribute to the continued turbulence that Dr. Rosenbloom discussed.

Asked whether the computing and semiconductor sectors would see continued rapid rates of innovation, Dr. Horn said that when it comes to lithography, he expected to see a slowing in the rate of advance in that field within 5 years, at least using traditional lithography technology. At the system level, however, there is room for substantial improvement in packaging technology, compilers, and additional software functions. This will more than compensate for the slowdown at the core silicon level. Thus, Dr. Horn said computing power would increase at a rate greater than exponential over the next 10 years.

Jim Turner of the House Science Committee suggested that the turbulence in the pharmaceutical industry that Dr. Rosenbloom predicted might not occur because the industry is subject to regulation by the Food and Drug Administration. Mr. Turner asked whether panelists had considered whether advances in computing power and bioinformatics might in fact lower the cost of regulation, and thereby create a climate in which more market entry into pharmaceuticals would be possible. If that came about, Mr. Turner suggested that perhaps venture capital would migrate more toward a less regulated biotech industry, and less so toward the unregulated Internet sector. The end result might be enormous payoffs to human welfare as new drugs and treatments come to market faster.

Dr. Horn discussed several technologies that might lessen regulatory burdens. Many pharmaceutical companies have very heterogeneous databases; it is costly to transfer information across them, and new database technology under development at IBM and elsewhere will address this. Emerging computing technologies will allow computational chemistry to estimate *in vivo* impacts of drugs. This will not obviate the need for clinical trials, but it will provide regulators excellent data on a drug's impact in a more timely way.

Dr. Alexander commented that the cost of bringing a new drug to market is between $150 million and $500 million—an enormous barrier to entry. One reason we need such large and costly human clinical trials for new drugs is the incredible genetic variability of human beings. As this variability is better understood through genomics research, it will be feasible to conduct trials on a smaller scale and in a far less costly manner. These advances will not be on-line for another 15 to 20 years, but when they are, the process will be shorter and cheaper.

Panel V:
Capturing New Opportunities

INTRODUCTION

Michael Borrus
Pektevich & Partners, LLC

Mr. Borrus observed that today's booming economy is due, in many ways, to the fruits of past innovations by the private sector and past government-industry partnerships. As Dr. Moore mentioned, many of today's innovations in electronics can be traced back 50 years to the invention of the transistor at Bell Labs. The conference's examination of the current state of research in biotechnology and computing identified innovations and partnerships between government, industry, and universities that stretch back over several decades.

In the first panel, the objective is to look at issues in biotechnology and computing that are presently subjects of intense scrutiny. The economic impacts of these topics are perhaps a decade or so away, but it is important nonetheless to make sure that current research needs are attended to, and that the public policy environment encourages further development. Each of the panelists, Mr. Borrus said, is uniquely qualified to provide insights into the possibilities of today's technologies, and what we need to do today to take full advantage of them in the future.

COMPUTING AND THE HUMAN GENOME

Mark Boguski
National Center for Biotechnology Information

Up until 5 or 10 years ago, Dr. Boguski said that biologists restricted their studies to *in vivo*—observing the body or other living organisms, or *in vitro*—the test tube. Increasingly, however, biologists are spending all of their time working *in silico*, that is, making discoveries entirely on computers using data from gene sequences.

A federal advisory committee in June 1999, urged the National Institutes of Health to train biologists in computing. The advisory panel, chaired by David Botstein of Stanford University and Larry Smarr of the University of Illinois at Urbana-Champaign, recommended that NIH establish up to 20 new centers at the cost of $8 million per center per year to teach computer-based biomedical research. This is $160 million per year out of NIH's budget for training computational biologists.

Citing a quote in *Nature* magazine that said, "It's sink or swim as a tidal wave of data approaches," Dr. Boguski said the need for computational biologists was urgent. The information landscape is vast in genomics research, and Dr. Boguski said he would describe that landscape and the staffing needs that must be fulfilled to negotiate it successfully.

One measure of the growth of scientific information is the growth in the MEDLINE database at the NIH's National Library of Medicine. The database contains over 10 million articles, and this is growing by 400,000 articles per year; these are peer-reviewed journal articles. If one looks at the subset of molecular biology and genetics articles, the growth rate is considerably faster than that for biomedical literature as a whole. Another measure of the growth in biological information is the growth of DNA sequences. Rapid DNA sequencing technology was invented in 1975, but until it was automated in 1985 and until the Human Genome Project took off, the growth of DNA sequences was modest. Since the early 1990s, however, the growth rate has been extremely steep.

Comparing the growth in articles on DNA sequences with the growth of DNA sequences, Dr. Boguski identified a serious gap for the biomedical research community. From 1975 through 1995, there were more papers published in DNA sequences than the number of sequences. Since 1995—roughly five years after the inception of the Human Genome Project—an enormous gap has opened up. There are now more genes than articles about those genes. The challenge for biology today is to find ways to bridge the gap between the number of genes being discovered and techniques to classify and understand their function.

One technology that seeks to bridge the gap is functional genomics. Functional genomics refers to

> The development and application of global (genome-wide or system-wide) experimental approaches to assess gene function by making use of the information and reagents provided by the genome projects. It is characterized by the high throughput or large-scale experimental methodologies combined with statistical and computational analysis of the results. The fundamental strategy in a functional genomics approach is to expand the scope of biological investigation from studying single genes or proteins to studying all genes or proteins in a systematic fashion.

The Human Genome Project was initially planned to last 15 years and conclude by 2005. Technological and competitive forces have accelerated that timetable, and by May 2000, a 90-percent draft sequence of the human genome will be in the GenBank database. To date, 800 million bases of DNA have been produced and are in GenBank; this is out of 3 billion bases to be sequenced, so the project is between 25 and 30 percent complete. This data is available—free of charge on the Internet—the taxpayers have paid for it, so Dr. Boguski's job is to ensure its availability—and anyone can access it.

Computational biology enables biologists to make new kinds of inferences about genes using computing power. In the past, biology has been largely an observational science. More recently, biologists have classified their observations and put them into databases. However, advances in computing capability allow scientists to take a gene sequence, infer the protein sequence using the genetic code, and then, using modeling techniques, infer the protein structure and function.

For the past 15 years, biologists have been engaged in the comparative analysis of genes. Approximately 40 years ago, scientists discovered that at the molecular level, life is very similar. When biologists discover a new gene and want to know its function, they search the genome database for homologues—genes whose function is known and from which the function of the newly discovered gene can be extrapolated. The basic insight is that what is true for yeast, whose gene has been sequenced, is true for humans. In fact, the cell nucleus for yeast and humans are the same, and one can take a gene out of yeast, replace it with a human gene, and the yeast will thrive as if nothing happened. Furthermore, biologists have advanced from comparing genes to comparing entire genomes of different organisms. In the coding sequence of the human, the human being is 85-percent similar to rodents; that is, to mice and rats.

The theory behind this is molecular evolution. When biologists discover a human gene, and then compare it to databases of mouse or yeast genes and discover that the genes are so similar that the similarity can be no accident, this is evidence of evolution at work. And the increasing processing power of modern computers enables these comparisons to be made.

The Power of the Internet

The Internet has also been a crucial tool for discovery in genomics. One researcher was involved with an 18-year research project whose final answer was compressed into a five-minute search on the Internet. The researcher was searching for a gene that causes a rare disease called ataxia telangiectasia. When the researcher finally discovered the sequence for the gene, he did not know what its role was in the body. After a five-minute Internet search of other organism databases, he was able to find out by comparison the cause of the disease.

Gene Mapping Milestones

In 1996, the first large-scale mapping of the human genome, containing over 15,000 genes, was published in an article by 105 authors from 18 different research institutions in five countries on three continents. This collaborative effort signals a sea change in how biological research is conducted, and indeed the paper resembles the type of collaboration usually seen in particle physics. Scientists have not yet identified many of the functions of these mapped genes. This is why functional genomics technologies are so important. Even with many gene functions unidentified, the current map of the human genome has been instrumental in accelerating the cloning of hundreds of genes for diseases prevalent in humans.

The so-called "obesity gene," leptin, is an example of the uses of genomic data. When this gene was first cloned, its sequence itself was uninformative; the sequence was not related to anything in existing databases. Therefore, the next level of sophisticated analysis, protein threading, was needed to discover its function. Using protein threading, a type of modeling technique, it was discovered that leptin was similar to a hormone called Interleuken-2. It was not at all expected that the obesity gene would function like this hormone, so this work initially was greeted with skepticism. Eventually, the structure of the protein was determined, and the notion that leptin functioned like Interleuken-2 was shown to be correct. The structure of the protein was determined by computing the structure completely *in silico*. Since this discovery, an entirely new pathway in obesity research has opened, one that would not have otherwise been undertaken.

An example of a functional genomics technology is DNA microarray analysis, which enables researchers to monitor the activity of approximately 20,000 genes on a single chip. For example, researchers can see how a gene's behavior changes when a drug is introduced—or perhaps how a gene does not respond to the drug. This would tell a doctor whether to modify dosage or change the drug entirely in order to ensure that the therapy targets what is intended. Computer-enabled functional genomics technology can enable such an experiment to consider tens of thousands of genes. Eventually, researchers want to consider millions of genes, and this will require new visualization and simulation technology.

Another example of functional genomics experiments is *in vitro* simulation of wound healing. Biologists take the cells responsible for healing, place them in a culture, and treat them with a serum. Biologists then observe which genes are "turned on or off" in response to a wound. That process is then studied and simulated in a test tube using functional genomics technologies. Because the government developed the techniques, databases, and other elements needed to conduct such experiments, it is all freely available on the World Wide Web. This has stimulated additional paths of inquiry in the scientific community.

Dr. Boguski pointed out that NIH did not start the human genome project, rather it was the Department of Energy (DOE) that did as part of its charge to study the biological effects of ionizing radiation. The idea was to sequence the human genome and then observe the changes in gene sequences that the introduction of radiation induced. One of the pioneers of the project in DOE, Charles DeLisi, said in 1988 that "the anticipated advances in computer speed will be unable to keep up with the growing [DNA] sequence databases and the demand for homology searches of the data." Luckily, said Dr. Boguski, he was wrong. Likening the growth of gene data to Moore's Law,[1] Dr. Boguski said the amount of gene data doubles about every 18 months. For the foreseeable future, there is sufficient computer power for the searches and structure predictions that he has discussed.

There are research areas in which supercomputing power will be needed. Dr. Ray Winslow at Johns Hopkins University's Department of Biomedical Engineering is simulating heart failure by studying five different genes. Using partial differential equations, Dr. Winslow simulated irregular electrical activity in the heart and represented it in a three-dimensional simulation of the beating heart. For five genes, this took two CPU days on a 16-processor Silicon Graphics computer. If one were to study 50 genes or more, the computational requirements would be dramatically higher. Biologists understand the growing need for high-performance computers. Dr. Boguski said that grant applications increasingly request funds for new computers.

In closing, Dr. Boguski said that with the completion of the genome mapping by next May, the ethical, legal, and social implications (ELSI) of these breakthroughs will grow. It is impossible at this time to anticipate all of these so-called ELSI issues, but the need to face them will grow with our understanding of the human genome.

Questions From the Audience

Dr. William Long asked Dr. Boguski whether software development was keeping pace with advances in understanding of the human genome. Dr. Boguski said that, even though the human genome has 3 billion base pairs, 97 percent of that is so-called "junk DNA," repetitive DNA that has no apparent biological function. This simplifies the computational problem in that the volume of data is not as vast as it seems. Dr. Boguski did emphasize, however, that improvements in software would be critical. New search algorithms will be necessary to search expanding databases filled with genomic information.

Dr. Boguski was asked how well biologists were able to model physically and chemically the role of gene expression in driving the "protein factories" within cells. Dr. Boguski responded that this has been accomplished for small

---

[1]See Gordon E. Moore, *op. cit.* 1965 and 1997.

pathways in bacteria. Modeling how a fertilized egg develops over 9 months into a baby is quite a ways off, but Dr. Boguski said it would be possible in the future. Today, biologists receive lots of data from functional genomic tools. The only way biologists will be able to make sense of all this data is to develop quantitative dynamic models of systems and simulate evolving biological systems in a computer.

Dr. Boguski was asked what his "wish list" to Congress would consist of in order to advance his field. Dr. Boguski responded that one colleague in biomedical engineering, Ray Winslow, said that his field needed high-performance computers, databases, and scientific visualization tools. Other colleagues said they need larger computer equipment budgets and larger grants. Dr. Boguski said it was a bit bothersome that on many grant applications, the salary for a computer consultant exceeds that of the principal investigator.

NANOFRONTIERS

Alton Romig
Sandia National Laboratories

Dr. Romig said that the driver behind nanoscience is miniaturization, and there are two schools of thought on this issue. One is that scientists should simply take components in use today and make them smaller and smaller. The other is that scientists should build from the micro-scale up; that is, build systems from the atomic level, perhaps using biological systems or materials, to create very small devices. Dr. Romig said that he thought that neither method would become dominant, but rather the two approaches would converge. At Sandia National Laboratories, the approach is to mix the two approaches in creating new materials and devices.

As background, Dr. Romig described Sandia as a multi-program national laboratory whose principle mission is to help protect the national security. This includes ensuring the reliability of the nation's nuclear weapons through the stockpile stewardship program, ensuring the efficiency and reliability of the nation's energy supply, protecting the environment through remediation programs, and developing sensing and monitoring devices to detect environmental problems. It is program need in each of these areas that drives Sandia's interest in developing nanotechnologies.

Dr. Romig said that the terms "nanoscience" and "nanotechnology" are often used interchangeably. In fact, "nanoscience" may be used in order to obtain money when the funding agency is interested in fundamental research, while "nanotechnology" may be used when a funding agency wants more applied research. To the extent that a distinction between the two terms exists, nanoscience often refers to nanocrystals or quantum dots, layered structures, and hierarchical assemblies. Nanotechnology refers to technologies such as photonics or magnetics, biomedical or bioremediation, and integrated microsystems. These technologies are integrated into small system-level applications.

When talking about the growing links between the biological and physical sciences, Dr. Romig said the discussion often focuses on how the physical sciences will help biology. There is an important flip side, perhaps further in the future, in which biology serves as a catalyst to advances in the physical sciences. Within the next 10 to 20 years, Dr. Romig said he is quite confident that a number of materials and devices in the market will be manufactured using biological methods, even though the final product is inorganic.

An important example of nanoscience in action, said Dr. Romig, is in building layered structures. Breakthroughs in layered structures in the 1980s using nanotechnology have had broad-based applications in gallium arsenide semiconductors, which in turn have had major impacts on the wireless communications industry. Nanotechnology has also contributed to enhancing magnetic behavior in materials and improving surface hardness. Using nanocrystals, scientists can finely target light emissions and create super-capacitors. Other "nano" applications include nanocomposites that enable the separation of membranes, thin film that can be used in sensors, and materials that can capture certain pollutants and impurities.

With respect to energy efficiency, Dr. Romig described two examples in which nanotechnology will make significant contributions. Displaying a graph showing magnetic response of samarium cobalt and iron multilayers, Dr. Romig said that the magnetic flux density changes as the microstructure of the material is changed. In practice this means that it is possible, for a given amount of material, to create equivalent magnetic effects with a lower volume of material and with greater strength. The net result is greater efficiency in electrical applications, such as in small motors.

At Sandia, work has been done involving implanting aluminum and oxygen in a nickel alloy that creates aluminum oxide precipitates at the surface. Alloys such as this do not usually have good "wear properties" on the surface, but the creation of nanometer-size aluminum oxides greatly strengthens the alloy. The result is greatly reduced friction and wear.

Turning to optical properties, Dr. Romig discussed vertical cavity surface emitting lasers (VCSELs). The typical laser—called a slab laser—consists of a piece of gallium arsenide that emits a single laser beam. In the case of a VCSEL, a quantum well can be built vertically into the surface of the gallium arsenide that enables lasers to be in very close proximity to one another. For example, a semiconductor chip of just a few millimeters square can be constructed that emits a number of laser beams. Among the applications is optical communication within electronic packages. This sort of optical communications enables higher-bandwidth communications and, in military systems, allows more reliable communications because the optical system is immune to electrical disruption.

Dr. Romig's second example was the photonic lattice, which emanates from work begun in the early 1980s at Bell Laboratories. At Bell, the technology was used to diffract microwaves; using nanotechnology, the same approach can be used to diffract light. In one case, a diffraction grid has been built to bend light 90 degrees with 100 percent efficiency in a silicon environment. An important application is development of optical interconnects on silicon chips. Within 10 years, Dr. Romig said that it should be possible to use photonic lattices to produce laser light from silicon. This could have revolutionary impacts in optical computing.

Semiconductor Nanocrystals in Biological Systems

At Lawrence Berkeley Laboratory, nanocrystalline materials have been injected into living cells and, under the right conditions, the cells have emitted light. So far this has been done only as a demonstration that nanocrystals can be injected into cells without damaging them. Applications are off in the distant future, but it opens up the possibility of tagging cells and watching where they go.

With respect to Sandia's work on microsystems, Dr. Romig said that with respect to microelectronics, miniaturization accompanied by expanding capability has been the hallmark of the industry for years. However, a microprocessor in principle is very much the same today as it was in the 1970s, only far more powerful. Over the course of the next decade, Sandia scientists believe that a revolution will occur that will enable more than just computation to be done on silicon. Silicon-based technologies will be able to sense physical things, such as acceleration and temperature, act on that data with micromechanical devices, and communicate to the outside world using optical signals. This is a "system on a chip" but a much broader notion of that concept than people discuss today (which usually refers to combining processing and memory or analog or digital functions on a chip).

To make this happen, nanotechnologies will be key, but a number of issues must be addressed to make nanotechnologies viable in practical applications. In microelectromechanical systems (MEMS), tiny gears may be used, but they must be able to withstand wear and tear. One must also be concerned about how these devices respond to humidity or other conditions in the environment. An important design issue to take into account is that devices at the micro-scale are affected by factors at the nano-scale. Gravity and inertia influence normal mechanical devices.

At the scale of a few microns, however, surface forces dominate inertial forces, so things such as "stiction" become more important, and therefore lubrication at a small scale is an issue. Finally, a single dislocation can affect performance at the nano-scale; in macrosystems, it is usually a network of dislocations that generally affects performance. Dr. Romig pointed out that computational tools could now model the impact of a single dislocation on a particular material under development.

Another application of nanoscale science on technological development at Sandia is the tunneling transistor. The CMOS technology presently used in semiconductor manufacturing, as one tries to shrink it down smaller and smaller, runs into a physics wall having to do with quantum mechanical tunneling—the walls of materials are so thin that electrons pass through them. There are ways, Dr. Romig said, that a structure can be built to take advantage of the tunneling, so that a transistor works because of the quantum tunneling, rather than malfunctioning. This technology may allow Moore's Law to be extended.

The Future: Nanotechnology in the Biosciences

Collaborative research between scientists at Sandia and Harvard University is developing ways to do rapid DNA sequencing. The technique uses an organic membrane, drills a hole through it using a virus, puts an electrolyte on either side, and then places DNA on one side. Scientists then put a potential between the two, which pulls strands of DNA through the hole, and the "correlation factor" enables the DNA sequence to be determined. Doing that process one at a time is not very efficient, but scaling up this process's speed could have exciting applications. Dr. Romig said that if one could sequence 1,000 base pairs of DNA per second through a pore developed using this technique, it would take about a month-and-a-half to sequence the entire human genome. If a structure could be built using 1,000 pores, the human genome could be sequenced in a day. Now, this technique would not be used for the human genome, since that sequence is nearly complete, but it could be helpful in countering biological warfare. This technique could very quickly identify the agent in a biological attack, which in turn could be used to rapidly develop a vaccine.

Future Challenges for Nanotechnology

Fabrication on a large scale will be one challenge for nanotechnology in the coming years. Many of the nanotechnologies Dr. Romig discussed have been fabricated once in a laboratory. To have widespread impact in the defense or commercial sectors, it will be necessary to develop methods to manufacture these technologies at a significant scale. Specifically, the challenges have to do with synthesizing, manipulating, and assembling devices at the atomic scale. In some cases, nanotechnology amounts to building an entire chemistry lab at the size of a chip. Another challenge is computational, both in aiding in the design of systems and then in processing all the data that nanotechnologies can collect.

To make further development of nanotechnologies a reality, advanced diagnostics and modeling are crucial. Within the Energy Department's complex of facilities, the national laboratories provide advanced photon and neutron sources, as well as electron microscopy centers, such as the one at Berkeley. Sandia has developed something called an "atom tracker" that enables measurements at very small scales with localized probes; the atom tracker actually allows a scientist to trace an atom on a particular surface. Finally, high-end computing helps scientists understand what nanotechnologies do. The passage of drugs through biomembranes calls for first principles models, models that scientists would like to link so that they can understand how an entire system works. Doing this is very computationally intensive.

As scientists and engineers accelerate research on nanotechnologies, Dr. Romig said that multidisciplinary partnerships would be needed. The federal government will have a prominent role, as a funding entity and as an owner of national labs. Universities and the private sector will play important roles as well. What will be most crucial is synergy across universities, the federal government, and the private sector.

Questions From the Audience

Dr. Romig was asked whether there were any commercial products on the market using nanotechnologies. In response, Dr. Romig said that the thin-layered structures in cell phones and pagers use nanotechnology. As for micromachine technologies—micron scale as opposed to atomic scale for nanotechnologies—air bags in cars and a number of engine controls are examples. Dr. Romig was asked about the status of nanotechnology research initiatives in other countries. Dr. Romig responded that the Europeans and the Japanese are every bit as aggressive in nanotechnologies as the United States. Most scientists would consider the United States to be slightly ahead of its competitors in nanotechnologies; with respect to micromachines, neither the Europeans nor the Japanese have shown a great deal of interest in this area.

DEFENSE INTERESTS AND APPLICATIONS

Timothy Coffey
Naval Research Laboratory

Dr. Coffey said that his focus would be on the military applications of nanotechnologies, which fall broadly into the following areas:

- Fires and Targeting—identifying where enemy fire is coming from and how to eliminate it;
- Command and Control—the systems needed to run military operations;
- Survivability and Sustaining Forces; and
- Chemical and Biological Defense.

Without focusing exclusively on biotechnology and computing, Dr. Coffey suggested that the following technologies over the next 10 or 15 years would have the most dramatic impacts on the military:

- Nanoelectronics and Microelectromechanical Systems (MEMS);
- Computers, Displays, Networks, and Advanced Algorithms;
- Engineered Biomolecular and Chemical Systems;
- High-Energy Density Batteries and Fuel Cells;
- Advanced Lightweight Materials, Structures, Composites;
- Micro-Air Vehicles and Systems; and
- Multi-Spectral and Hyper-Spectral Sensors and Processors.

In thinking about the biotechnology and computing industries, Dr. Coffey pointed out that a variety of scientific communities, from the biosciences, to materials, to computers and information technologies, to electronics are increasingly interconnected. From the Department of Defense's (DoD) point of view, the confluence of these technologies holds the most promise.

Micro-Air Vehicles

To illustrate this confluence and what it means to DoD, Dr. Coffey used micro-air vehicles and systems as examples. First, the vehicle is not a traditional airplane and indeed looks more like a biological system than an airplane. It is also small—only 15 centimeters across with a weight of 100 grams and a payload of 15 grams. Its range is 10 kilometers at the most and its flight endurance is between 20 minutes and one hour. The types of missions for these vehicles would include radar jammers, small visual cameras (now fully capable at the one-ounce level to take and transmit video), uncooled infrared cameras, and chemical/biological sensors. The key technologies for such vehicles are MEMS, ultra-low Reynolds number aerodynamics, wing-flapping aerodynamics, advanced materials, and artificial intelligence.

One prototype vehicle is the Micro Tactical Expendable (MITE), whose mission is radar jamming. Because disabling radar takes very little power, this small vehicle flies out to the radar tower, lands on it, and emits enough power to jam the radar. New materials make such a vehicle possible, as well as advances in motor technology that enable small and sufficiently powerful motors for such a mission. The MITE is currently functioning and it today disables radar in military missions.

The next evolution of the MITE will allow a number of these devices to work in a swarm—to act in concert as they enter the battlefield and identify and label a number of tanks. The vehicles need enough intelligence to tag vehicles spectrally and communicate among them. Using today's technology, such devices would have a range of only one kilometer.

A more advanced version of this technology would involve teaching the devices to behave collectively, which would require advanced communications and sensor technology. Individually, the vehicles may be less sophisticated than those used today, but collectively comprise far more sophisticated systems than we presently have. Even with a communications range of a kilometer, a swarm of such things could transmit a large amount of information. But they would have to be very intelligent, able, for instance, to regroup and complete the mission should a substantial number be destroyed by the enemy.

With respect to detecting biological or chemical warfare hazards, a sensor is placed on the small vehicle that then flies into the hazardous area and transmits back what it has found. In this example, biology, electronics, computers, and materials all come together to make the sensor work. A biologically based sensor on a small vehicle will detect the agent, become fluorescent in a specific way depending on what the agent is, and transmit the identified agent using a fiber optic link to the command center. This yields very rapid detection of a particular agent. MEMS will enable this technology to be scaled down even further.

In biosensors, the military has developed sensors that are based on neurons. Neurons have been taught to grow on electronic substrates, and their vitality is monitored. Essentially, this system is analogous to the canary in the mineshaft, in that when these sensors detect certain things, their electrical patterns change in accordance to what the sensors have found.

Another application of biology in military systems is in "biofouling control." In one example, Dr. Coffey said, the Navy has developed a material that allows a coating that might be used on amphibious vehicles to repel growths that would be harmful to the coating. The result is an environmentally sound way to lengthen the life of coatings used on ships and submarines.

In summary, Dr. Coffey said that as the battlefield demands on the military grow, it will increasingly turn to biological systems for solutions, but that these systems will rely on advances in computing and electronics technologies. Such technologies give military systems the lightweight characteristics that are increasingly important, as well as the communications capability that is also required.

DISCUSSION

A questioner asked Dr. Romig about the intellectual property regime with respect to nanotechnologies being developed at Sandia. Dr. Romig said that Sandia recognizes the value of intellectual property in terms of possible royalties. Sandia and other DOE labs are also carefully thinking through the right intellectual property arrangements in partnerships. If the intellectual property rights for products from DOE labs are not carefully specified, then partnerships with industry can be poisoned, even though such partnerships can be very mutually beneficial.

Charles Wessner asked panelists to discuss what their agencies need from policymakers to better fulfill their missions. Making an analogy with the Semiconductor Industry Association's technology roadmap, which identifies technology "show stoppers," Dr. Wessner asked panelists to point out any "show stoppers" in their fields. Dr. Coffey said that in terms of facilities, the DoD is in reasonably good shape. Computer technology must advance, he added, to enable the sorts of things he discussed with respect to micro-air vehicles. The Defense Department will go to any source, public or private, to find the necessary computing capability. In terms of DoD's technical contributions, Dr. Coffey said that the Department could probably be most helpful in the interface between the physical sciences and biology.

Dr. Romig said that the DOE labs are resource-constrained at present. The Defense program budget, which is a funding source for the labs, is always under close scrutiny. The labs themselves have adopted a dual production and laboratory role, and production requirements have placed great strain on research budgets. While noting that most laboratories will always make a case for more money, Dr. Romig said that the best way, in his experience, to stimulate collaboration across disciplines is through research money. If the goal is to encourage collaboration between biology and the physical sciences, there must be some targeted funding toward this end, regardless of which agency funds it.

In closing, Mr. Borrus said that at the frontiers of science and technology, whether it is in biosciences or nanotechnology, scientific inquiry is a highly social enterprise increasingly enabled by cross-disciplinary fertilization and applications that permit technology to flow in many directions. Cross-institutional collaborations—among government, industry, and universities—are necessary. For the broader project on Government-Industry Partnerships for the Development of New Technologies, it is important to understand which partnerships work, which do not, and why. This should be of help to the entire process of discovery and breakthrough, one that is always surprisingly fragile. From a policy perspective, this means support for fundamental science, and development of the enabling technologies, tools, methods, and partnerships that create products and processes that impact our economy and society. If past generations had been more timid in support for computing and biotechnology, Dr. Borrus said he doubted very much that the United States would be enjoying the economic boom it does today. Dr. Borrus said that we should be no less timid moving ahead.

Panel VI: Intellectual Property and the Public Domain: Sectoral Perspectives

INTRODUCTION

Jorge Goldstein
Stern, Kessler, Goldstein, & Fox

Mr. Goldstein opened the panel by commenting that patents reflect a "grand compromise" in giving inventors incentives to create new technologies—which it is hoped will widely benefit society—by allowing them to privatize the economic returns of an innovation's benefits. There are questions about what we should privatize, for what length of time privatization should be ensured, and which institutions should benefit from privatization and which should not. The presentations in the panel would, Mr. Goldstein said, reflect the competing forces that run through the patent issue. Dr. Wes Cohen's presentation would consider how different sectors view patents. Dr. Maryann Feldman's presentation would focus on patents and universities, and the government-university relationship with respect to patents. Finally, Mr. Robert Blackburn of Chiron Corporation would provide the perspective of an in-house practitioner of patent law who has done both litigation and scholarly writing on patent law.

Dr. Goldstein reminded the audience of a recent *Washington Post* article about Craig Venter of Celera Genomics. Venter, according to the article, is filing more patent applications than he originally promised as his company sequences the human genome. Congress and other entities have criticized him for this. Although his company is filing thousands of patent applications, Dr. Venter, in defending Celera against critics, said that he anticipated his company having no more than a few hundred patents at the end of the day. Dr. Goldstein wondered whether Celera would in fact have just a few hundred patents by the time it finishes sequencing the human genome.

In turning over the podium to the panelists, Mr. Goldstein said that the patent system reminded him of one of the animals from Dr. Dolittle, the Push-Me-Pull-You, the two-headed llama that wanted to go in two directions at once. The patent system, Dr. Goldstein said, may be a multi-headed llama, with a number of conflicting tensions in constant search of a grand compromise to determine in what direction it will finally go.

SECTORAL VARIATIONS IN THE ROLE OF INTELLECTUAL PROPERTY

Wesley Cohen
Carnegie Mellon University

Dr. Cohen began by saying that the title of the paper he was going to discuss is "Patents, Public Research, and Implications for Industrial Innovation in the Drug, Biotechnology, Semiconductor, and Computer Industries" and that his co-author is Dr. John Walsh of the University of Illinois at Chicago. The paper is part of a larger project being conducted in collaboration with Richard Nelson of Columbia University.

Dr. Cohen said he would try to examine some of the assumptions underpinning the patent system, especially when it comes to the public-private interface and public research. By public research, Dr. Cohen said he meant research conducted by universities using public funds, and research conducted in federal laboratories. He also said that he would look closely at the assumptions underlying the Bayh-Dole Act. The rationale for Bayh-Dole was that there was an "urn full of untapped possibilities in universities and other public research institutions" and that universities needed the ability to patent this knowledge to fully exploit their R&D potential. This would serve as an incentive to universities to commercialize and provide the complementary R&D that would take these ideas into the commercial marketplace. Another part of this rationale was to give faculty members extra incentive to push their innovations from the university laboratory into the commercial arena. Dr. Cohen said that most of the circulation of university research is done through the traditional channel of publication in journals.

The questions Dr. Cohen planned to address included the following:

- Do patents really protect industrial R&D?
- Does public research need to be privatized to ensure its commercialization?

The industries to be examined are drugs, biotechnology, semiconductors, and computers. The data used in the analysis comes from the 1994 Carnegie

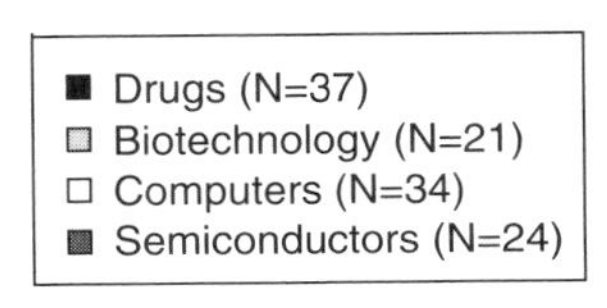

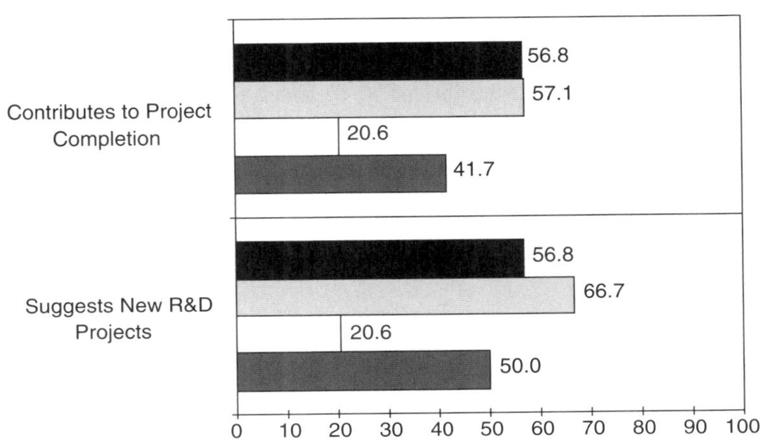

FIGURE 14 Importance of public research for industrial R&D.

Mellon Survey of Industrial R&D. The respondents were R&D managers or unit heads in U.S. manufacturing industries. Overall, the survey received 1,478 responses, reflecting a 54 percent response rate. The respondent firms include the very large ones as well as those with just a handful of employees. For the firms Dr. Cohen discussed, the sample size was: drugs, 37; biotechnology, 21; computers, 34; and semiconductors, 24. One thing that Dr. Cohen would not discuss today, given the 1994 date of the collection of the data, is the growth of software patenting and the patenting of business methods, both of which are prominent topics today.

As background, Dr. Cohen discussed the importance of public research to industrial R&D. The traditional academic wisdom on public research is that it is commercially important in a small handful of industries. In the full Carnegie Mellon sample, industrial research managers in all sectors reported that public R&D was important to their companies. For the four sectors analyzed in his paper, Dr. Cohen said that when asked whether research from universities or government labs suggested an R&D project or contributed to the completion of a project, the positive response was quite high. Figure 14 shows that for the category "Contributes to Project Completion" or "Suggests New R&D Project," more than 50 percent of drug and biotech firms responded positively, while a large portion of semiconductor firms did so. Only for the computer sector was the positive response relatively modest. These findings suggest the stakes involved with policies such as Bayh-Dole. If, for example, an industry reports that publicly funded R&D is of little commercial relevance, then policies such as Bayh-Dole may be of modest consequence.

Dr. Cohen and his colleagues then asked industrial R&D managers about the channels through which they learned about public research. The channels of knowledge flow that managers were asked to consider were patents, publications, meetings or conferences, informal exchange, new hires, licenses, joint ventures, contracts, consulting, and temporary personnel exchanges. Across the entire sample of 1,478 firms, by far the most important channels were publications and meetings or conferences. Informal exchanges were also very important. In sum, the traditional channels of public science—publications and conferences—are the main sources of information for industrial R&D managers. For the four industries focused on for his research, Dr. Cohen said that papers and publications and meetings or conferences served as the primary means of communicating public research to industry.

Looking more closely at drugs, biotech, computing, and semiconductors, Dr. Cohen said that "formal, market-mediated" channels, that is, "privatizable" mechanisms, are most important in drugs and biotech. Recalling an earlier figure, Dr. Cohen said that these two sectors are where public research is most important; in those industries, the public channels (i.e., papers and conferences) are most important. For semiconductors, the public channels, such as papers, are dominant, although informal exchanges and consulting arrangements are quite prominent in moving public research into the commercial arena. Although "privatizable" mechanisms play a role, Dr. Cohen said that given the weight placed by respondents on public channels, it is fair to conclude that it is not necessary for channels of information transfer to be private in order for public research to finds its way into commercial use. Even when considering the relative importance of private versus public channels, Dr. Cohen said that much of public research flows readily to the commercial sector using public channels.

Dr. Cohen turned to the issue of how firms use patents to protect inventions, that is, the effectiveness of appropriability mechanisms for product innovations. Firms use a variety of ways beyond patenting to protect their innovations, such as secrecy, lead time, complementary sales and service arrangements, and complementary manufacturing. In the survey, respondents were asked to give the mean percentage of product innovations for which each mechanism was considered effective. The notion of effectiveness was defined as protecting the commercialization or licensing of an innovation that was protected using a particular mechanism, e.g., patenting or secrecy.

As shown in Figure 15, there are differences across sectors in which mechanisms are more effective. In drugs and biotech, patents, secrecy, and lead time were the leading mechanisms. In semiconductors and computers, patents are

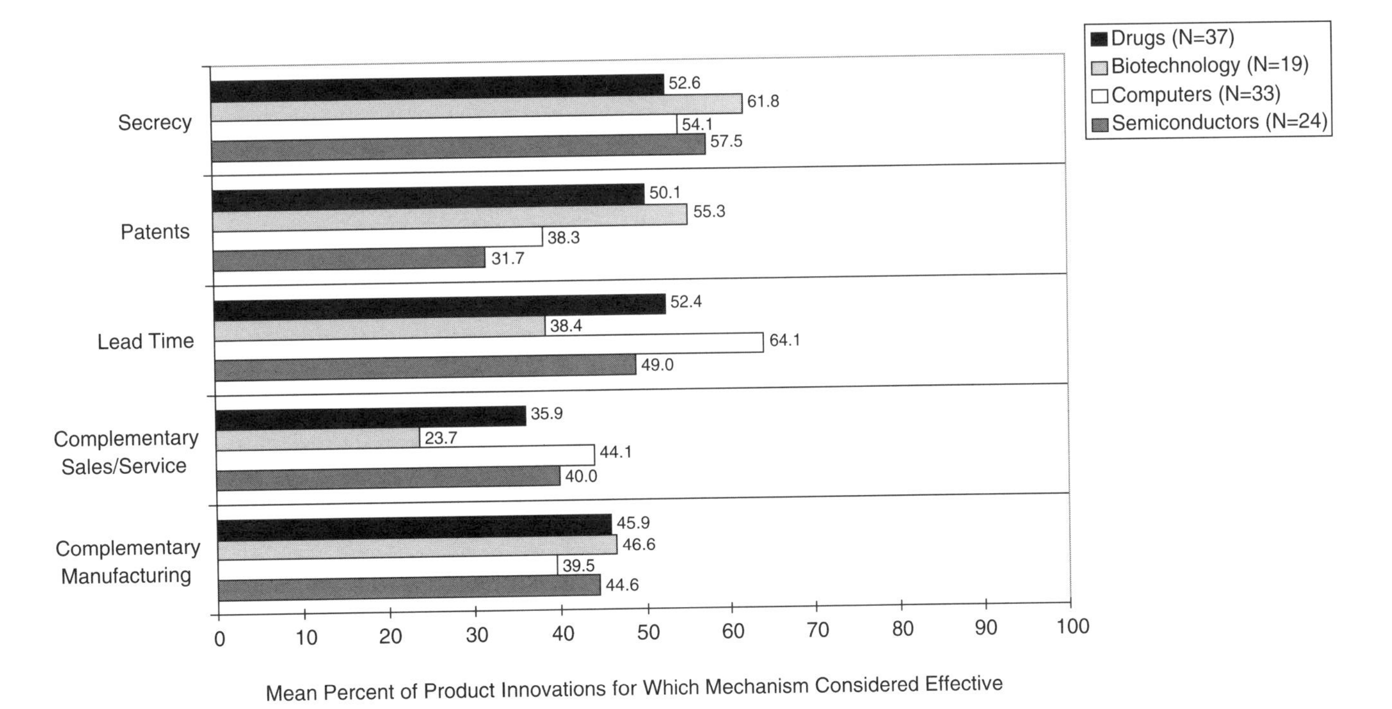

FIGURE 15 Effectiveness of appropriability mechanisms for product innovations.

less important, with secrecy, lead time, and complementary sales and service being more useful mechanisms. Looking at patents alone, it is clear that biotechnology and drugs are two industries—either in comparison with semiconductors and computers or with all manufacturing sectors—in which patenting matters quite a bit.

With the varying effectiveness of patents across sectors, one might ask why firms patent at all. In the case of biotechnology and drugs, Dr. Cohen said that indeed firms use patents for traditional reasons, namely to prevent other firms from gaining commercial advantage from their innovation. In semiconductors, where patenting is less effective, firms patent as part of a strategy to build a portfolio of knowledge in a particular industry to build "player strategy" in the field. Patents are also used in cross-licensing negotiations. In computers, standard licensing is more prevalent than in semiconductors, and patents are used in cross-licensing negotiations.

In conclusion, Dr. Cohen said that most of the public research—that which takes place in universities or government laboratories—in the drugs, biotechnology, semiconductor, and computer industries is made available to these industries through public channels. This does not mean that privatization of some of these knowledge flows is a bad idea. These data suggest, however, that policymakers should carefully consider how essential privatization is in encouraging the commercial exploitation of public research. A key point is that importance of public research and the way in which it is transmitted to the commercial arena *varies across industries*. Therefore, what is at stake with respect to the Bayh-Dole Act also varies across industries. In biotech and drugs, Bayh-Dole may have the intended effect of promoting innovation by providing the legal framework for exclusive rights to intellectual property. But the importance of public channels in these industries (and public channels dominate private ones) suggests that exclusive rights are not critical to commercialization. Looking to the semiconductor industry, where patents are not terribly important, granting exclusive rights to public institutions for R&D is not likely to have much effect. Dr. Cohen sounded a cautionary note, saying that exclusive rights for public institutions may inhibit public-private partnerships. At Carnegie Mellon, Dr. Cohen recalled the story told to him by an electrical engineer, who said that he would gladly patent his research, but not press for license fees from corporate partners. If that happened, the corporate partners would not talk to him at all, fearing that each communication might result in a license fee or a legal dispute. The net result would be to inhibit the free flow of information between universities and industry.

Questions From the Audience

Dr. Cohen was asked whether the size of the firm affected the importance of patents. Dr. Cohen responded that he and his colleagues looked carefully at the role of firm size, and found that large firms tend to view patents as being much more effective. The primary reason is that large firms have the legal resources to defend patents. Dr. Cohen added that the role of size with respect to patents varies across industries. In the biotechnology and drug sectors, small firms find patents to be very valuable; indeed, patents are absolutely essential to obtaining venture capital financing.

POST-BAYH-DOLE UNIVERSITY-INDUSTRY RELATIONSHIPS

Maryann Feldman
Johns Hopkins University

In talking about university-industry relationships after passage of the Bayh-Dole Act, Dr. Feldman said that she would synthesize a multi-year research project funded by the Andrew Mellon Foundation. These studies have explored universities' responses to the Bayh-Dole Act. David Mowery of Berkeley and Richard Nelson of Columbia conducted the first set of Mellon studies on this topic; they examined the University of California at Berkeley, Stanford University, and Columbia University. The second set has looked at Johns Hopkins University, Duke University, and Penn State. The objective is to increase the number of university intellectual property databases that are available for analysis. The studies also aim to study the relationship between industry and universities and understand that relationship in an evolutionary framework.

Dr. Feldman said her presentation would attempt to accomplish the following:

- Interpret some of the changes in university-industry relationships post-Bayh-Dole;
- Demonstrate the variations in which universities have responded to the new opportunities presented by Bayh-Dole; and
- Offer an interpretative framework to understand the evolving industry-university relationship.

The post-Bayh-Dole era for the university-industry relationship has been marked not just by a change in the intellectual property regime, but also by a number of other changes in the environment. In many sectors, such as biotechnology and computing, universities have been sources of "high-opportunity" technologies over the past 15 to 20 years, and industry has sought access to these technologies. The new technological opportunities available from universities have been accompanied by a growing need for universities to find new sources of research funding. Competition from researchers has also grown; more researchers are out there seeking grants to conduct research.

Although Bayh-Dole's passage in 1982 serves as an important policy watermark, the six universities under study were patenting prior to that and, since its passage, in a changing environment. Nonetheless, Bayh-Dole deepened universities' existing patenting activities, while also expanding patenting to universities that were inactive in patenting in the past.

Under Bayh-Dole, all universities acquired broader intellectual property rights, but the Mellon studies found great diversity across universities with respect to institutional policies. For example, universities place different weights on how founding a company is credited; some may look very positively on this when evaluating professors' performance, others may place little weight on it. Different universities may look differently on a professor funded by industry versus one being funded by the National Science Foundation or the National Institutes of Health. In many universities, commercial activities are still regarded as "second tier" sorts of undertakings. Universities also vary in the amount of funding delay they will tolerate as a consequence of working with industry. Finally, the amount of authority—in terms of funding and charters—accorded to university technology transfer offices varies tremendously across universities. Many of these variations simply reflect the different cultures in the universities studied.

With respect to research results, Dr. Feldman displayed a table showing total research expenditures among selected universities and research royalties (Table 3). As the table shows, there is not a perfect match between overall university research expenditures and royalties received. The University of California system spent close to $1.6 billion in 1997 on research and also received the most dollars back in royalties, $61 million. The second largest royalty generator, however, was Columbia University, which received $46 million in royalty revenue, and it was last in total research expenditures among universities listed. Johns Hopkins University, in contrast, spent close to $1 billion on research in 1997, but had only a meager $4.6 million in royalty revenues.

Exploring the data further, Dr. Feldman displayed a table (Table 4) showing measures of the productivity of university research spending, using invention disclosures and licensing revenues as metrics. For a university's invention ratio—the number of invention disclosures per $1 million in research expenditures—Stanford, Wisconsin, the Massachusetts Institute of Technology, and Columbia had the highest ratios. In terms of royalty per license, Columbia topped the list, followed by Stanford and Wisconsin.

The patents and licensing revenues on which the economics literature has focused with respect to innovation give only a very narrow snapshot of what has been going on with universities in the post Bayh-Dole era. Universities have been undergoing a search process as they try to adapt to Bayh-Dole. Should universities retain their traditional role as generators and disseminators of knowledge? Or is revenue generation now a legitimate goal?

Dr. Feldman and her colleagues have been conducting a series of interviews on these topics at universities, and they have discovered a great deal of conflict

TABLE 3 Differences in University Technology Transfer Activities

University	Total Research Expenditures	Adjusted Royalties Received	Licenses Generating Royalties	Start-ups Formed
UC System	$1,586,533,000	$612,800,000	528	13
Johns Hopkins	$942,439,696	$4,686,519	103	3
MIT	$713,600,000	$19,860,549	255	17
WRF	$528,602,441	$11,478,605	142	25
Michigan	$458,500,000	$1,708,939	83	6
Stanford	$391,141,224	$34,014,090	272	15
WARF	$379,600,000	$17,172,808	133	2
Harvard	$366,710,262	$13,402,273	232	1
North Carolina	$263,517,405	$1,665,909	61	2
Columbia	$244,100,000	$46,105,192	201	4

SOURCE: Association of University Technology Managers, 1997

on these questions. In the years immediately after Bayh-Dole, universities often granted exclusive licenses to industry for their innovations; now universities are quite selective when it comes to conferring exclusive licenses. Universities have also developed new mechanisms for commercialization; in 1997, there were 248 university-generated start-up companies, and universities took an equity position in 70 percent of them.

In thinking through a framework for understanding industry-university relationships, Dr. Feldman and her colleagues have found that commercializing university technology will involve multiple types of transactions. Multiple licenses for a technology or family of technologies are one approach that universities may employ. In exchange for a license from a university, companies will often sponsor a research project at the university. This helps the company build exper-

TABLE 4 Productivity of University Research Spending

University	Invention Ratio	Royalty Per License
UC System	4.51	$116,061
Johns Hopkins	2.43	$45,500
MIT	5.04	$77,885
WRF	5.30	$80,835
Michigan	3.66	$20,590
Stanford	6.34	$125,052
WARF	5.24	$129,119
Harvard	3.25	$57,768
North Carolina	3.57	$27,310
Columbia	6.02	$229,379

tise in a technology; a company may also do this through hiring students that also work with the faculty principally involved with the technology. Dr. Feldman and her colleague have also observed companies making gifts and endowments to universities as the relationship between a company and a university has developed.

An important point to keep in mind, Dr. Feldman said, is that the technology transfer transaction is not the most important issue, but rather it is the relationship between a university and private sector partners that develops over time. Understanding those relationships is the best way to appreciate the evolution of university-industry ties in the post-Bayh-Dole era. Universities and companies have, therefore, engaged in a very active exchange of information, using transactions such as licenses, sponsored research, consulting arrangements, recruitment, equity arrangements, support for start-ups, and others. Precisely what transactions dominate will vary depending on the sector, the technology, and the size of the company, as Dr. Cohen said in his presentation.

The formal and informal rules and norms of universities will also shape the evolution of their relationships with companies. In terms of exogenous policy factors that will influence these interactions, Dr. Feldman pointed to the intellectual property regime and funding availability. Dr. Feldman and her colleagues have started interviews to develop case studies of particular firms, universities, and technology areas to gain a greater sense of the factors that affect the evolution of university-industry relations in the post-Bayh-Dole era.

In conducting archival research on the history of patenting at Johns Hopkins University, Dr. Feldman uncovered correspondence between the National Research Council and the dean of the medical school from 1949. At the time, NRC was conducting a survey of university intellectual property policy, and it was inquiring about the possible participation of personnel from Johns Hopkins in a conference on the topic. The response from the dean of the medical school stated that individuals in the school would neither be interested in patenting innovations that may impact the public health nor participating in the NRC conference. In fact, the medical school viewed it as "undesirable" for doctors associated with Hopkins to patent their inventions or discoveries as they pertain to public health. As of 1949, no member of Hopkins' medical school faculty had applied for a patent.

Based on her interviews, Dr. Feldman said that Johns Hopkins has a legacy of discouraging patenting. Faculty members hired in the 1970s, coming from places such as Stanford that encouraged patenting, were concerned about a culture at Hopkins that discouraged such activity. In recent years, Johns Hopkins University has been trying to change this culture.

Conclusions

Dr. Feldman raised the following issues in conclusion:

- What are the most effective means to organize and conduct technology creation and diffusion?
- Are the financial and transaction costs of doing science increasing in the post Bayh-Dole era?
- Does emphasis on patenting create divided loyalties for faculty and does this hurt students?

Questions From the Audience

Dr. Kathy Behrens asked whether Dr. Feldman and her colleagues looked at a) the number of inventions that created the royalties at universities and b) the year in which the intellectual property was created that generated the revenue shown for 1997. Dr. Feldman responded that the data does not yet exist that permit these important questions to be answered. One of the motivations of the Mellon studies is to gain access to offices within universities that will enable the necessary data to be assembled. From her research, Dr. Feldman said that it appears that certain licenses generate a great deal of revenue and that there is a gestation period until licenses begin to generate returns. It seems clear, said Dr. Feldman, that there are just "a few big hits" in the university-patenting world.

Mr. Goldstein asked whether Dr. Feldman's research is consistent with Dr. Cohen's that licenses in the biotechnology and drug area are more valuable than those in engineering or electronics. Dr. Feldman responded affirmatively, noting that at Johns Hopkins approximately 90 percent of the university's intellectual property portfolio is in the biosciences. That figure is very similar for Duke University and Penn State University. At Johns Hopkins, Dr. Feldman continued, the largest revenue source from a license is for software from the School of Public Health; this raises classification issues for license revenues.

INTELLECTUAL PROPERTY AND BIOTECHNOLOGY

Robert Blackburn
Chiron Corporation

Mr. Blackburn began by saying that the number of biological and medical threats that we face today is growing, if we take seriously what epidemiologists tell us about the advance of HIV in the Third World and the growth of new drug-resistant strains of tuberculosis. In a very real sense, Mr. Blackburn said, we are in a biological war. The U.S. patent system is designed to encourage progress in the "useful arts and sciences," and the need for such encouragement has never been so urgent.

For the biotechnology industry, the patent system has been superb at protecting products sold in the marketplace and methods of manufacturing. Where there has been significant controversy is in the area of research tools. Research tools, in the eyes of some industry observers, should not be patentable. Others believe that they should be patentable because research tools represent something of value. Mr. Blackburn said that this proposition should be closely scrutinized, and he suggested that it might not be true. If it is not true, the concern arises that venture capitalists may not have sufficient incentives to invest in such tools.

The debate over the patentability of research tools is complex. It includes issues such as the statute of limitations, the clinical trial exemption under the Hatch-Waxman Act, and reach-through royalties (i.e., whether they can be awarded by a court). If one is not cognizant of these complexities, then one is debating with only 10 percent of the relevant information at hand. Mr. Blackburn said he would discuss these complicating issues, with the hope of moving beyond some of the more emotional elements that sometime arise in debates about the biotechnology industry.

The Evolution of Drug Discovery

Mr. Blackburn first discussed some of the differences between the traditional pharmaceutical industry and the biotechnology industry. The process of discovery in the traditional pharmaceuticals industry involves finding a lead chemical compound that has a desirable biological property and that can be administered orally. Often these chemicals were found by happenstance, and the real science begins when a pharmaceutical chemist optimizes it into a compound that has less toxicity, more activity, and is more biologically available. This is often a slow process, as each version of the drug has to be synthesized and tested.

Biotechnology is very different, in that it involves finding a disease with a large protein and then finding the gene for that protein. The next challenge is developing a way to make the protein in a reasonably cheap manner by recombinant DNA expression. These proteins, however, are generally expensive to make, although it is less expensive to make them using recombinant DNA than by finding them in nature. Proteins made by recombinant DNA are also not usually orally available, so there are limitations to this approach. Examples of these types of inventions are Factor Eight for hemophilia, Interlukin 2 for cancer treatment (which also has promise for AIDS), a hepatitis B vaccine that is a recombinant yeast expression of the vaccine, and human growth hormone. All of these are large proteins made by recombinant DNA technology.

Today's multidisciplinary biopharmaceutical research has brought the tools of chemistry and biotechnology together. This interaction is driving modern pharmaceutical science into an era of unprecedented drug development, and it represents nothing short of a revolution. A technique called combinatorial chemistry, for example, allows a chemist to make a library—essentially a test tube—with 100,000 randomly generated chemical structures. This is in contrast to traditional methods that require synthesis of individual chemical compounds. With the invention of targets, which is the site of action for particular drugs, chemists test the randomly generated compounds against the targets. Other scientists have invented high throughput assays that automate this process. These targets are often proteins that have been discovered using recombinant DNA techniques or genes themselves.

When an individual compound is found in a library, it is known as a "hit," but scientists will still not know what it is, as it is in the complex mixture. So-called "deconvolution technology" pulls the compound out of the mixture and gives scientists its structure. This allows them to go back to the laboratory to make the compound in large quantities. This amalgam of technologies can create a number of lead compounds in a short period of time. From that point, more traditional pharmaceutical chemistry, such as optimization, can take over, as well as newer tools of "rational drug design."

The power of this technology is truly remarkable. For diseases for which no drug therapy had been found for decades, scientist are finding multiple lead compounds, sometimes in a matter of weeks. In the coming decades, we will have, in all likelihood, multiple drugs for diseases for which we have had either poor treatments or no treatments.

New Developments in Biotechnology and Patenting

There is, however, a problem with this amalgam of technology. The current patent system may not provide the right balance of protection and freedom of operation that will allow these technologies to flourish. These technologies are not the product being sold in the marketplace; they are used in the research phase of drug development. The patent laws we have today are superb at protecting final products, but are untested when it comes to patents covering research tools. The problem is very significant because it relates to the most critical technology that we have to benefit human health.

To understand fully why patent law has been untested on research tools, it is necessary to understand what the scope of a patent is and what a patent covers. If you invent a research tool, the Patent Office does not grant a patent to you on the things that people invent using the tool. The inventor receives a patent only on the research tool, and perhaps the method of using the tool, as well as the reagents and machinery involved with the tool's use. The patent will not cover the downstream product.

Possible infringement of the patent will occur in the research phase, prior to a drug entering the marketplace. Detecting infringement will therefore be very difficult, because companies do not usually publish their lead compounds or optimization research used in drug development. This is where secrecy enters the picture. Recalling Dr. Cohen's finding that secrecy is important to drug companies, Mr. Blackburn said that keeping research methods secret is key to pharmaceutical companies having quick time-to-market.

By the time a patent-holder discovers that a research tool has been improperly used, one usual remedy, an injunction against use, is ineffective. Granting damages is another possible remedy, but there is a six-year statute of limitations in the United States on filing an infringement suit. However, because of the time it takes to develop drugs, it would not be unusual for the infringement to be discovered after the statute of limitations has expired.

The Hatch-Waxman Act

Even if a holder of a research tool patent clears these hurdles in infringement litigation, the Hatch-Waxman Act presents another barrier. This legislation was enacted to permit generic drug companies to run clinical trials on a patented drug before the patent on that drug expires so that generic drug companies would be ready to enter the market as soon as the patent expired. A court had found that running a clinical trial was a patent infringement, and Congress overturned the ruling on the rationale that the court ruling amounted to an unwarranted patent extension. Mr. Blackburn said the act exempted "from infringement those activities reasonably related to submitting data to the FDA [Food and Drug Administration]." The Act does not limit the types of patents to which the legislation could be applied.

Because the Hatch-Waxman Act was passed before the invention of research tools, Congress passed the law only with patents for drugs in the marketplace in mind. It was impossible to consider the act's effect on research tools. Parties that have allegedly infringed research tool patents have used the act as a defense, arguing that Hatch-Waxman exempts them from infringement suits. If that argument stands, the end result is that it may be impossible to infringe patents on research tools.

Recent Court Decisions

Assuming further that a research tool patent holder has cleared the Hatch-Waxman hurdle and is in court, problems arise from recent court decisions that apply patent law doctrine developed for the synthetic chemicals industry to biotechnology. Patents in biotechnology typically involve discoveries of things found in nature—a DNA sequence for example—which are subject to a wide degree of variation in specific applications, and the variations are relatively obvious. In applying patent law developed for the synthetic chemical industry, courts have rightly recognized as a significant invention the compound that will be sold in the market. But in controversial cases, such as *The University of California v. Lilly*, courts have extended this doctrine, and this has resulted in a significant reduction in the scope of patents for biotechnology inventions.

Mr. Blackburn said that the doctrine made sense when the invention was making the new structure, as in synthetic chemicals, where the structures simply did not exist before. It probably does not make sense, argued Mr. Blackburn, to apply that doctrine when the invention involves finding preexisting structures in nature and making that information available so that it can be turned into products.

Those in favor of the *Lilly* decision might argue that broad patents might be just as harmful to innovation in the biotechnology industry as the unavailability of patents. Mr. Blackburn said he did not envy the Federal Circuit Court that would have to ultimately strike the right balance, and he pointed out that the courts are the only governmental bodies presently addressing these basic policy issues. A problem is that the legal process limits the amount of input the court considers. In the *Lilly* case, the court had to address about a dozen technical issues, but the petitioners had to limit their briefs to no more than 50 pages. That is a small information base with which to adjudicate very technical scientific and legal issues.

Delay is another problem. In the *University of California v. Lilly* case, this decision—which now serves as the guide on how to make a patent application in biotechnology—was handed down 20 years after the invention in dispute was made. Two decades is simply too slow in an industry as dynamic as biotechnology. On average, it takes 10 to 15 years from "benchtop to marketplace" for drugs, and under the U.S. patent system, it is very difficult to challenge patents prior to the date of award.

A Scenario of New Drug Development

To elaborate on this point, Mr. Blackburn presented a hypothetical situation about the development of a drug in a biopharmaceutical company. Suppose you are the director of research at this biopharmaceutical company. You ask the firm's patent attorney whether a particular patent will block the development of a drug you would like to develop. The patent in question covers a final product, and your company's patent attorney says that you will probably infringe upon the patent when the drug comes on the market. In the meantime, however, Hatch-Waxman will probably protect research you conduct. The patent attorney adds that, in her opinion, the patent is invalid; it involves a doctrine that the courts have yet to address, and in her reading of current trends, she estimates a 60 percent chance that the patent would be defeated in federal circuit court—in 15 years.

As research director, you then turn to your business development people, who cannot procure a license to the technology on reasonable terms. The 60 percent chance that the patent will be thrown out is causing you great pause. The chances of any one drug making it through to the market are very low; to further discount the chances by 40 percent makes it an unacceptable expenditure of money. The cost of a successful research program to develop the drug is likely to approach $200 million. In this case, a 40 percent chance exists that a product could not be sold. However, the potential upside of the drug is so high, several hundred thousand dollars in research expenditures today seems worthwhile to test the patent.

To obtain a declaratory judgment that the patent is invalid, however, requires a case and controversy; in this case, this means that a reasonable chance exists that the patent holder will litigate. The patent holder, not wanting a court to possibly invalidate his patent, declines to provide a letter or any evidence to suggest that litigation is likely. Having a patent declared invalid, your patent attorney informs you, does not invoke the already weak reexamination procedures available from the Patent Office. A federal court would really have to make a ruling. That sort of litigation would cost approximately $5 million, which is far more than the several hundred thousand dollars that you wanted to invest in exploratory research.

In the end you, as research director, kill the project, even though further development seems technically feasible and promising, and there is a 60 percent chance that the existing patent would not block you. Biopharmaceutical companies face this sort of decision every day. Some attorneys might argue, said Mr. Blackburn, that this is how our patent system works and it has worked well for several centuries. While this may be a good patent system, argued Mr. Blackburn, "it is a lousy industrial policy."

Patenting Policy and the Climate for Innovation

One possible solution to these problems is to discontinue patenting research tools. This, of course, lessens incentives to innovation in this field. What is needed, Mr. Blackburn argued, is a system in which drug developers can obtain reasonable certainty about the climate for innovation in the industry. Such certainty must be attainable at a low cost and in a timeframe so that investment decisions are not unduly hindered. Legislation could perhaps address some of these problems. For instance, legislation could make it easier to obtain a declaratory judgment by defining a "case and controversy" to include the existence of a patent that might block the development of a new drug. Legislation could also address the current application of chemical patent doctrine—developed over 50 years ago—to biotechnology. Biotechnology is, after all, based on completely new science, and thus calls for new rules.

The Japanese and German models of patent enforcement, said Mr. Blackburn, should be closely considered as solutions to U.S. problems. In those countries, expert tribunals within their patent offices hear all challenges to patent validity. This is much cheaper and faster than federal court litigation in the United States. Federal courts would still hear cases on infringement and damages, which are reasonably predictable. Mr. Blackburn did not advocate wholesale adoption of the German or Japanese systems, because in some respects they run counter to U.S. notions of due process. But those systems offer useful guidance in developing alternatives to an American system that currently restricts the subject matter on which patents could be granted. The stakes are important, Mr. Blackburn concluded, because "getting it [the patent regime] wrong" could severely suppress the entrepreneurial spirit that drives innovation in biotechnology. In today's highly competitive climate, "the inventions you don't make could kill you."

DISCUSSION

Mr. Goldstein commented that he sees two important issues coming from the presentations. With respect to the papers by Dr. Cohen and Dr. Feldman, he wondered whether there were differences in the computer industry between hardware and software in terms of the importance of patents. Given the changes in the past five years in the legal regime governing software patents, Mr. Goldstein said that software patents are likely to become much more important. Labeling them as "computer patents" may thus be somewhat misleading. Mr. Goldstein suggested a fertile area of research would be exploring how legal decisions affirming software's patentability would affect industry behavior.

Mr. Blackburn's discussion of research tools brings a number of important themes together. Research tools, said Mr. Goldstein, will not only be one of the great high-technology developments in finding new drugs in the next 10 to 15 years, but will also cause the most friction between industry and universities. In Mr. Goldstein's law practice, he has had more problems with either academics or industrial researchers being irritated with having requested a research tool, and receiving a lengthy letter from a lawyer with terms about "reach throughs," future intellectual property rights, and other legal conditions. The NIH has issued guidelines on this, but the legal problems surrounding research tools are greatly straining relations between universities and industry.

Dr. Stephen Merrill asked Mr. Goldstein to comment further on his earlier statement about patents as a public source of technical information. Mr. Goldstein responded by saying that it is clearly in the best interests of everyone who files for patent applications to know what the best "prior art" is. It does no one any good to receive patents that are invalid; they are a real burden on the system. Cynics might say that an invalid patent in hand is better than no patent, but in Mr. Goldstein's experience, he does not see applicants "hiding their head in the sand" by ignoring prior art in order to obtain a patent at all costs that will then "hold up the industry."

In the software area, critics say that the Patent and Trademark Office issues many invalid patents in software; indeed, Mr. Goldstein said that some software patents may be declared invalid for "obviousness reasons." The Patent Office lacks the manpower and databases to understand the range of products and innovations in software. However, it is not an industry-wide phenomenon where software firms file for patents, which they suspect are invalid, as a strategy to hold back technical progress by other firms.

Concluding Remarks

Gordon Moore
Intel Corporation

Dr. Moore observed that the conference had been full and informative, and he thanked each of the speakers and discussants for making their time and expertise available. Leaders in government, industry, and universities identified key future challenges in two of our most vital industries. The intellectual stimulation of Dan Goldin's vision of the future inspired us all to plan today for the things we hope to accomplish 25 years from now. From Dan Goldin's speech, and the presentations of many others, a number of important themes have figured prominently in the discussions of the past two days.

Interdisciplinary Challenges

It is impossible to make progress in one branch of science while neglecting other branches of science. Scientific advance is an "interconnected whole" that must make broad progress on all fronts. This is not something that appears to be fully understood by some of the parties that fund research. There appears to be a natural constituency for biotechnology in Congress, given that everyone grows old, contracts diseases, and wants new remedies and treatments. In the physical sciences, the audience is less focused in Congress and there is a less vocal lobbying effort than with biotechnology. This imbalance should be addressed.

One point that was not raised prominently in the conference, Dr. Moore said, was that many of the challenges in biotechnology and computing are technical in nature, and not just about basic science. Disguising nanotechnology as nanoscience because the latter is easier to fund may be a useful short-term expedient. However, we are on the brink of developing some very exciting technologies, and we should be explicit in generating a constituency in the technology community that will actively support the necessary technical research.

We also learned that partnerships are not easy. The language and culture of the various disciplines do not mix easily. For example, it is clear that biology will increasingly rely on mathematics and computing in the coming years; both disciplines must make efforts to understand each other's culture and languages.

Trends in Government Support for R&D

Dr. Moore said that he "was appalled" to see the dropping percentage of Gross National Product that is going to R&D in the past several years. The trend is not encouraging. The problem cannot be addressed overnight, but those of us who feel strongly about our technical future must address this trend. Government research is extremely important; the private sector is increasingly funding short-term R&D because it can capture the results of such research fairly easily. It is a legitimate function of government to fund the longer-term, fundamental science that is so vital to our economy and society.

The point was raised several times that high-performance computing capability is important for a variety of reasons, and that the government is the sole market for high-performance computing these days. Efforts to extend computing power must recognize that fact and take it into account. In the field of computing, it is unique that the government plays such a key role in terms of demand for sophisticated high-end products.

Intellectual Property

Intellectual property is another topic that will shape innovation in biotechnology and computing in the coming years. The legal regime that governs intellectual property will greatly affect the rate of progress, and policymakers face significant challenges in this arena. It is a policy field that evolves slowly; rapid change is difficult to implement. But intellectual property policy will have to be addressed to provide a climate in which innovators can capture value, and in which all participants are well served.

The steering committee on Government-Industry Partnerships for the Development of New Technologies is now faced with the task of writing a report that includes findings and recommendations based on the input from this conference. In laying the basis for the report, Dr. Moore thanked each of the speakers and discussants for making available their time and expertise for the conference, as well as everyone in the audience for their attentiveness and contributions. Finally, Dr. Moore thanked the staff of the STEP Board for making this conference as informative and enjoyable as it has been—especially Chuck Wessner, who directs the program on Government-Industry Partnerships, as well as John Horrigan and McAlister Clabaugh.

V

RESEARCH PAPERS

The Federal Partnership with U.S. Industry in U.S. Computer Research: History and Recent Concerns

Kenneth Flamm
The University of Texas at Austin

In 1999, two distinguished advisory panels, a congressional budget analysis, and an administration budget proposal[1] either explicitly or implicitly focused on recent concerns that declines in federal and industry support for computer research pose potential problems for the long-term health of the information technology sector in the United States. This paper analyzes extant data on changing patterns of computer R&D funding, and how these changes are correlated with changes in the internal structure of the U.S. computer industry. We begin by briefly reviewing the history of federal support for computer R&D in the United States, drawing largely on the author's 1987 analysis of the development of government-supported computer research programs worldwide,[2] and update that

[1]I thank Jongwoo Kim for his outstanding research assistance in undertaking the work described in this paper. See President's Information Technology Advisory Committee (PITAC), *Information Technology Research: Investing in Our Future*, Washington, D.C.: Government Printing Office, 1999; National Science and Technology Council, Subcommittee on Computing, Information, and Communications R&D, Committee on Technology, *Information Technology: Frontiers for a New Millennium*, Washington, D.C.: Government Printing Office, 1999; National Research Council, *Making IT Better: Expanding Information Technology Research to Meet Society's Needs,* Washington, D.C.: National Academy Press, 2000; Congressional Budget Office, *Current Investments in Innovation in the Information Technology Sector: Statistical Background,* Washington, D.C.: Congressional Budget Office, 1999.

[2]See Kenneth Flamm, *Targeting the Computer*, Washington, D.C.: The Brookings Institution, 1987. The companion volume, *Creating the Computer,* Washington, D.C.: The Brookings Institution, 1988, focuses on the economic history of the industry.

analysis to the present. Disaggregated estimates of R&D funding, over time, are then constructed and presented. The proposition that much R&D activity has migrated out of computer hardware production and into semiconductor production is measured against available data. A final section produces a net assessment of recent trends, summarizes recent proposals that address these trends, and offers some concluding observations on the problems that these proposals are likely to encounter.

The United States government has been an important patron of information technology research throughout its history, beginning with the development of the first electronic digital computers during the Second World War, and continuing up through the present.[3] Much has changed since the birth of this technology in the 1940s, however, and in the last several years prominent voices have articulated new concerns over changing relationships between government, industry, and information technology research. This paper surveys available empirical data on computer research activity with an eye to the legitimacy of these worries. Before examining the present, we briefly survey the past.

THE BIG PICTURE[4]

In the first decade after the war, most significant computer research and development projects in the United States depended—directly or indirectly—on funding from the United States government, mainly from the U.S. military. The second decade of the computer, from the mid-1950s through the mid-1960s, saw rapid growth in commercial applications and utilization of computers. By the mid-1960s, business applications accounted for a vastly larger share of computer use. However, the government continued to dominate the market for high-performance computers, and government-funded technology projects pushed much of the continuing advance at the frontiers of information technology over this period.

From the mid-1960s through the mid-1970s, continued growth in commercial applications of vastly cheaper computing power exploded. By the middle of the decade of the seventies, the commercial market had become the dominant force driving the technological development of the U.S. computer industry. The government role, increasingly, was focused on the very high-end, most advanced computers, in funding basic research, and in bankrolling long-term/leading-edge technology projects.

From the mid-1970s to the beginning of the 1990s, the commercial market continued to balloon in size relative to government markets for computers, leav-

---

[3]For more details, see Flamm, *Targeting the Computer, op. cit.*; Flamm, *Creating the Computer, op. cit.*; National Research Council, *Funding a Revolution: Government Support for Computer Research,* Washington, D.C.: National Academy Press, 1999.

[4]Unless otherwise noted, our summary history is based on the sources cited in footnote 2.

ing the government an increasingly niche segment in a huge, wide market for computers of all shapes and sizes. Only in small niches—particularly in the very-highest-performance computers and data communications networks—did government continue to play a role as a quantitatively or qualitatively dominant customer for leading-edge technology. The government also remained the primary funder of basic research, and long term and long-range computer technology projects.

The 1990s saw some significant changes in the pattern of funding for computer research in the United States, and a sharp turn in its R&D relationships with industry. Indeed, the industry itself seems to have undergone some abrupt changes in its patterns of R&D funding. Both sets of changes, at some fundamental level, are what have motivated recent expressions of concern over the magnitude and direction of information technology R&D in the United States. Before examining the basis for these concerns, we will briefly mention some of the major players within the U.S. government responsible for bankrolling the IT revolution.

DEFENSE

What was arguably the first stored-program, electronic computer was built in Britain during World War II to assist in cracking German codes, and it was the intelligence community that dominated the development of the very fastest computers, and the components needed to make those computers, in the United States in the early years. Through the 1950s and 1960s, the National Security Agency was the primary funder of what came to be called supercomputers.

From the early 1960s on, a unique new U.S. defense agency came to dominate the U.S. computer R&D agenda. The Advanced Research Projects Agency (ARPA, later renamed DARPA with a "Defense" prefix added on), under a succession of technology leaders rotated in from the U.S. computer community, devised a strategy of focusing on long-term, innovative, and highly experimental research projects that ultimately generated an outstanding record of accomplishment. The DARPA-funded projects of the 1960s and 1970s are the building blocks of today's bread-and-butter computing. Such advances as time sharing and distributed computing, computer networks, the ARPAnet (the precursor of today's Internet), expert systems, speech recognition, advanced computer interfaces (like the mouse), computer graphics, computer-based engineering design and productivity tools, advanced microelectronics design, manufacturing and test tools—all drew significantly on results of an astounding portfolio of people and projects.

That is not to say that everything the Department of Defense touched turned to gold. DARPA occasionally burned serious resources on large-scale, "political" projects that didn't seem to produce much in the way of results—the so-called Strategic Computing Initiative, the centerpiece of DARPA computer research in the late 1980s, is a prime example of a massively expensive program that seems to have ultimately slipped beneath the waves with barely a ripple. Other Defense customers have funded even larger, marquee R&D programs with what have been decidedly mixed results—the Office of the Secretary of Defense did not have a lot to show for its billion-dollar VHSIC (Very High Speed Integrated Circuit) program of the seventies, or its various ADA programming language initiatives of the 1980s and 1990s. Indeed, the Defense-defined ADA language seems destined to become the Latin of computer languages—used only by the most faithful within the inner sancta of the Pentagon.

Overall, DoD's role in federal funding of computer research has declined significantly over the years. Detailed NSF data on federal funding of math and computer science research (that is, excluding development expenditure) begin in the 1970s. Figure 1 shows the growth of federal spending in this area over 30 years, expressed in 1996 dollars (using the fiscal year GDP deflators compiled by NSF). Note that overall, federal spending on basic research in the area has gone from exceeding 50 percent of the total in the mid-1970s, to a one-third/two-thirds split today. The growth in basic research funding has certainly been considerably less impressive than growth in applied research.[5]

Within the relatively slow-growing federal basic computer research budget, DoD's role has shrunk even more sharply over the decades. Figure 2 shows that in the early 1970s, over 50 percent of federally funded basic computer research came from DoD. Today, that figure is about 20 percent.[6]

Similarly, Figure 3 shows that DoD's share of applied federal computer research funds has fallen from as much as 80 percent in the early seventies, to under 50 percent today.[7] Interestingly, even as DoD's overall share of applied federal research funding outside of universities was falling—mainly because other agencies' funds for applied computer research in industrial and nonacademic settings exploded (see Figure 1)—DoD remained the clearly dominant funder of applied research in universities. Better than 90 percent of federally funded applied math and computer science research in universities for 1999 came from DoD, preserving its role as the primary funder of computer research (aggregating basic and applied) in universities.

In short, the major trend affecting DoD in recent years has been a dramatic shrinkage in its relative importance as a funder of computer research within the federal government. This has happened at a time when the overall federal research portfolio was shifting away from basic and toward applied research, and a dramatic surge in recent years in applied computer research by agencies other than DoD, channeled mainly to industrial and nonacademic performers. Next, we briefly consider the other federal patrons of computer research.

---

[5]This Figure is based on official NSF data on federal obligations for research, by field of science and engineering, by fiscal year.

[6]The data used here is the same as in the previous Figure.

[7]This data, too, is from the same sources as the previous two Figures.

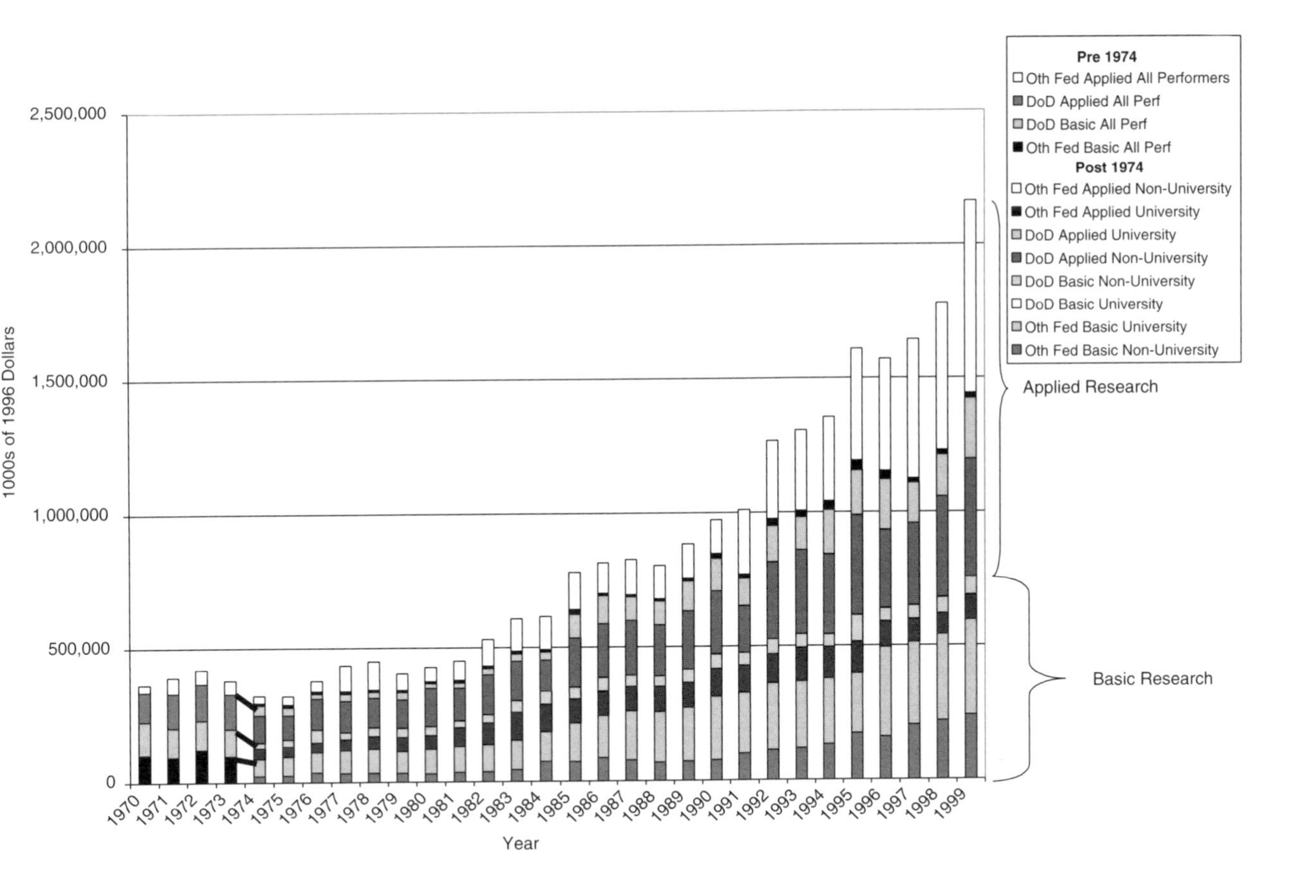

FIGURE 1 30 Years of federal math & CS research funding.

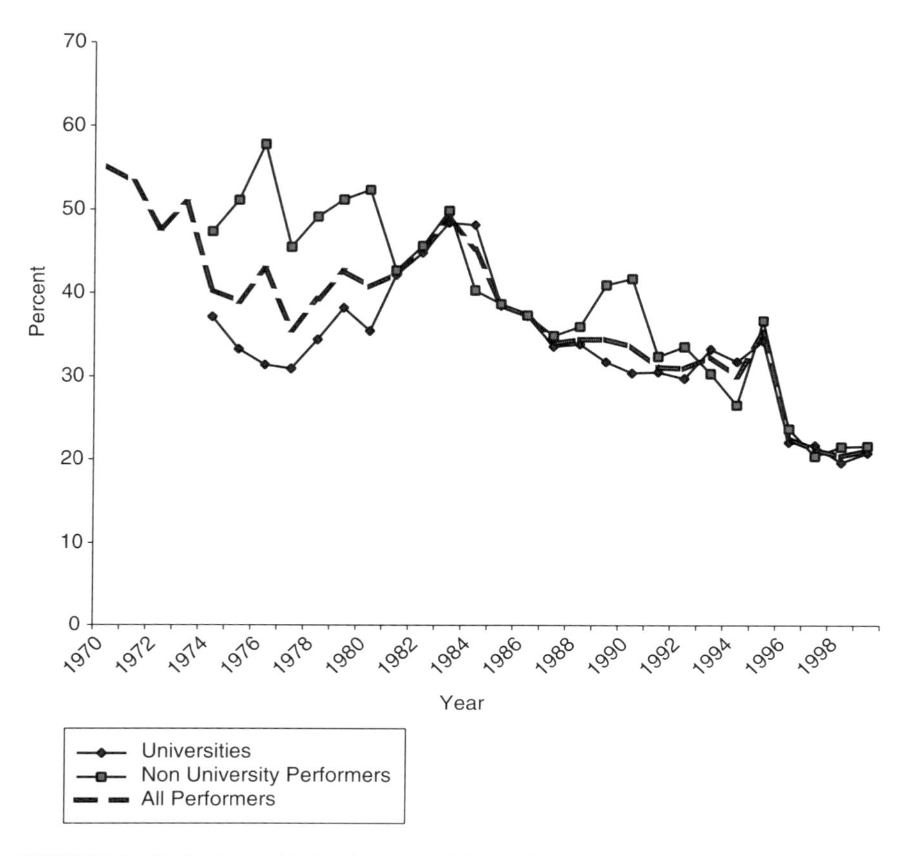

FIGURE 2 DoD share of federal math and CS basic research dollars.

THE OTHERS

Department of Energy/Atomic Energy Commission

Historically, the second-largest federal funder of computer research and development has been the Department of Energy, and its predecessor agency, the Atomic Energy Commission. This was true 50 years ago, and it remains true today. Energy's special interest has been in funding the development and use of supercomputers, and the modeling and operating system software for nuclear weapons design that runs on these ultra-high-performance computers. In the 1950s and 1960s, these machines bore the names of now defunct companies—Engineering Research Associates, Philco, Control Data, Cray Research—and IBM. In the eighties and nineties now familiar new names were added to the Energy roster (along with IBM)— Intel, Silicon Graphics, and Sun Microsys-

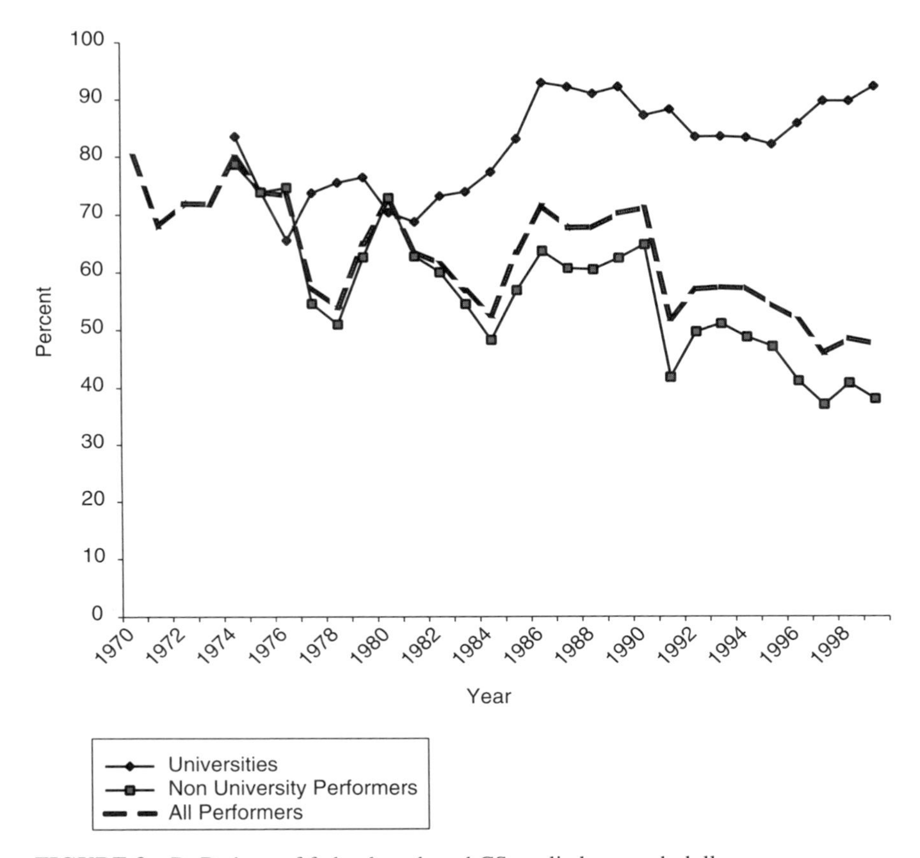

FIGURE 3 DoD share of federal math and CS applied research dollars.

tems, and the focus shifted toward massively parallel supercomputing using large numbers of processors lashed together.

In addition to funding the second-largest slug of computer research funding in the federal budget, Energy was particularly creative in supporting development and production of horrendously expensive new hardware. A variety of means—advance purchase agreements, prepayment for hardware, direct subsidy of development expenditure—have been used to underwrite creation of these top-of-the-line machines.

NASA

In the sixties and seventies, NASA was probably the third-most significant federal player in computers. While the scale of its efforts was significantly smaller than that of DoD and even Energy, it was highly focused and had great impact on a number of areas. Principal among these were artificial intelligence, computer simulation of physical design (the NASTRAN finite-element structural modeling software that is today an industry standard was originally developed for NASA), image processing, large-scale system software development (the Space Shuttle was a major landmark for large-scale complex systems development projects), and radiation-hardened, reliable computing hardware. Today, however, NASA influence has dwindled along with its budget, and it is now the fifth-largest player in funding computer research.

NSF

The National Science Foundation was a late arrival on the computer scene. Computer science did not emerge as a separate academic discipline until the 1960s, and with its orientation toward academically defined disciplines, the NSF initially only funded problem-oriented computer applications in traditional academic disciplines. The first organized thrust into computing came in 1967, with a large facilities investment program designed to strengthen computing hardware resources available in the nations' universities and colleges. In the mid-1970s, computer science finally took off as a supported discipline. From 1985 on, another large infrastructure investment—this time in making supercomputers available to academics, outside of the traditional defense and nuclear weapons communities—was launched. Finally, in the 1990s, NSF was given the leadership role in the new High Performance Computing and Communications Initiative, which was merged into a single, larger interagency IT R&D budget for FY 2001.

Today, interagency briefings typically show NSF with the largest share of IT R&D in the federal budget,[8] but this is an arguably misleading assertion. (For one thing, substantial computer R&D support by the individual services within DoD, and lower-profile computer R&D efforts embedded within other programs elsewhere within the federal budget are clearly being omitted from such IT "crosscuts.") A safer claim is that NSF is today the dominant force in basic computer research (with close to 60 percent of federal basic computer research funds, and close to half of all basic math and computer science research funding). Within research more broadly writ—i.e., including applied research—NSF today has a solid hold on the number-three spot, after DoD and Energy.

NIH

The National Institutes of Health in recent years have come to occupy an increasingly prominent role as a supporter of computer research, in no small part

[8]See, for example, Dr. Ruzen Bajcsy, "Information Technology R&D Crosscut for the President's FY2001 Budget," presentation to the President's Information Technology Advisory Committee, February 25, 2000.

because of the growing role of computers and computing technology in automated genetic sequencing and analysis. It is by no means a new role, however. In earlier decades, NIH had played a small but focused role in supporting the development of artificial intelligence and expert systems for interpreting data and diagnostics. In a sense, the recent upsurge in the NIH presence in computers represents a continuation of this activity in a new and promising area of technology. Today, NIH is the fourth largest player in federal computer research, basic and applied. With roughly two-thirds of its funding going into basic research, it is number three in basic research, after NSF and DoD.

THE U.S. GOVERNMENT AND ACADEMIC COMPUTER RESEARCH

Thus far, we have seen that there has been significant growth in federal funds for math and computer research over the last several decades, more than doubling in real terms from 1991 to 1999 alone. Federal math and computer science research obligations approached $2.5 billion in 1999. We did, however, note that the agencies involved in providing these funds have altered their relative shares of this pie over the years, and that overall, there has been a marked shift away from basic research and toward more applied projects.

Does this mean that the federal role in academia, which we tend to identify with more basic research activities, has declined? Briefly, the answer is no. Based on NSF data, Figure 4 shows the share of federal funds in math and computer science research and development undertaken in universities, and in university-administered federally funded research and development centers (FFRDCs).[9] Though there has been some downward trend most recently, the federal share of all math and computer R&D funds in universities has fluctuated between 65 and 75 percent over the last fifteen years, and shows no sign of fundamental change.

In short, federal and non-federal funding of computer R&D in universities have increased at roughly similar rates over the long haul. Given the continuing predominance of federal funding in universities and their FFRDCs, however, the data in Figures 1 and 4 together suggest that the math and computer science R&D undertaken in academic institutions has shifted away from more basic projects and toward more applied research and development.

---

[9]In contrast to the previous three Figures, these data show both research and development expenditure at universities, by field of science and engineering, by calendar year.

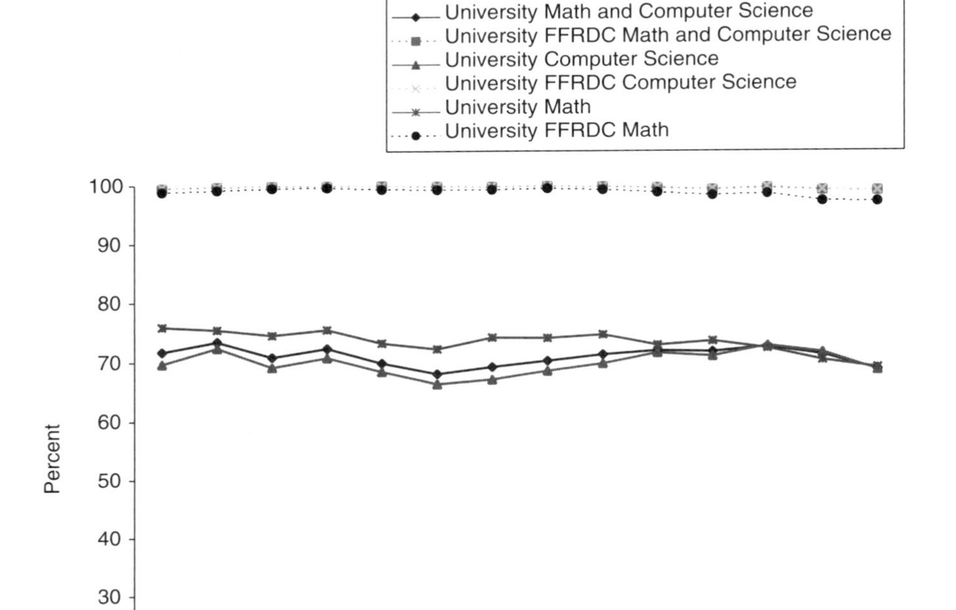

FIGURE 4 Federal share of university math & CS funding.

THE U.S. GOVERNMENT AND INDUSTRIAL COMPUTER RESEARCH

The same sort of assertion of continuity does not apply to federal computer research funds going to the industrial sector. In the 1950s, the federal government paid for perhaps 60 percent of research and development undertaken by firms producing computer hardware. By the mid-1960s, the federal share had declined to roughly a third of total R&D.[10] By 1975, NSF data show that firms

[10]See Flamm, *Targeting the Computer, op. cit.*, for details.

loosely classified as being computer hardware producers ("office, computing, and accounting machines," or OCAM) received federal R&D funds accounting for 22 percent of their total effort, and by 1980 that share had further declined to 13 percent.[11] (See Figure 5.) After increasing slightly, to 15 percent of OCAM R&D in the mid-1980s, the federal share plunged to 6 percent by 1990, plunged again to 0.6 percent by 1995, and had dropped by another third, to 0.4 percent, by 1997.

Clearly, after years of slow, gradual erosion, the role of federal funds in computer hardware industry R&D plunged to virtually nil in the nineties. Further, since the industry R&D effort is so much larger in the aggregate than research undertaken in universities, even lumping robust federal support for university R&D efforts with industry R&D would not arrest the precipitous decline of the nineties in an overall U.S. aggregate. Adding university math and computer science research to computer hardware industry R&D, the federal share of the total went from 26 percent in 1975, hovered in the 16 to 22 percent range from 1980 through 1995, then dropped to 10 percent in 1997 (Figure 5). Beginning in 1995, industrial R&D by computer software and service firms has been measured in NSF statistics, but folding in that, too, does not arrest a sharp decline from 1995 to 1997.

A positive interpretation of these changes might be to suggest that while federal funding grew at a robust enough rate, the computer industry grew at a truly phenomenal rate in the nineties, resulting in a federal share of computer R&D that dropped sharply in the late nineties. In other words, while the federal slice of the pie increased, the overall R&D pie grew vastly—much faster with the explosive computer industry growth of the 1990s.

[11]In 1995, an estimate of $30 million in federal funds for office, computing, and accounting industry R&D was derived from NSF data and used in this graph. There were 49 firms with greater than 500 employees, of which 4 received federal R&D funds; 377 with less than 500 employees in 1997, of which 2 received federal R&D funds. We also know the four firms receiving the largest amounts of federal R&D funding received $23 million, and that the two firms with less than 500 employees received $7 million in federal R&D funds. Therefore, if the 4 federal fund-receiving firms with more than 500 employees coincided with the four largest federal R&D fund recipients, total federal funds to all in the industry would have been $30 million.

An identical but completely separate estimate of $30 million in total federal funds to computer hardware producers in 1995 is obtained by adding up the total R&D funds of all firms with R&D programs exceeding $200,000, then adding on an estimate for total R&D of firms with R&D funds less than $200,000 equal to 133 (the number of such firms) times $129,000 (average R&D in 1994 for firms in this category). We therefore conclude this is a fairly accurate estimate.

A similar process was used to estimate federal funds going to industry in 1990. Only 5 firms received federal funds that year, of which 4 were in the largest 17 firms by employment, and 1in the next 27 firms sized by employment. Assuming that the 20 firms with the largest total R&D programs also were responsible for the 20 largest company-funded R&D programs, these 20 firms accounted for $760 million in federal R&D funds.

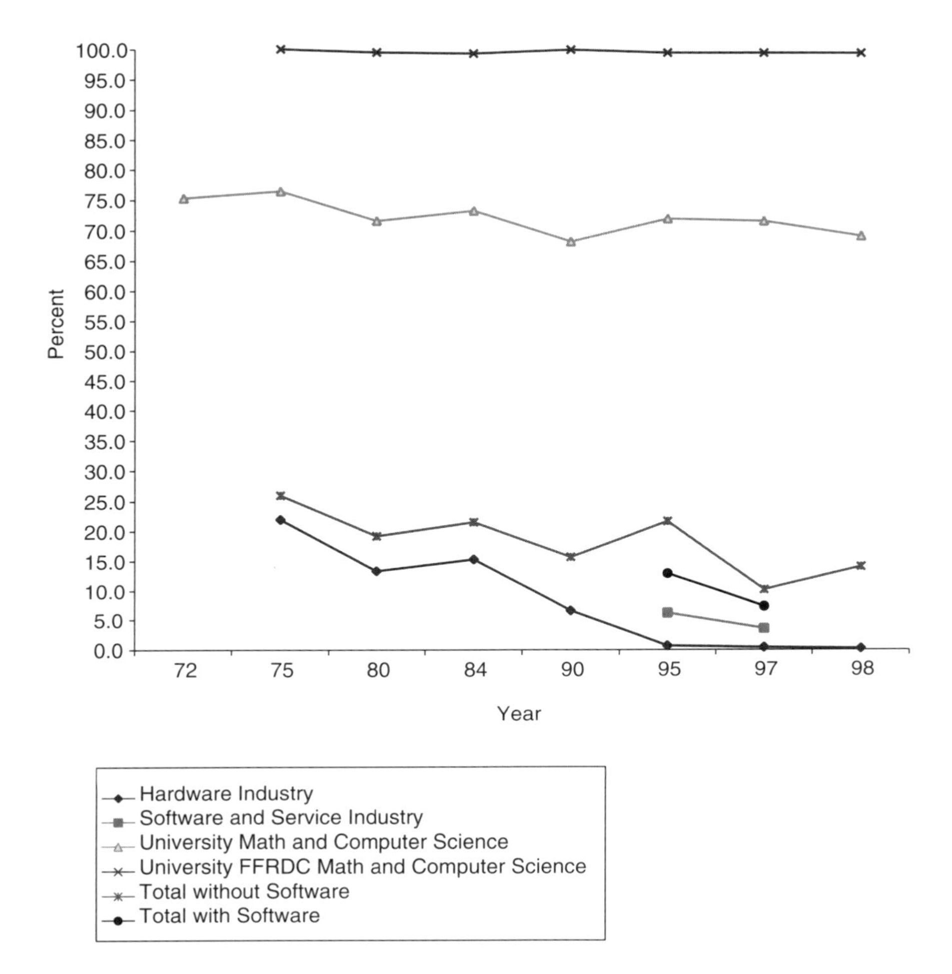

FIGURE 5 Federal share of computer R&D.

COMPUTER R&D UNDER THE MICROSCOPE

Unfortunately, the latter suggestion turns out to be untrue. Indeed, surprisingly and perhaps even shockingly, U.S. computer hardware industry R&D seems to have actually declined in real terms during the 1990s, according to NSF statistics! Rather than being a bigger federal slice of an even bigger industry pie, federal funds seem to have become a truly tiny piece of a much smaller hardware industry pie.

Figure 6 tells this surprising story. NSF estimates (in constant 1992 dollars) of industry-funded R&D in the computer hardware (and software, when available) sec-

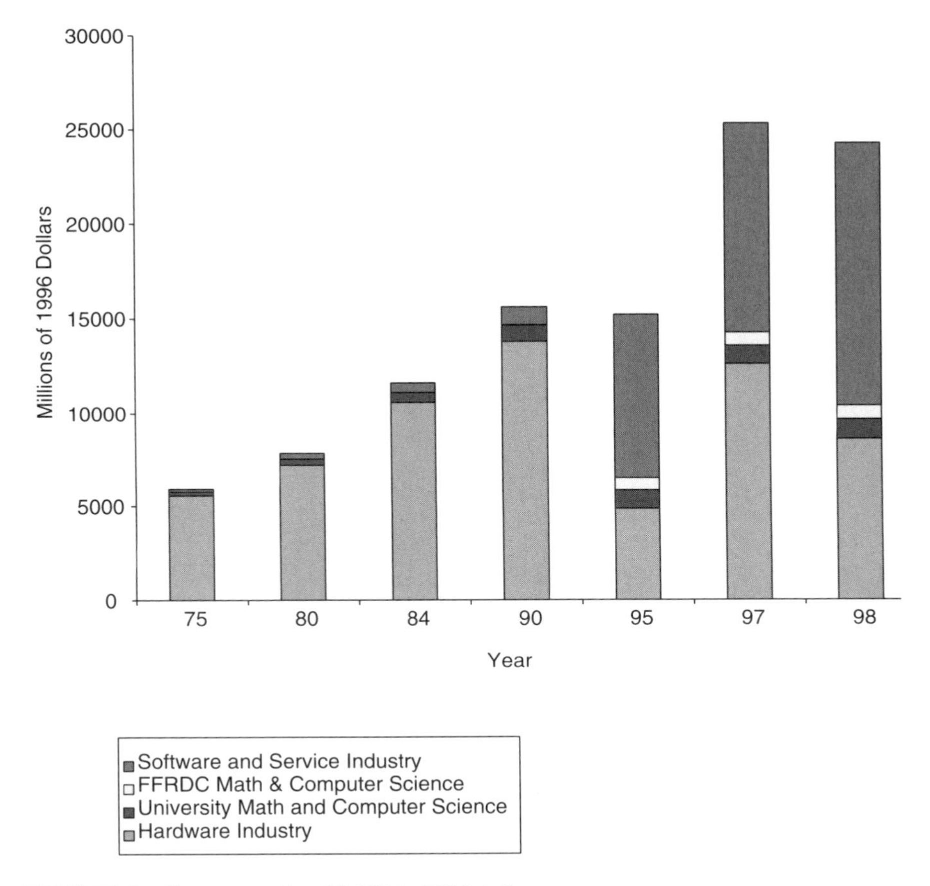

FIGURE 6 Computer-related R&D in 1996 dollars.

tors are shown in comparison with each other and with the size of math and computer science R&D undertaken in universities and university-administered FFRDCs. (Data is from the same NSF sources as in earlier figures.) Both OCAM industry R&D and university math and computer science research grow in real terms through 1990. But from 1990 through 1995, OCAM industry R&D drops by about two-thirds. Even if software and service industry R&D is added on to hardware (OCAM) for 1995, total industry-funded R&D drops from 1990 to 1995.

After some recovery in 1997, computer hardware industry R&D drops again in 1998. Even with increasing software and services industry R&D added on, the total again declines from 1997 to 1998. By 1998, OCAM total R&D has still not recovered to its 1990 level, while software industry R&D has increased by al-

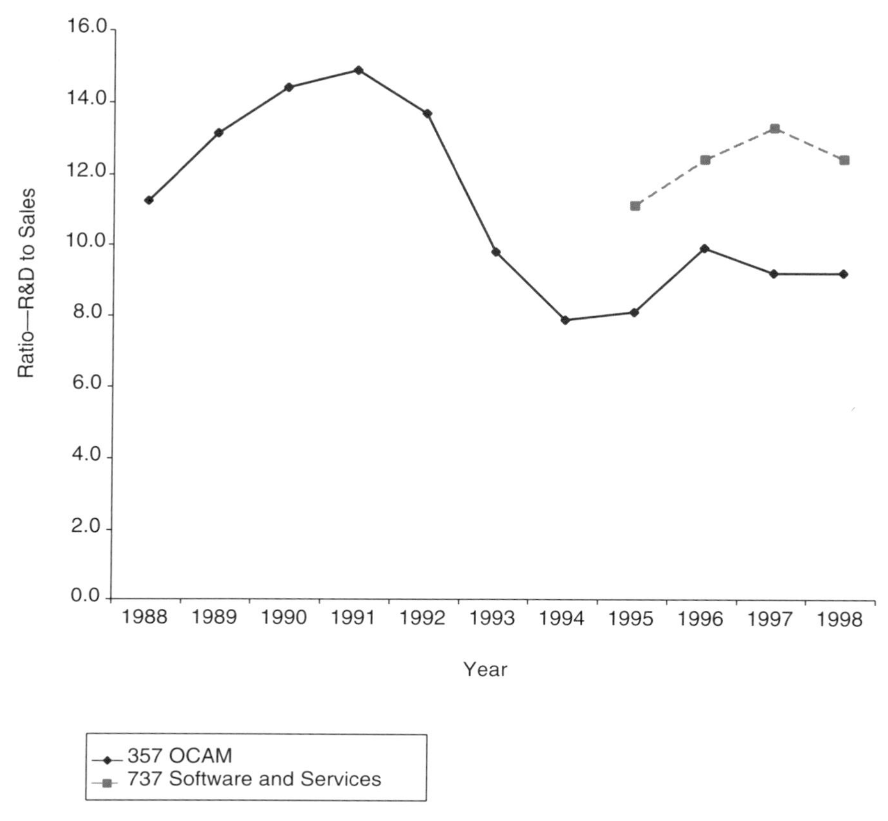

FIGURE 7 Ratio of R&D to sales for computer firms per NSF statistics.

most one-half of its 1995 level. Math and computer science research in universities and their FFRDCs, by contrast, is relatively stable in the 1990s, as was the federal share of the total.

Industrial computer R&D seems to have not only fallen in absolute terms in the 1990s, but also relative to sales by computer hardware producers. Figure 7 shows that the research intensity of computer hardware makers tracked by NSF statistics fell sharply in the 1990s.

At least superficially, then, we are left with a somewhat troubling picture of a high-tech computer industry that cut back its investment in R&D over most of what we know of the decade of the 1990s. How and why did this happen? Before exploring this in greater detail, it is helpful to discuss whether this trend might possibly be dismissed as a statistical artifact.

THE IBM EFFECT

The NSF statistics on industry R&D we have been examining classify R&D at the company level, imputing all R&D to the primary industry of the company. Industrial classification of a company is generally based on employment, and it is therefore possible for companies to switch industries from one year to the next if one activity of the company increases and surpasses another that previously might have been the company's largest.

NSF officials have in the past suggested that this was going on in statistics for the computer hardware sector in the 1990s, and consideration of the case of IBM makes concrete how this could cause large shifts in reported R&D. Investment analysts covering IBM typically reported in their market research that in the mid-1990s, IBM's largest revenue-producing sector shifted from hardware to software and services. Corporate R&D figures reported in IBM's annual reports suggest that this could have large consequences for NSF industry R&D estimates. In 1975, IBM's R&D amounted to about 55 percent of total company funds for OCAM R&D in NSF statistics. By 1985, the relative share of IBM R&D had fallen slightly, to 50 percent, and by a decade later, in 1990, was still about 45 percent. Thus, a transfer of IBM out of hardware and into software in the NSF statistics could very well have reduced OCAM R&D significantly in 1995 purely as a result of the classification change.

However, even if IBM and other computer hardware firms had shifted out of OCAM and into software and services in the NSF classifications, the data would still seem to show an absolute decline from 1990 to 1995, even if a very large share of software and service R&D in 1995 had been coming from such "transferees." In order to not have declining R&D from 1990 to 1995 among firms classified as computer hardware producers in 1990, there would had to have been large-scale movement out of OCAM and into still other sectors, like communications equipment or scientific instruments. While not impossible, this seems unlikely.

Furthermore, those companies that continued to be classified as OCAM producers clearly underwent a decline in their R&D-intensity in the 1990s, compared with prior decades. It is *not* true that the firms remained as research intensive as before, but that there were merely less of them, thus reducing the aggregate investment by this sector in R&D. In fact, there were very large declines in R&D intensity for all types of OCAM producers, large and small. Figure 8 shows the extraordinary declines in R&D intensity when OCAM companies are sorted into groups according to the size of their absolute R&D effort—the four largest, the next 4, the next 12, and the rest of the industry. All of these groupings of firms, using the same NSF data described earlier, showed large declines in research intensity in the 1990s.

The particularly sharp declines in research intensity within the largest play-

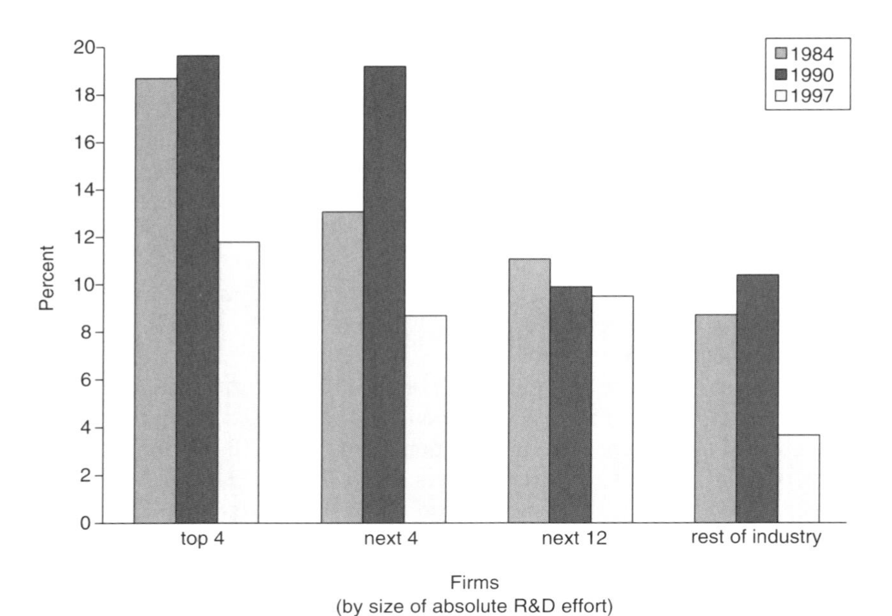

FIGURE 8 Total R&D as a percentage of OCAM industry sales.

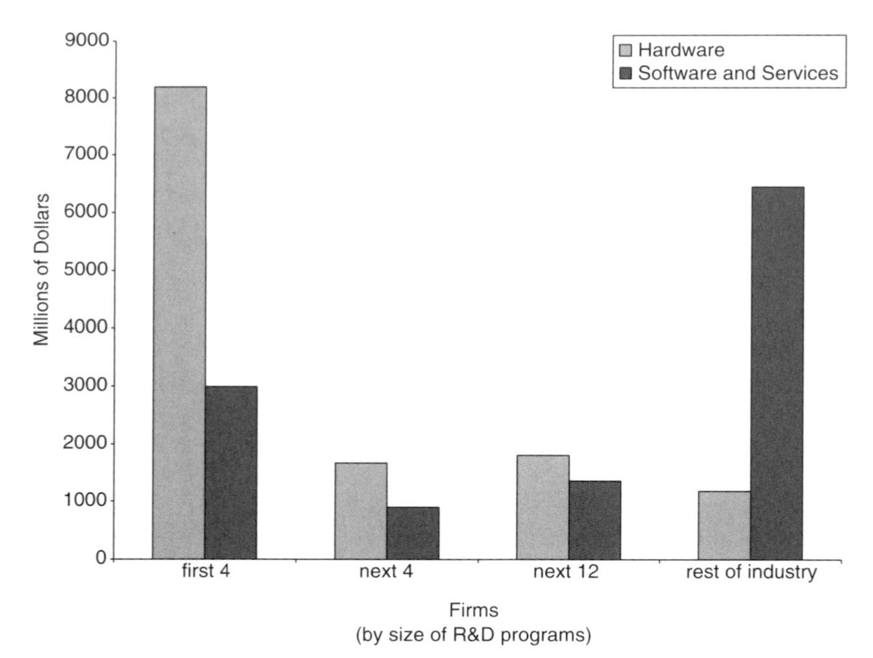

FIGURE 9 1997 Total R&D funds.

ers are notable because the biggest players are responsible for most of the R&D in computer hardware. Figure 9 shows, for 1997, the distribution of R&D across firms, ranked by the size of their R&D programs. The situation, as can be seen, is quite different in software and services, where smaller players, in the aggregate, account for most of the overall R&D activity. A similar Figure would be obtained analyzing sales for the same groupings of firms, and would tell the same story about differences between hardware, and software and services.

In short, it is not true that the decline in computer hardware industry R&D depicted in NSF data might purely reflect classification changes that reduced the number of firms in the industry, thus lowering total R&D while possibly leaving R&D intensity unchanged. The R&D intensity of those who continued to be classified as computer producers clearly declined in the 1990s, and played a role in the overall decline in R&D undertaken by the hardware-producing sector.

Still, it would be useful to have a more direct way of analyzing changes in research intensity among computer industry firms due to classification changes, versus changes within firms. To examine this question, we need data that does not suffer from the NSF data's abrupt classification shifts.

CHANGES IN COMPUTER INDUSTRY RESEARCH INTENSITY

For this purpose, we turn to corporate financial data collected in Standard and Poor's *Compustat* database. Unlike the NSF data, only publicly traded company-funded R&D is included in these statistics, and foreign companies with ADR shares sold on American stock markets are included. However, as we have already seen, federal funding was minimal by the 1990s, and therefore the first of these differences would little affect our examination of industry R&D activity over this period. Since we have access to individual company data from this source, we can—unlike the NSF data—examine the composition of some sample of computer industry firms directly, and eliminate the potential impacts of classification changes on R&D trends.

To isolate the potential impacts of classification changes, I used a 1999 edition of the Compustat database to construct a sample of 396 firms classified in OCAM (Standard Industrial Classification Code 357) over 1979-1998 when last assigned an industry (some of these firms exited the sample—they failed or were acquired by other companies—in which case their classification at the time of exit was used). I remark again that the Compustat sample includes firms no longer currently in the industry, and foreign firms, which are not included in the NSF sample, and excludes privately held firms, which are. Therefore, the number of U.S. firms tracked in the Compustat data is certainly less than the number of firms in the NSF survey, which we know exceeded 400 in recent years (see above). Of the Compustat firms, 100 were computer producers (SIC 3571), 64 computer storage makers (SIC 3572), 34 makers of computer terminals (SIC 3575), 131 producers of I/O peripherals (SIC 3577), 40 makers of accounting machinery (SIC 3578), and 27 producers of office machinery (SIC 3579). In addition, historical data was also available for 118 producers of computer networking and communications equipment, for whom Compustat created a classification (SIC 3576) not found in official government statistics for SIC 357. This latter group includes firms like Cisco, 3Com, and Adaptec, and was excluded from our analysis of SIC 357, in order to ensure greater comparability to NSF statistics. It is also important to reaffirm that the NSF data, collected by the Census, cover all U.S. firms, while Compustat data cover all publicly held firms listed on U.S. stock exchanges.

The Compustat data, in short, is a sample of firms classified in SIC 357 in 1998, if currently in business, or when last in business, if they exited the industry prior to 1998. If there were missing values for either sales or R&D in any year, that observation was dropped from that sample. Note that there are missing values for a number of Compustat firms in 1998 that significantly affect totals for that year.[12] Like the NSF data, firms entering an industry push up totals; unlike the NSF data, a firm in the industry in 1998 (or when last classified) is being counted in that industry in all previous years.

Figure 10 shows the distribution of sales within SIC 357 across product segments for our sample of Compustat firms over the years 1988-1998. The surge in computer sales from 1992-1994 is clearly shown. Much of the choppiness in Figure 10, however, comes from firms exiting the industry (or being acquired), or new firms entering the industry.

To make this point, I have constructed a subsample of the Compustat SIC 357 data showing only data for those firms with both R&D and sales data for every year over the 1979-1997 period. Out of our original sample of almost 400 companies, there were only 65 that met this test! Figure 11 compares total sales for the firms in my Compustat SIC 357 sample (which covers worldwide sales by these companies) with net domestic (i.e., excluding foreign) sales for the U.S. firms undertaking R&D classified in the NSF SIC 357 sample over these same years, and to my subset of 65 firms with continuous R&D and sales data. Both full-sample Compustat sales and NSF sales jump abruptly from year to year, while my continuous data subset of 65 firms, which accounts for a majority of the sales over this period, increases quite smoothly.

My conclusion is that entry and exit of firms in both Compustat and NSF samples seems to account for an important portion of the ups and downs in sales. When downturns hit, some incumbent players exit. When the computer industry turns back up, new and different players come in and account for an important

---

[12]Perhaps the most extreme example is that of SIC 3579, office machines, where a single firm—Japanese firm Sanyo, whose ADR shares are sold on a U.S. stock exchange and therefore is included in the Compustat sample—accounted for better than 80 percent of the sector's total R&D in 1997, and U.S. producer Pitney Bowes most of the remainder. With Sanyo's number missing for 1998, the sector's R&D falls sharply in the data pictured in Figure 10.

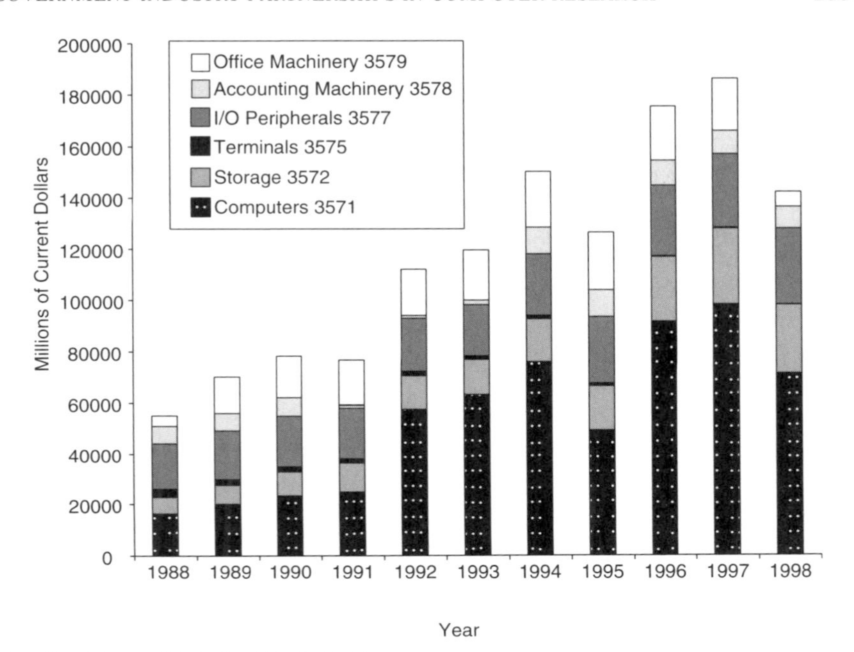

FIGURE 10 COMPUSTAT SIC 357 sales.

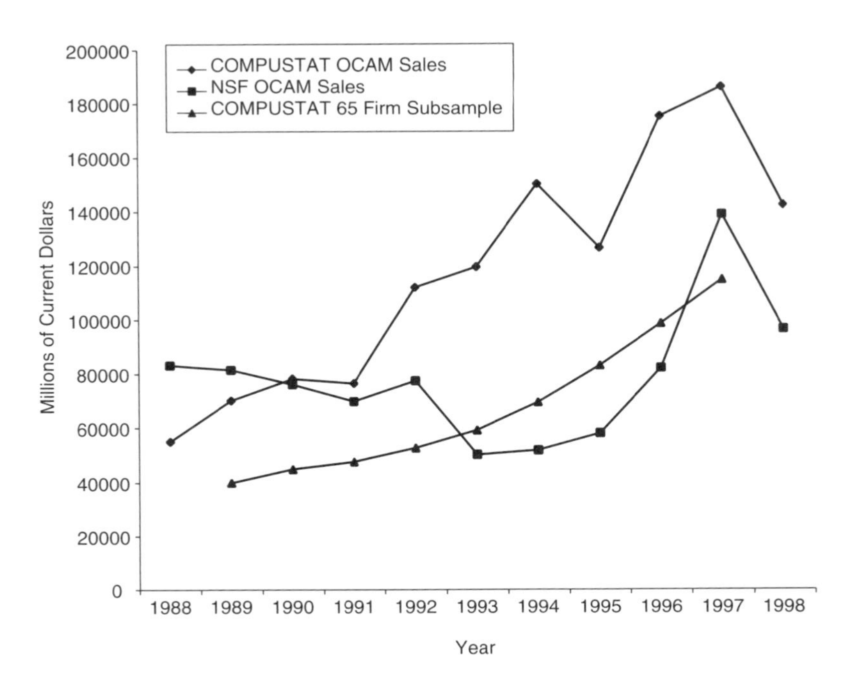

FIGURE 11 NSF vs. COMPUSTAT SIC 357 sales.

component of increased growth. Since the NSF data exclude foreign sales, they would normally be substantially less than the Compustat data (which are worldwide sales) for any given group of firms. The expected relation held, and the two series moved in a generally similar manner from 1995 on. Prior to 1995, however, the NSF data generally are nonincreasing, and even exceed the Compustat totals in the late 1980s. This suggests that the NSF data are capturing firms which either exited the computer industry, or shifted the bulk of their sales into other industries, over the 1988-1994 period, while these same firms are never included in the 1998 Compustat sample (which is based on the firm's activity in 1998 or its last activity prior to acquisition or failure in other cases). Thus, it appears that the NSF data include substantial sales by computer hardware firms that either failed or switched their main line of business to other activities over the 1988-1994 period.

Putting the above facts together, both flat or declining sales by firms focused on computer hardware, and declines in the research intensity of those sales, combined to produce the declining computer hardware R&D figures in the early 1990s in the earlier Figures showing the NSF data. In the Compustat data, by way of contrast, there are increasing sales over this period. Both NSF and Compustat, on the other hand, show increasing sales after 1995.

Figure 12 shows company-funded R&D, measured in 1996 dollars (using the GDP implicit price deflator), over the period 1988-1998 for both NSF and Compustat samples. Comparison of the full Compustat sample with the "continuous" 65-firm subsample strongly suggests again that entry and exit in the computer hardware business in the early 1990s must be the explanation for the zigs and zags in total company-funded R&D. As with sales, the NSF R&D data show further evidence of major classification changes in companies assigned to OCAM in the early 1990s.

Figure 13 compares company-funded R&D intensity (R&D as a percent of sales) for firms in the NSF SIC 357 OCAM sector with the aggregate R&D intensity for the overall Compustat SIC 357 sample, as well as the 65 firms in our "continuous" Compustat sub-sample for SIC 357. The NSF OCAM R&D intensity (which takes R&D as a fraction of domestic U.S. sales only) falls sharply in the early 1990s, as do the Compustat figures for computer (SIC 3571) R&D intensity (which take worldwide corporate sales as the denominator). Exports are a significant share of U.S. computer companies' sales, so we would expect the NSF figure to be substantially greater, as it generally is. Therefore, this figure shows three things: that the movement of NSF OCAM 357 R&D intensity over time parallels that of SIC 3571 in both Compustat samples; that the SIC 3571 R&D intensity converges to SIC 357 R&D intensity in both Compustat samples; and that SIC 357 OCAM R&D intensity, in all three data sets, is down somewhat relative to the early 1990s by the late 1990s—by a little in the Compustat data; by a lot in the NSF data.

All the above is consistent with a story in which many highly R&D-inten-

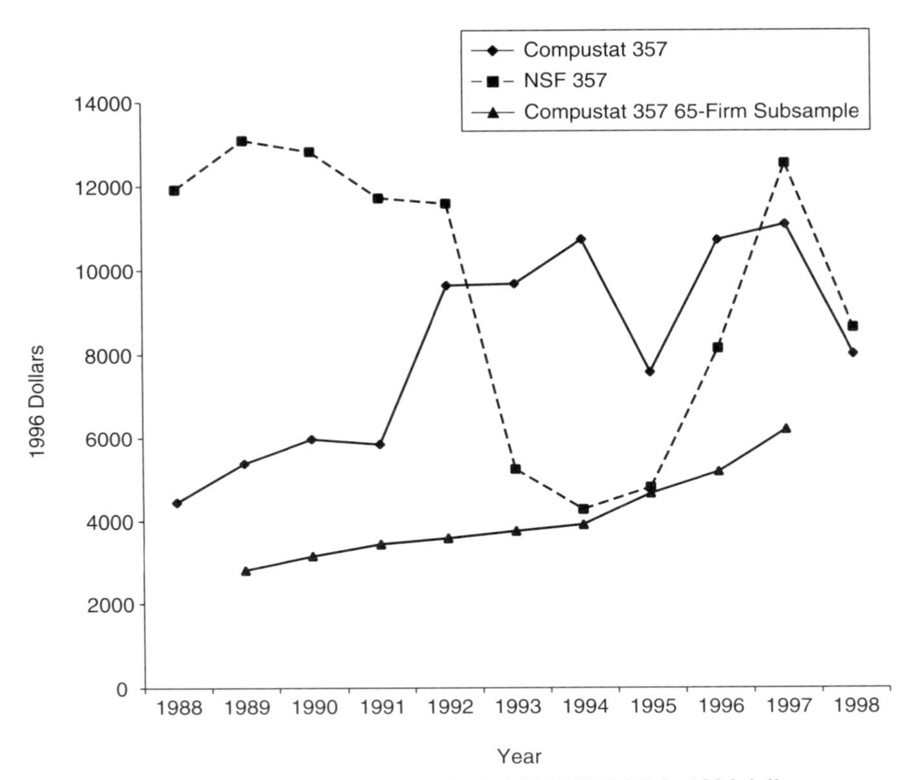

FIGURE 12 COMPUSTAT vs. NSF co-funded SIC 357 R&D in 1996 dollars.

sive computer firms either failed, shifted into other lines of business, or were acquired by others in other lines of business in the 1990s. (Fortunately, this matches up with actual events for this decade.) All these much more R&D-intensive firms—whether they failed, merged with other computer firms, or merged with non-computer firms— are picked up in the early part of the 1990s in the NSF sample. The subset that was acquired by other computer companies remaining in the business, or that failed, is picked up in the broad Compustat sample over the entire period. The subset that was acquired by other computer companies remaining in the business is picked up in the small (65-firm) continuous Compustat sample after the period of acquisition only, while failed enterprises and new entrants in the late 1990s are omitted in the continuous subsample.

In short, the decline in R&D intensity appears to be real, but mainly in computer systems (SIC 3571), where it plunged precipitously. The above data can be read as suggesting that the NSF survey procedures may give much greater relative weight to these exiting, high R&D computer companies, in constructing

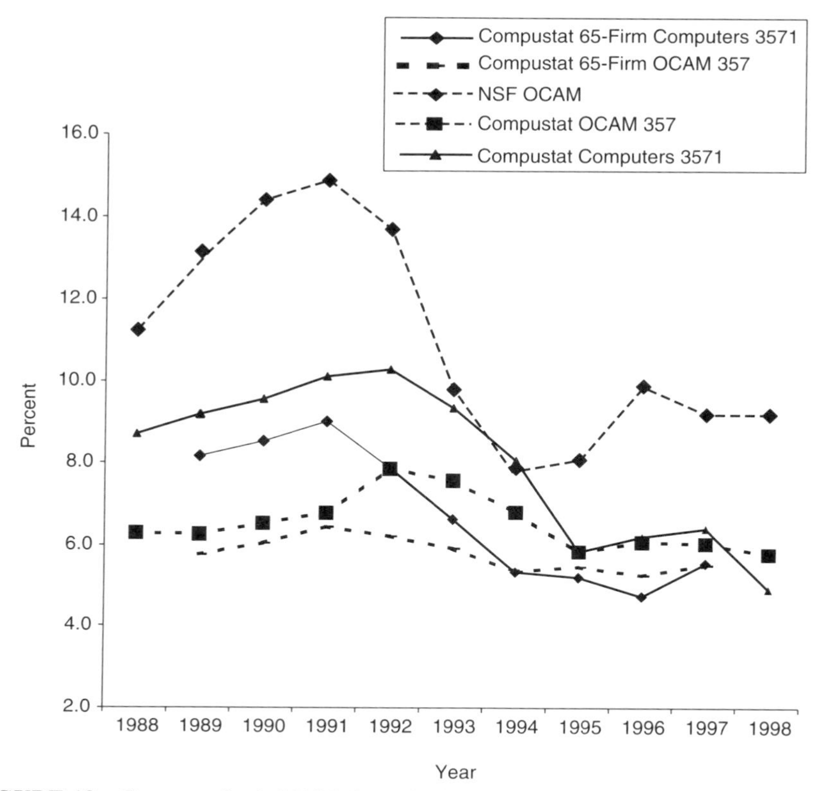

FIGURE 13 Company funded R&D intensity in SIC 357.

an estimate for all OCAM, than the smaller subset of firms tracked in the Compustat classification scheme.

Figure 14 displays R&D intensity, by subsector, for the full Compustat sample. Declines in R&D-intensity in computers appear to be offset to a significant extent by increasing R&D intensity in I/O peripherals. More interestingly, this raises the issue of whether the decline in R&D intensity in computers, and the relative stagnation of computer hardware R&D in the late 1990s, really masks the fact that the leading edge of computer technology has moved out of traditional computer companies and into new and different kinds of firms.

MIGRATION OF COMPUTER R&D INTO OTHER SECTORS

In many respects, it is obvious that much R&D that used to be done by computer firms is now being done elsewhere. The classic example is the person-

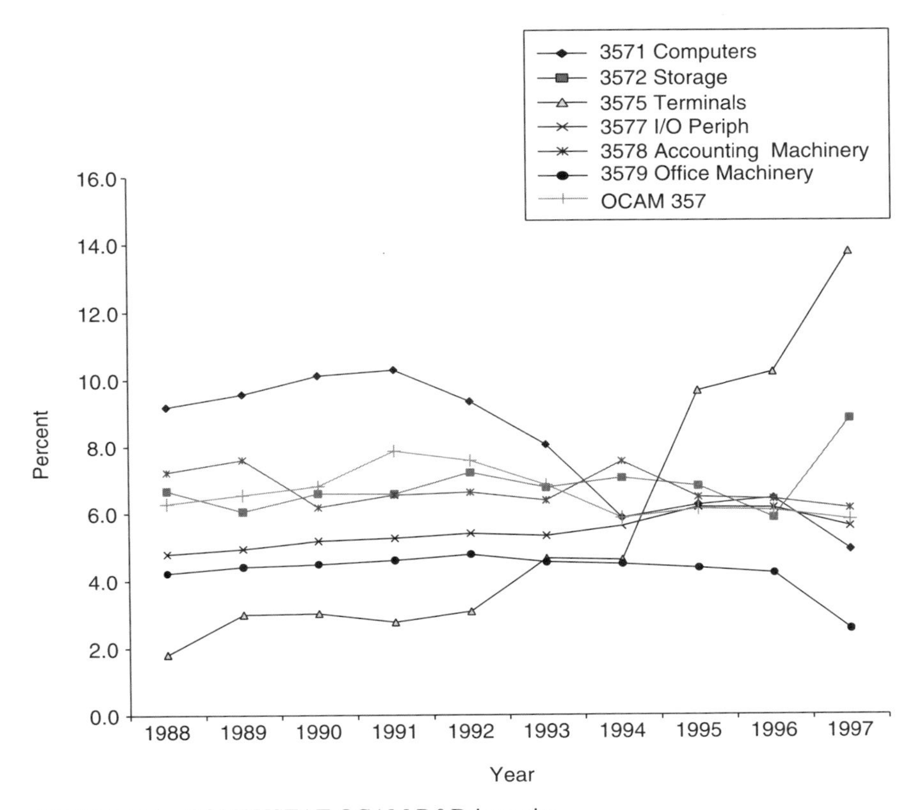

FIGURE 14 COMPUSTAT OCAM R&D intensity.

al computer (PC) industry. Where computer firms once did most of the R&D, design, and engineering work needed for a new model of PC, virtually all of the design is now incorporated into highly integrated, standardized semiconductor components now designed, developed, and shipped by Intel and a handful of other specialized chip firms. Today, a builder of "commodity" PCs, like Dell or Gateway, typically spends something like 1 percent of revenue on R&D, compared with the 10 percent or more that was the case for computer makers in the 1980s. The "technology" is largely in the chips and other components, not the "box" that is the PC maker's responsibility.

Figure 15 makes this point graphically by showing the split of value added between U.S. computer hardware manufacturers (SIC 357) and semiconductor producers (SIC 3674) over the last 40 years. The changes are striking. Semiconductors and computers accounted for roughly equal shares of value added in their sectors' output in the early 1960s. Today, the share of value added in

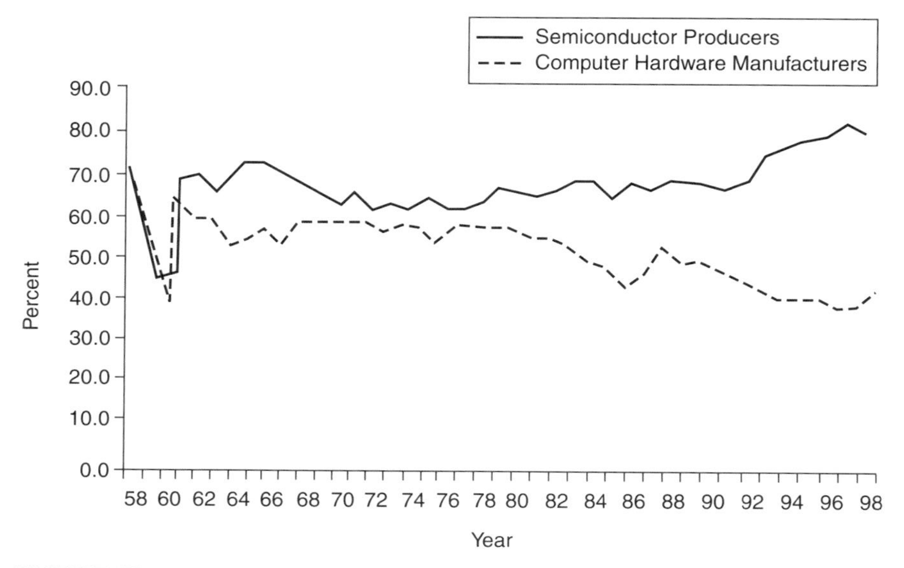

FIGURE 15 Value added/shipments in OCAM and semiconductors.

computer shipments is roughly half the share of value added in semiconductor shipments.[13]

Alternatively, Figure 16 shows the split of value added between these two sectors. In the early 1960s, computers accounted for 80 percent of the value added in the two industries. By the 1990s, computers' share of the two industries' value added was halved, to about 40 percent.

Semiconductors were not the only IT-related sector benefiting from relatively higher levels of R&D investment in the 1990s. Figure 17 shows that in addition to semiconductors (contained in NSF statistics within SIC 367, "electronic components"), communications hardware expanded sharply in both real terms, and relative to OCAM R&D. With the rise of the Internet, the World Wide Web, and e-commerce, this is not surprising. It is borne out by the available data.

[13]These data are from the 1997 U.S. Census of Manufactures, and the U.S. Census Bureau's Annual Survey of Manufactures, for various years. Census shifted from the SIC classification system to the NAICS classification system in 1997, and data for the old SIC 357 are unavailable after 1997. Therefore, the 1998 numbers for the elements of SIC 357 have been estimated from NAICS-classified data for 1998, based on the relative 1997 NAICS to SIC correspondence of shipments.

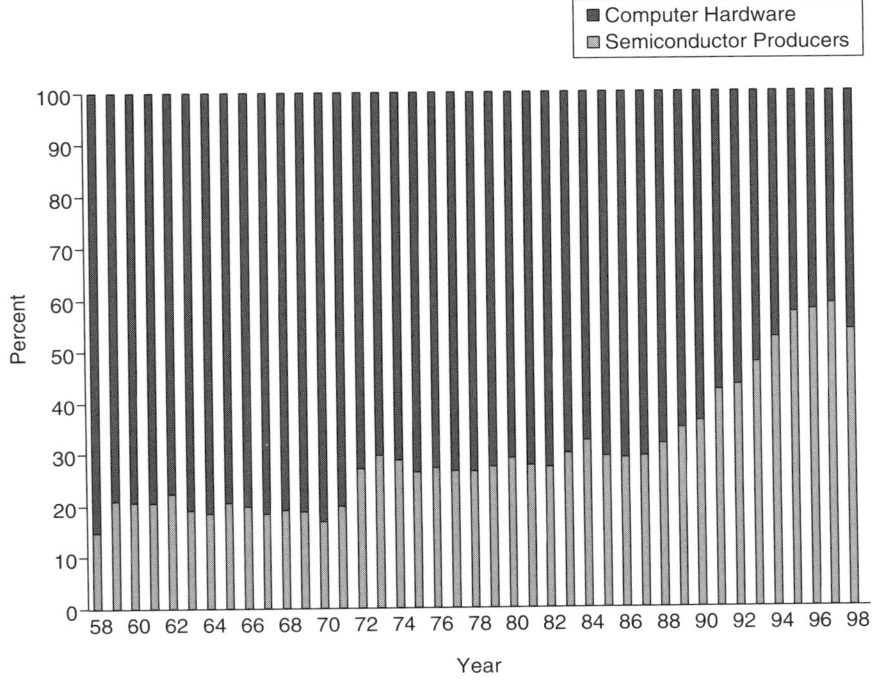

FIGURE 16 Distribution of U.S. manufacturing value added between OCAM and semiconductors.

SUMMARY AND CONCLUSIONS

Government's role in the funding of computer R&D has declined greatly, in both relative and absolute terms. Virtually no government R&D funds now support projects at computer hardware companies, in marked contrast to the situation just two decades ago.

The decline in support for basic research is particularly striking, and perhaps even troubling. Even more disturbing are various anecdotal reports (since statistics on basic research in industry are virtually nonexistent) suggesting that within industry's domain, it has turned even more sharply away, in relative terms, from long-term, basic, and fundamental research on computing technology.

Furthermore, considering just the federal investment in computer research, informed observers suggest that even the federal investment has increasingly been funneled into "targeted" research, and away from "fundamental" projects.[14]

[14]See PITAC, *Information Technology Research: Investing in Our Future, op. cit.*; and National Research Council, *Making IT Better: Expanding Information Technology Research to Meet Society's Needs, op. cit.*

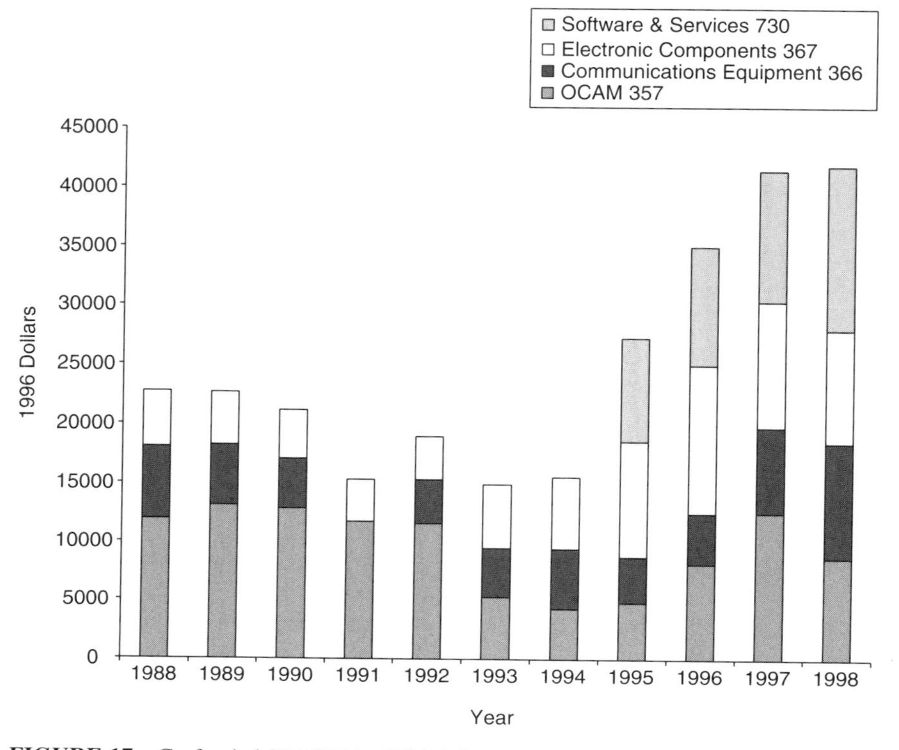

FIGURE 17 Co-funded IT R&D in 1996 dollars.

Certainly, within the Department of Defense, there has been considerable pressure on DARPA to shift from more basic, fundamental projects, toward near-term efforts with expected impacts on short-term military requirements. Given DARPA's enormous historical importance in this area, this raises important questions about whether federal funding of long-term, fundamental research on information technology is at optimal levels.

After looking at a variety of data sets, we have tentatively concluded that the apparent decline in industrial R&D intensity in computer hardware is real. It is particularly pronounced in computer systems; less so in storage and peripherals.

Relatively stagnant or even declining levels of R&D in computer systems have to some extent been offset by increasing R&D intensity in areas such as networking, communications, and peripherals, and by the shift of considerable computer architecture R&D into semiconductor producers.

The statistics collected by Census and NSF in their current form are not well suited to tracking trends in R&D. Given the growing importance of these sectors in overall output, and the widespread belief that they are associated with recent large improvements in productivity, investment of reasonable resources in tracking the U.S. investment in these technologies would seem uncontroversial. A serious review of the objectives and conceptual basis for surveys and statistics of R&D would seem useful given the apparent structural changes affecting these high-tech industries.

In short, the scale of R&D investments in computers and computer architectures has dropped in both absolute and relative terms recently, and longer-term investments in basic and fundamental research are clearly falling short by historical standards. This would not be a source of concern if we were convinced that computing technology had "matured," and was no longer an area with a high social payoff for the U.S. economy (with substantial excess over private returns). However, given a growing economic literature suggesting an important link between information technology and the recent performance of the U.S. economy, the widening role of high-performance computing as a complement to continued technical advance in other high payoff areas, like biotechnology, and a widespread technical belief that much remains to be done and many high payoff areas remain in utilizing computers, it would be prudent for the United States, as a whole, to plant more seed corn in this particular field.

Indeed, three recent reports have called for precisely this sort of action.[15] The response to these reports has been lukewarm, however, with only partial funding to date for the increases in R&D called for in these reports. Perhaps this is because the question of what is actually being spent is so muddy in the current statistics, and perhaps because the focus for a new wave of long-term R&D projects is as yet undetermined.

There clearly is a consensus that this has been a productive area for government-industry collaboration, and remains so. The history of government and private R&D investments in IT has been a highly complementary one, with each bringing unique and different strengths and focus to the enterprise. The biggest challenge is likely to be how to move beyond a vague approval of greater collaborative public/private investments in information technology, and on to concrete particulars: How many dollars are justified, in what sorts of projects and topics, and organized how?

The greatest contribution of the economics community to this debate may be to better define a framework in which to collect and analyze data on these technology investments and their payoff. The greatest contribution of the science and technology community may be to better specify what concrete research themes and areas are worthy of greatly increased support, and collaboration between public and private sectors. If we can execute both of these agendas competently,

---

[15]PITAC, *Information Technology Research: Investing in Our Future, op.cit*; National Science and Technology Council, *Information Technology: Frontiers for a New Millennium, op.cit.*; National Research Council, *Making IT Better: Expanding Information Technology Research to Meet Society's Needs, op.cit.*

we may yet manage to extend the impact of this extraordinary set of innovations well into the twenty-first century.

REFERENCES

Congressional Budget Office. 1999. *Current Investments in Innovation in the Information Technology Sector: Statistical Background.* Washington, D.C.: Government Printing Office.

Flamm, K. 1987. *Targeting the Computer.* Washington, D.C.: The Brookings Institution.

Flamm, K. 1998. *Creating the Computer.* Washington, D.C.: The Brookings Institution.

National Research Council. 1999. *Funding a Revolution: Government Support for Computing Research.* Washington, D.C.: National Academy Press.

National Research Council. 2000. *Making IT Better: Expanding Information Technology Research to Meet Society's Needs.* Washington, D.C.: National Academy Press.

National Science and Technology Council. 1999. *Information Technology: Frontiers for a New Millennium.* Committee on Technology, Subcommittee on Computing, Information, and Communications R&D. Washington, D.C.: Government Printing Office.

President's Information Technology Advisory Committee (PITAC). 1999. *Information Technology Research: Investing in Our Future.* Washington, D.C.: Government Printing Office.

Public Research, Patents and Implications for Industrial R&D in the Drug, Biotechnology, Semiconductor and Computer Industries

Wesley M. Cohen
Carnegie Mellon University
and
John Walsh
University of Illinois at Chicago

INTRODUCTION

The links between industry and academia have deepened over the past two decades. For example, of the 1,056 estimated university-industry R&D centers existing as of 1990, almost 60 percent were established in the prior decade alone. Academic patenting activity has also increased. In 1974, 177 patents were awarded to the top 100 research performing universities. By 1995, there were 1,561. In 1980, 25 American universities had offices administering technology transfer and licensing. By 1990, the number had grown to 200. The share of academic R&D supported by industry has also increased from 2.6 percent in 1970 to 6.9 percent in 1990. Although no systematic data exist on either spin-offs or faculty participation in new firms, anecdotal evidence suggests an increase over the past twenty years, especially in biotechnology and software.[1] In addition to such closer ties between industry and academia, Narin et al. have claimed that there has been a parallel increase in the commercial impact of university research, at least in selected domains.[2]

These deepening ties and impacts likely emerge from several sources. Academic research in a number of areas, such as in the biomedical sciences, has moved in directions that offer more immediate commercial potential. Such move-

[1]See Cohen, et. al, "Industry and the Academy: Uneasy Partners in the Cause of Technological Advance," in Noll, R. ed., *Challenges to Research Universities*, Brookings Institution, 1998.

[2]Narin, et. al., "The Increasing Link Between U.S. Technology and Public Science," *Research Policy*, Vol 26, No. 3, 1997, pp. 317-330.

ments are surely associated with policy decisions to provide key financial support. There are also numerous other policies and features of the policy environment that have pushed universities and industry closer together. For example, stimulated by the broad expectation dating from the late 1970s that federal support for academic research would fall, universities and particularly their faculty began to seek out more industry support for their work. At the same time, there was a shift in government attitudes toward collaborations between industry and universities. Prompted largely by growing international competition, legislative changes have encouraged academics to solicit support from industry and have also given industry an incentive to be more forthcoming. Specifically, the Economic Recovery Tax Act of 1981 extended industrial R&D tax breaks to support research at universities. In addition, since the 1970s there has been substantial growth in government programs (such as the NSF's Science and Technology Centers and Engineering Research Centers) that tie government support for university research to industry participation.

Policy has also changed regarding the ability of universities to profit from their research and allow others to profit from it as well. The Patent and Trademark Act of 1980, known as Bayh-Dole, permits universities and other nonprofit institutions to obtain patent rights to the output of federally sponsored research. Part of the impetus behind Bayh-Dole and related legislation was the assumption that there was some stock of underexploited, commercially applicable knowledge residing in universities and other research institutions receiving federal funding. The problem, as seen by policy rnakers at the time, was a lack of incentive to transfer this knowledge to the private sector and subsequently embody it in product and process innovation. So, the idea was to use the award of patents to "incentivize" the private sector to undertake the downstream R&D and related investments necessary for commercialization.[3] Although perhaps less critical due to the already existing incentives for faculty to publish, some suggested that patents could also benefit the public by similarly providing universities with an economic incentive to increase their technology transfer efforts.[4]

The conventional wisdom is that Bayh-Dole has indeed induced an outflow of commercially fruitful technology as reflected in recent increases in university patenting and licensing. Mowery et al. have, however, voiced skepticism. Reflecting on their analysis of the patenting and licensing behavior of the University of California, and Columbia and Stanford Universities, Mowery et al. conclude: "An array of developments in academic research, industry and policy . . . combined to increase U.S. universities' activities in technology licensing, and Bayh-Dole, while important, was not determinative. . . . Even in the absence of

[3]See R. Mazzoleni and R. Nelson, "Economic Theories About the Benefits and Costs of Patents," *Journal of Economic Issues*, Vol 32, No. 4, 1998, pp. 1031-1052.

[4]*Ibid.*

Bayh-Dole, we believe that all of these three universities would have expanded their patenting and licensing of faculty inventions during the 1980s."[5]

At this juncture, it is difficult to test directly whether Bayh-Dole has indeed stimulated the commercial application of public research, no less promoted technical advance more generally. In this paper, we will exploit the 1994 Carnegie Mellon Survey (CMS) on the Nature and Determinants of Industrial R&D at least to probe some of the key assumptions of policies underpinning the privatization of public research.[6] Assuming that firms need to be able to capture some significant share of the returns to their innovations to have the incentive to invest in R&D to begin with, is it reasonable to assume that patents are central to that capture? Second, should we assume that the knowledge flowing into the firm (as opposed to out of the firm) also has to be protected in some way for the firm to be willing to devote the complementary efforts required to bring it to market?

One of the themes of this paper and the prior research upon which it builds is that the answers to most of these questions will differ across industries. The CMS data allow us to address many of these issues on an industry-by-industry basis. We will exploit these data to consider the experience of four industries. We will examine computers, semiconductors and drugs. For the purpose of this analysis we win also break out our drug industry respondents into drugs and biotechnology on the basis of their self-reported R&D and market activities. Although not a distinct industry, we will consider biotechnology as such for this paper.

Section 2 briefly describes the CMS data. Section 3 provides background on the impact of public research on industrial R&D. Section 4 considers the channels through which public research impacts industrial R&D and the role that patents might play in affecting those knowledge flows. Section 5 considers the importance of the different mechanisms that firms employ to protect their inventions and, in that context, the particular role of patents. In Section 6, we examine the reasons why firms patent, and implications of those motives for the interface between public research and industry. We conclude the paper in Section 7.

DATA AND METHOD

The data come from a survey of R&D managers administered in 1994. The population sampled includes all the R&D labs or units located in the U.S. con-

[5]See Mowery et. al., "The Effects of the Bayh-Dole Act on U.S. Univeristy Research and Technology Transfer: An Analysis of Data from Columbia University, the University of California, and Stanford University," 1999.

[6]While perhaps the clearest exemplar of such policies is Bayh-Dole, that is not the only one. Indeed, others might be quite subtle, For example, there are no federal strictures against the disclosure restrictions that are now sometimes associated with the research outputs originating from federally supported university-industry research collaborations. See Cohen, et. al, "Industry and the Academy: Uneasy Partners in the Cause of Technological Advance," *op cit.*

ducting R&D in manufacturing industries as part of a manufacturing firm. The sample was randomly drawn from the eligible labs listed in the Directory of American Research and Technology or belonging to firms listed in Standard and Poor's COMPUSTAT, stratified by 3-digit SIC industry.[7] In our survey, we asked R&D unit or lab managers to answer questions with reference to the "focus industry" of their R&D lab or unit, defined as the principal industry for which the unit was conducting its R&D. We sampled 3,240 labs,[8] and received 1,478 responses, yielding an unadjusted response rate of 46 percent and an adjusted response rate of 54 percent.[9] Our survey data are supplemented with data on firm sales and employees from COMPUSTAT, Dun and Bradstreet, Moodys, Ward's and similar sources.

We have 40 observations in our drug industry sample, 21 in biotechnology, 34 in computers, and 25 in semiconductors. The average overall firm sales for our respondents is almost $3 billion in the drug industry, $1.8 billion in biotechnology, $4.4 billion in computers, and $6.3 billion in semiconductors. The average R&D professional employment in the focus industry for each of our respondents is 577 in drugs, 234 in biotechnology, 1580 in computers and 302 in semiconductors.

IMPORTANCE OF PUBLIC RESEARCH TO INDUSTRIAL R&D

As background to our discussion of the impacts of intellectual property policy on the interface between public research and industrial R&D, it is useful to consider the importance of public research to industrial R&D more generally. As noted in Cohen et al., one widely accepted view holds that the short-term impact of university research is negligible except in a few industries.[10] Accumulating evidence suggests, however, that we may wish to revise this view. Studies published since 1989, as well as the results of the 1994 Carnegie Mellon Survey, suggest that university research provides critical short-term payoffs in some in-

---

[7]We also oversampled Fortune 500 firms.

[8]For each case, we verified the contact information by telephone before mailing the survey. Data were collected from May to December, 1994. We mailed a questionnaire to the contact person at each lab with a cover letter describing the purpose of the research and ensuring confidentiality. Follow-ups were conducted following Dillman's method. (Dillman [1978]).

[9]A nonrespondent survey allowed us to determine what percent of nonrespondents were not in our target population. The results showed that 28 percent of nonrespondents were ineligible for the survey because they either did no manufacturing or did no R&D. Excluding these from our denominator, as well as respondents who should not have been sampled, yields an adjusted response rate of 54 percent of eligible respondents.

[10]Cohen et al., "Industry and the Academy: Uneasy Partners in the Cause of Technological Advance," *op cit.* See also Rosenberg and Nelson, "American Univerisities and Technical Advance in Industry," *Research Policy*, vol. 23, no. 3, 1994, pp. 323-348 and Klevorick et al. "On the Sources and Significance of Interindsutry Differences in Technological Opportunities," *Research Policy*, vol. 24, no. 2, 1994, pp. 195-206.

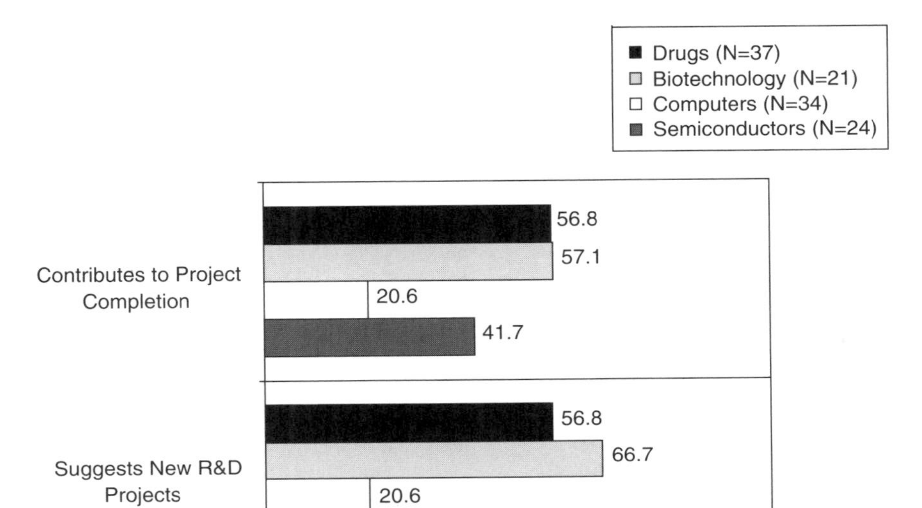

FIGURE 1 Importance of public research for industrial R&D.

dustries (such as drugs), and is broadly important in numerous industries.[11] A key result that should not be revised, however, is that the impacts of university research vary substantially across industries, and we see this variation across drugs, biotechnology, computers and semiconductors.

The CMS data provide a number of different measures of the impact of public research on 'industrial R&D. For brevity, we will only focus on a couple of these. The results reported here are, however, qualitatively similar across all the measures. We begin our analysis by examining responses to a key question in the survey that considers the importance to each respondent's R&D activities of information originating from a broad range of information sources, of which university and government R&D—collectively labeled "public research"—are one. In order to make the notion of the importance of information from these sources tangible, we asked respondents to indicate whether, over the prior three years, information from a source either suggested new R&D projects or contributed to the completion of existing projects over the prior three years. Presented in Figure 1, the results show that the majority of respondents for both drug and

[11]Cohen et al., "Industry and the Academy: Uneasy Partners in the Cause of Technological Advance," *op. cit.*

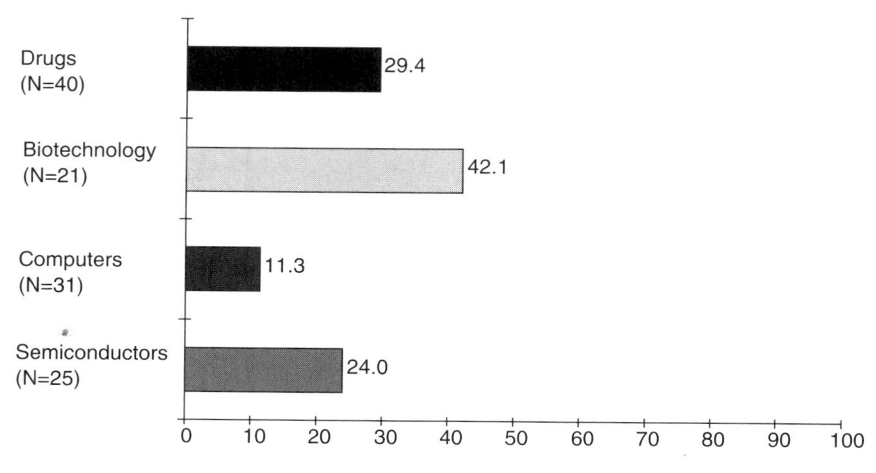

FIGURE 2 Importance of public research findings for industrial R&D.

biotechnology companies report that public research both suggested new R&D projects and contributed to the completion of R&D projects. Relative to the results for the full sample spanning all of manufacturing reported by Cohen et al., these two industries score among the highest.[12] Respondents from the semiconductor industry evaluate the contribution of public research to be almost as high as that for drugs and biotechnology in both categories of contribution. In contrast, public research has relatively little impact on the R&D of our computer industry respondents. One point to recognize here is that the conventional view of public research is that it principally suggests new ideas, presumably spawning new projects. That is not the case. Our data indicate that in the three industries where its contribution is substantial, it contributes both by suggesting new R&D projects and by increasing the efficiency of existing R&D projects.

The same cross-industry qualitative pattern of importance of public research is observed in Figure 2 that presents the percentage of firms' R&D projects reported to make use of the findings of public research over the prior three years. Public research was obviously a central source of information for biotechnology firms, important for drug and semiconductor firms, and considerably less important for computer firms.[13]

---

[12]*Ibid.*

[13]Although not displayed, we considered not only firms' use of research findings originating from the institutions of public research, but also their use of the prototypes and instrumentation originating from public research. Only drug and biotechnology companies reported substantial use of these last two categories of public research output in their R&D operations, though not comparable to their use of research findings.

Thus, we conclude that public research is most central to the R&D activities of drug and biotechnology firms, perhaps a bit less important for the R&D of semiconductor firms (though still quite important), and least central to the R&D of computer firms.

INFORMATION CHANNELS

In this section, we examine the findings from the CMS on the importance of the different channels through which public research findings might flow to industrial R&D labs. By examining the importance of the different channels, we hope to arrive at some idea of the extent to which firms might rely on channels of information from public research institutions that are public, private or lend themselves to privatization.

We asked our respondents to report on a four point Likert scale the importance to a recently completed major R&D project of each of ten possible sources of information on the research findings or R&D activities of universities or government R&D labs or institutes. The information sources considered include patents, informal information exchange, publications and reports, public meetings and conferences, recently hired graduates, licenses, joint or cooperative ventures, contract research, consulting and temporary personnel exchanges. For each of the four industries, Figure 3 shows the percent of respondents reporting that a given source was at least "moderately important." We observe that the most important channels across all four industries are the public ones, namely publications and reports and meetings or conferences. Informal information exchange is also quite important. The results reported here are similar to those for the manufacturing sector as a whole as reported in Cohen et al.[14] There, the four dominant channels of communication between public research and industrial R&D are, in order of importance, publications and reports, public meetings and conferences, informal information channels and consulting. A factor analysis also indicated that these four information channels tend to be used together. Thus, person-to-person interactions, such as informal information exchange or consulting, tend to be used in conjunction with more public channels such as publications or conferences.[15] Although we do not know the reasons for the apparent complementarity across these channels, one might imagine, for example, that firms contact faculty or engage them as consultants in response to the publication or public presentation of research judged to offer commercial potential.

---

[14]Cohen et al., "Industry and the Academy: Uneasy Partners in the Cause of Technological Advance," *op. cit.*

[15]In an earlier study of information usage in technological innovation, Gibbons and Johnston [1975] similarly found that the technical literature and "personal contact" were most beneficial as information sources when used together. Faulkner and Senker [1995] confirm this finding in a recent study of the effects of public sector research on industrial innovation in the areas of biotechnology, engineering ceramics and parallel computing.

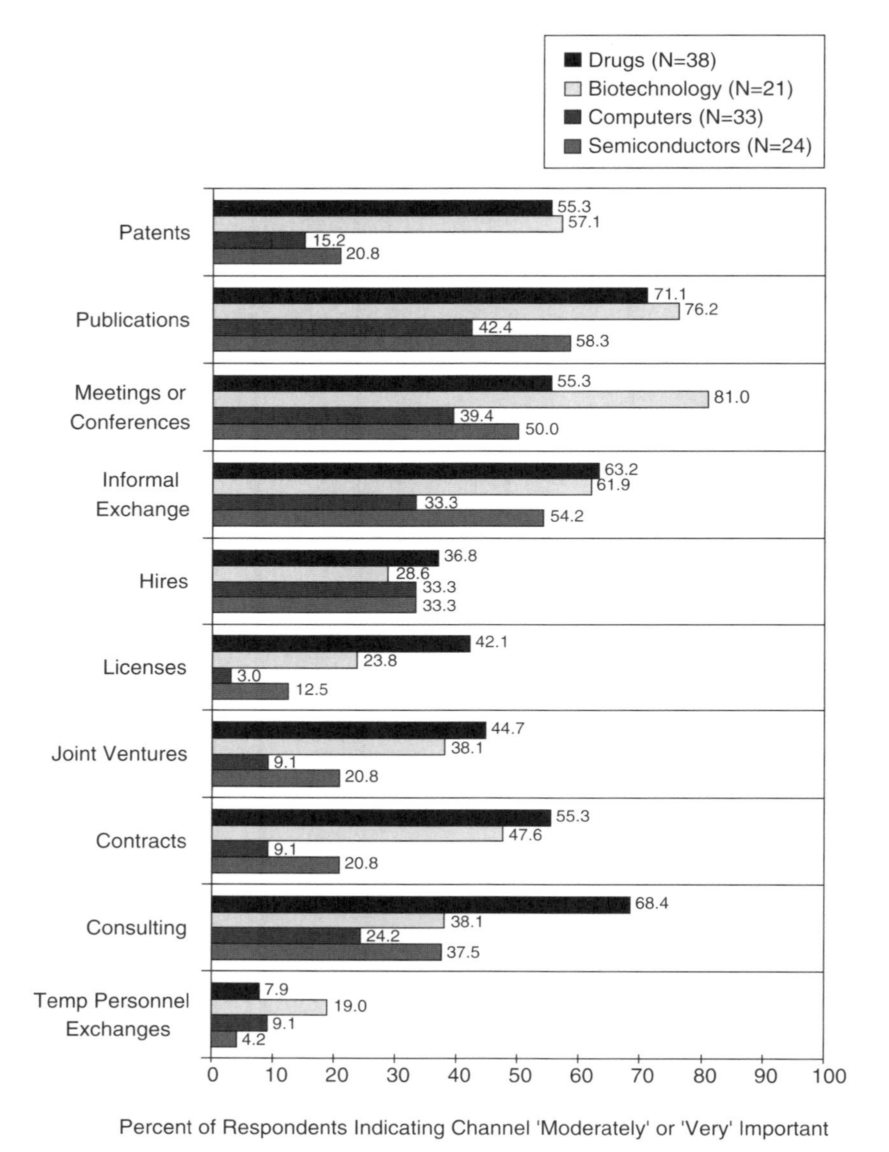

FIGURE 3 Importance of channels of knowledge flow from public research.

The results presented in Figure 3 also reflect some clear differences in the employment of information channels across the four industries. First, a comparison of this figure with Figures 1 and 2 suggests that all or most of the channels tend to be more important to the extent that public research is more generally reported to be more relevant to each industry's R&D. Thus, we observe that almost all the channels (except for recent hires and temporary personnel exchanges) are reported to be less important by our computer industry respondents. We also see differences in the importance of channels across the three remaining industries reporting the contribution of public research to be relatively high. First, we see a sharp difference between drug and biotechnology industry respondents versus semiconductor industry respondents. Specifically, most of the formal and market-mediated channels, including, patents (which may convey information via licenses or freely through patent disclosures), licenses, joint ventures and contracts are much less important for the semiconductor respondents than they are for the drug and biotechnology respondents. The importance of communication channels differ even between drugs and biotechnology. For the former, consulting is far more important. For biotechnology, public meetings and conferences are more important, as are licenses to some extent.[16]

Our results offer several implications and raise numerous questions. First, the dominant importance of public channels for disseminating public research across all four industries suggest that public research findings disseminated via the standard channels of "open science" are commonly used by industrial R&D labs. In areas, however, where public research has greater impacts such as drugs or biotechnology, channels that are private and more restricted, such as consulting, contract research and licenses, appear to be more important. Even where, however, such channels are important, observe that the public channels are more important still.

Although we cannot yet explain the cross-industry differences in the relative importance of the different channels, it is possible that market mediated channels, notably licenses, contracts and consulting, are used more intensively to the extent that patents offer more protection, which, as shown below, they do in the biotechnology and drug industries. This raises the question of the extent to which the overall impact of public research on industrial R&D depends on the effectiveness of specific channels, and particularly private channels. A different possibility—and one that is not particularly consistent with the rationale for Bayh-Dole-is that the overall impact of public research on industrial R&D may have little to do with the strength of any specific or set of channels, but rather simply

[16]Note that movement of people, either through recent hires of graduates or even through temporary personnel exchanges, are not considered to be as important as either the public channels or even the more market mediated channels for biotechnology and drugs. Moreover, the results on the importance of recent hires are quite similar across all four industries, notwithstanding the overall importance of public research.

reflects the overall relevance of public research. In this view, if the research is relevant and useful to industry, it will be used somehow. There is some support for this position in that the three industries reporting public research to be important overall (per Figures 1 and 2) rate most of the channels to be more important than does the one industry, computers, reporting public research to be relatively unimportant.

APPROPRIABILITY MECHANISMS

In this section, we review the CMS results on how firms in the four industries protect their innovations, and address the question of the role of patents.[17] For the sake of brevity, we focus on the effectiveness of the different mechanisms that firms use to appropriate the returns to their product innovations, including patents, secrecy, lead time, complementary sales and service and complementary manufacturing facilities and know how.[18] The role of these mechanisms in protecting process innovations is reported in Cohen et al.[19]

To measure the effectiveness of appropriability mechanisms, we asked respondents to report the percentage of their product and process innovations for which each appropriability mechanism had been effective in protecting their firm's competitive advantage from those innovations during the prior 3 years. The five response categories were: 1) less than 10 percent, 2) 10 percent through 40 percent, 3) 41 percent through 60 percent, 4) 61 percent through 90 percent, and 5) greater than 90 percent. This response scale reflects how central a mechanism is to firms' strategies of appropriating rents to their innovations in the sense that it reflects both the frequency with which a mechanism is employed and the effectiveness of that mechanism given its use, and we interpret effectiveness as reflecting the ability of a mechanism to protect the profits accruing to the commercialization or licensing of the protected invention.[20]

For each of the four industries, Figure 4 presents the percentage of respondents reporting a given mechanism to beat least "moderately effective." Figure 4 shows that patents are reported to be among the most effective mechanisms for

---

[17]See Cohen et al. "Protecting Their Intellectual Assets: Appropriability Conditions and Why U.S. Manufacturing Firms Patent (or Not)," National Bureau of Economic Research, Working Paper No. 7552, 2000, for a discussion of the ways that firms protect their innovations for the U.S. manufacturing sector as a whole.

[18]We also collected data, on the effectiveness of the different mechanisms firms use to protect process innovations.

[19]See Cohen et al. "Protecting Their Intellectual Assets: Appropriability Conditions and Why U.S. Manufacturing Firms Patent (or Not)," *op. cit.*

[20]One limitation of this scale is that it may not index the return to using any particular appropriability mechanism due to the skewness in the distribution of the value of innovations. For example, patents or secrecy may effectively protect only ten percent of a firm's innovations, but that ten percent may account for 90 percent of the value of all of its innovations.

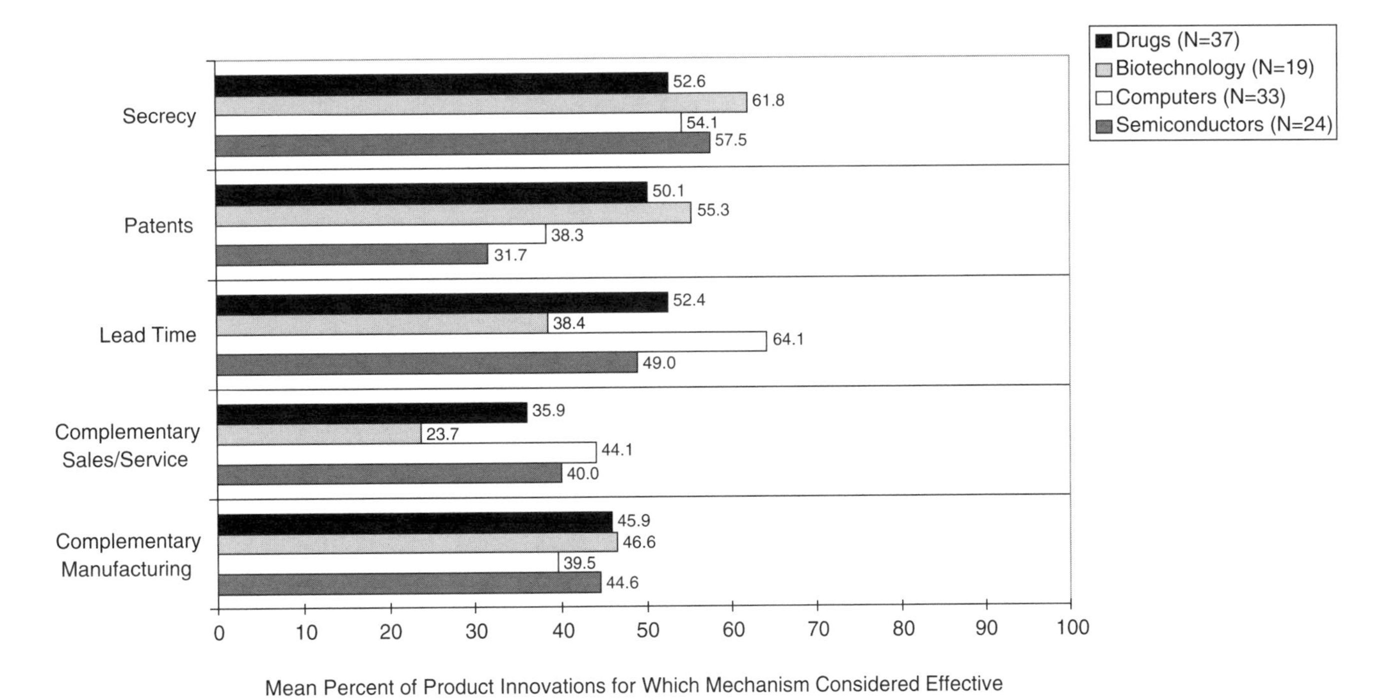

FIGURE 4 Effectiveness of appropriability mechanisms for product innovations.

drug and biotechnology firms. For these two industries, secrecy is comparably important to patents, and lead time is almost as important as patents and secrecy for drug companies. In contrast to the drug and biotechnology industries, we observe that patents are not among the key means used to protect innovations in either the computer or semiconductor industries. In those two industries, firms rely more heavily on secrecy, lead time and complementary capabilities to protect their inventions. A comparison of these results to those for the manufacturing sector as a whole reported in Cohen et al. suggests that the semiconductor and computer industries are not unusual.[21] Similar to the earlier findings of Scherer et al., Mansfield et al., and Levin et al., patents are reported to be the least effective of the various means employed by firms to protect their inventions in the preponderance of manufacturing industries, with secrecy, lead time and the use of complementary capabilities dominant.[22] This implies that the high effectiveness score of patents observed in the drug and biotechnology industries is unusual. In fact, in Cohen et al., where biotechnology is included in the drug industry category, patents are reported to be more effective in drugs than in any other industry.[23]

Although patents are not considered to be among the most effective mechanisms for protecting the profits due to the commercialization or licensing of invention in the computer or semiconductor industries—or for manufacturing firms in general—that does not mean that they are not used or are inessential. Cohen et al.'s analysis of the relationships across the appropriability mechanism effectiveness scores for the manufacturing sector as a whole indicates that appropriability mechanisms are not mutually exclusive in the ways that they are employed.[24] Firms will often use more than one to protect even the same invention. In drugs, for example, firms may well use a combination of secrecy, patents and lead time advantage to protect the same product. Thus, to say that patents are less effective relative to other mechanisms does not mean they do not confer value. Rather, it suggests that they are less central to the methods firms use to protect their inventions.

Aside from firms' judgments of patent effectiveness, the CMS data also provide estimates of the relative frequency with which firms apply for patents. CMS data include measures of patent application propensities, defined as the percentage of each respondent's innovations for which they apply for a patent. Figure 5 presents the average product (and process) patent propensities by indus-

---

[21] *Ibid.*

[22] See F.M Scherer et al., *Patents and the Corporation*. 2nd edn. Boston, privately published, 1959, E. Mansfield, "Patents and Innovation: An Empirical Study," *Management Science*, 1986, and Levin et al., "Appropriating the Returns from Industrial R&D," Brookings Papers on Economic Activity, 1987, 783-820.

[23] See Cohen et al. "Protecting Their Intellectual Assets: Appropriability Conditions and Why U.S. Manufacturing Firms Patent (or Not)," *op. cit.*

[24] *Ibid.*

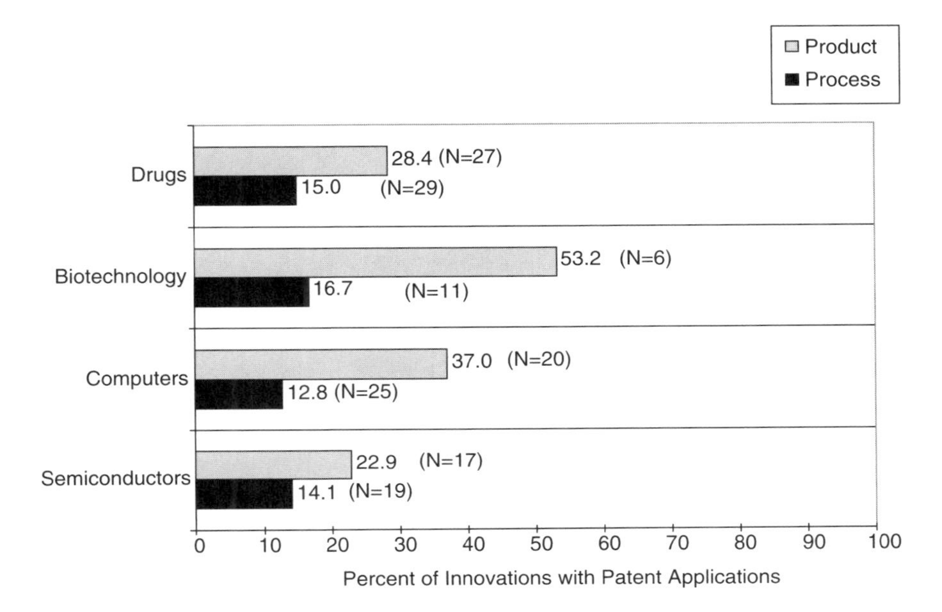

FIGURE 5 Propensities to patent product and process innovations.

try. The results sensibly suggest that the industry, namely biotechnology, which rates patent effectiveness the highest has the highest patent propensity. It is surprising, however, that computer industry respondents that rate patents to be less effective than drug industry respondents, report an average patent propensity of 37 percent, which is higher than the average propensity of 28 percent reported by drug industry respondents. This result changes, however, once we focus on more R&D active firms. For drug industry respondents with greater than the median R&D intensity, product patent propensity is 47 percent as compared to a comparably computed 40 percent for the computer industry. The average patent propensity for the semiconductor industry is the lowest of the four industries, which is consistent with their low rating of patent effectiveness. Thus, product patent propensities among more R&D intensive firms tend to be higher in industries where patents are reported to be more effective.

These results lead us to the question of why firms do not patent; what are the limits on their effectiveness? The CMS data provide some insight here. In our survey, we asked firms to report the reasons contributing to their most recent decision not to apply for a patent. The reasons for not applying for a patent considered in our survey include: 1) Difficulty in demonstrating the novelty of an invention; 2) the amount of information disclosed in a patent application; 3) the cost of applying; 4) the cost of defending a patent in court; 5) the ease of legally inventing around a patent. Figure 6 indicates the percentage of respon-

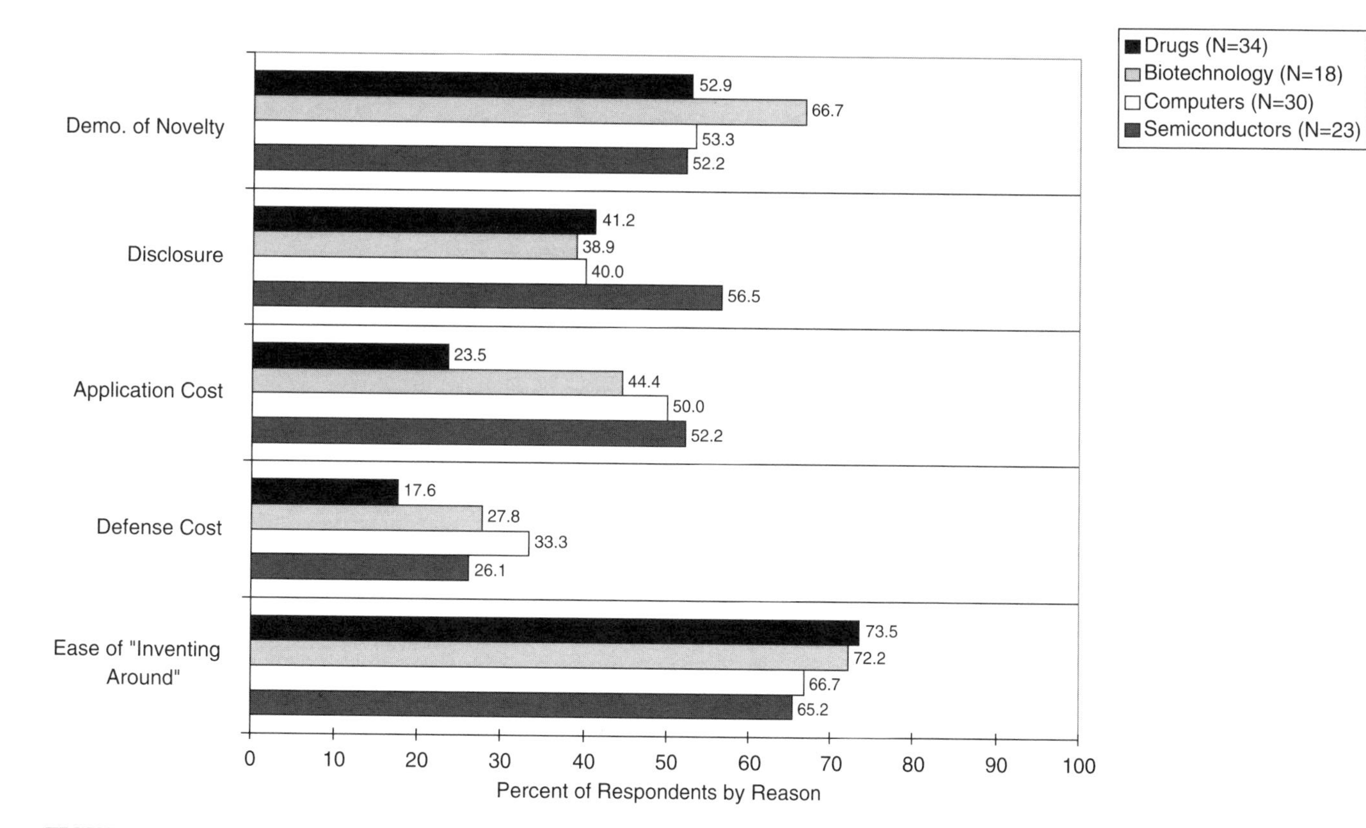

FIGURE 6 Reasons not to patent.

dents indicating whether a particular reason figured into their decision not to patent.[25] Our results in Figure 6 show that ease of inventing around is the most cited reason, and is of greater concern to the more R&D active firms in drugs and biotechnology than in computers and semiconductors. Aside from concerns over the demonstration of novelty—and hence whether an invention is patentable to begin with—the disclosure of information that comes with patenting is also of concern. The only cross-industry differences of note is that concern over the disclosure of information appears to be of less concern to the semiconductor respondents than others, perhaps because the technology is sufficiently complex and know-how based such that a patent discloses less of real value to rivals. Also, concern over the cost of either the patent application or subsequent legal defense is less of an impediment to patenting in the drug industry.[26]

REASONS TO PATENT

In this section, we will try to develop a better understanding of how firms use patents, particularly when patenting does not appear to protect the commercialization or licensing of their patented inventions. We will also speculate on the implications of the uses of patents for the exploitation of public research.

In our survey, we asked respondents to indicate which of seven possible reasons motivated their most recent decisions to apply for a patent. The reasons for patenting considered in our survey include the prevention of copying, the prevention of other firm's attempts to patent a related invention (which we call "patent blocking"), the earning of licensing revenue, use to strengthen the firm's position in negotiations with other firms (as in cross-licensing agreements), the prevention of infringement suits, use as a measure of internal performance of a firms' technologists, and the enhancement of the firm's reputation. Figure 7 presents the percent of respondents by industry indicating a given reason applied to their most recent decision to patent.

Aside from the prevention of copying, the most prominent reasons for patenting across all four industries include patent blocking, prevention of suits and for use in negotiations. This result resembles that for the manufacturing sector as a whole reported in Cohen et al.[27] One cross-industry difference observed in

---

[25]To save space in the survey questionnaire, we did not distinguish between process and product innovations for this question. We did, however, draw this distinction when we considered the reasons to patent.

[26]There is also an interesting difference between the full sample responses for biotechnology versus the responses from the subsample of the more R&D active firms in the industry (i.e., those respondents with greater than the median R&D intensity). As compared to the results for the full sample, the cost of application and defense drops considerably as a reason not to patent for the more R&D active biotechnology firms.

[27]See Cohen et al. "Protecting Their Intellectual Assets: Appropriability Conditions and Why U.S. Manufacturing Firms Patent (or Not)," *op. cit.*

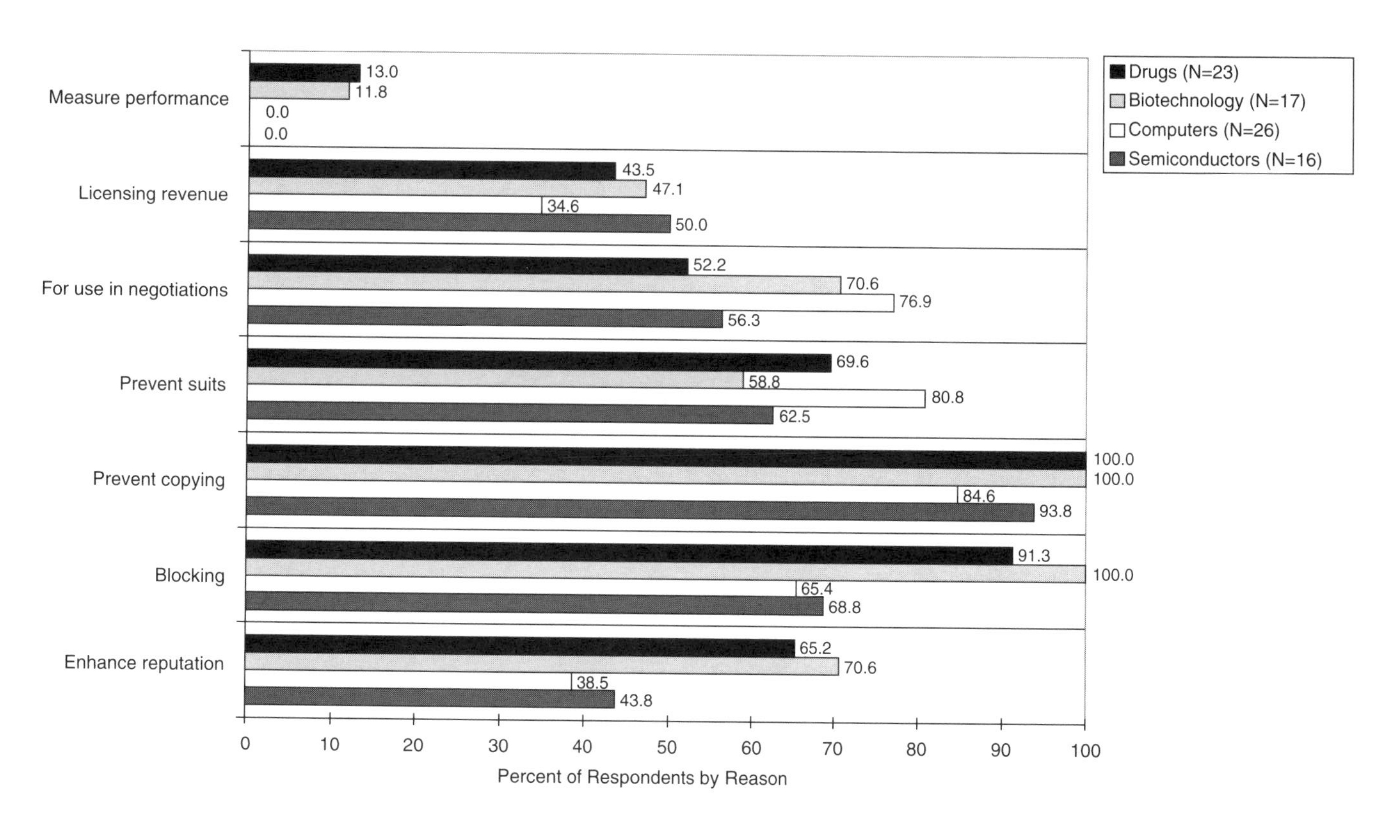

FIGURE 7 Reasons to patent product innovations.

Figure 7 is that licensing revenue is a less pervasive motive for patenting among firms in the computer industry, but using patents for negotiations is somewhat more pervasive, implying that most of those negotiations are associated with cross-licensing. Patent blocking appears to be more pervasive in biotechnology and drugs than in computers or semiconductors, though high in all four industries. Patenting to enhance the firm's reputation is less evident in the semiconductor and computer industries than in drugs or biotechnology.

In the more comprehensive analysis of the reasons to patent for the manufacturing sector as a whole, Cohen et al. highlight other cross-industry differences in the reasons to patent.[28] They find that once one focuses on firms that patent more aggressively (by weighting responses by each respondent's number of patent applications), a sharp difference becomes apparent between semiconductor and drug industry respondents (which include biotechnology firms in the Cohen et al analysis).[29] Firms in the drug industry appear to use patents in the way they are typically thought to be used, namely to protect the commercialization or licensing of the patented inventions. Firms also use patents to block the patenting of substitute or competing inventions, but rarely use patents for cross-licensing negotiations. In contrast, in semiconductors, the same patents that are used for blocking are also commonly used for strengthening the firm's position in crosslicensing negotiations, suggesting that patent blocking tends to be directed against not substitute but complementary inventions. This makes sense in a complex product industry like semiconductors where commercialized innovations are often comprised of numerous separately patentable elements and incumbent manufacturing firms consequently either require access to one another's technology, or at least the freedom to pursue projects that are similar to those of rivals without being sued.[30] Indeed, on the basis of both the survey data and field interviews, Cohen et al.[31] conclude that firms will often patent in semiconductors not to profit directly from a particular patented invention through either its commercialization or licensing, but to build portfolios to compel their inclusion in cross-licensing negotiations or at least provide some protection against suits by other manufacturers.[32]

The use of patent portfolios to achieve such "player" status in the semiconductor industry suggests that patents will tend to be most useful to firms that can generate or acquire rights over numerous patents and have access to the considerable legal resources essential to such a strategy. Where firms are both active in

---

[28] *Ibid.*

[29] *Ibid.*

[30] *Ibid.*

[31] *Ibid.*

[32] For the case of semiconductors in particular, this conclusion is also supported by the interview findings of Hall and Ham, "The Patent Paradox Revisited: Firm Strategy and Patenting in the U.S. Semiconductor Industry," NBER Conference on Patent System and Innovation, Jan. 8, 9, Santa Barbara, 1999, Working Paper, 1998.

the manufacture of semiconductors and conduct R&D, they will tend to cross-license among themselves in recognition of their mutual dependence rather than extract licensing revenue. The example of Texas Instruments (T.I.) suggests, however, that where firms do not hold a large stake in manufacturing the product but nonetheless possess patents on key technology, there is little benefit to be realized from cooperative cross-licensing agreements. Rather, since firms in T.I.'s position do not depend on access to other firms' technologies and thus cannot be hurt al.1 that much by a countersuit, they have a lot to gain and risk losing little by aggressively defending their patents and pursuing licensing revenue.[33]

Use of patents in the computer industry appears to differ from their use in either the drug or semiconductor industry, and has some similarities to both. First, patenting for use in cross-licensing negotiations is common in the computer industry, and, to this extent, is similar to the practice in the semiconductor industry. Yet, using these same patents to block appears to be less common. Patenting to earn licensing revenue is more pervasive in the computer industry among the larger, more patent intensive firms.

The contrast between the motives for patenting among drug and biotechnology firms, on the one hand, and particularly semiconductor firms on the other, challenges the rationale behind Bayh-Dole and similar policies. For drug and biotechnology firms, patents indeed appear to play a role in stimulating the commercialization of invention, as Bayh-Dole assumes. On the other hand, in semiconductors, patents play a different role. They are not used by firms to protect the commercialization of invention. Rather, the accumulation of patents into large portfolios confers player status in the industry, and, in turn enables access to the technology of rivals and the ability to pursue R&D and commercialization without fear of suits from other incumbents. In this latter setting, what is the use of a patent on an invention originating from public research laboratories? It is unlikely that such patents provide protection that induces the follow-on R&D and other investments necessary for commercialization. Thus, patents on public research in areas such as semiconductors may be of limited relevance to the enterprise of commercialization, and consequently play little of the role envisioned by the framers of the Bayh-Dole Amendment.

It does appear, however, that patents on public research outputs have stimulated universities and other publicly funded institutions to attempt to garner licensing revenues from firms. Although benefitting universities, such actions also raise the cost of innovation and may dampen firms' incentives to conduct follow-on complementary R&D in the absence of offsetting benefits. And, from the discussion above, we conjecture that the offsetting benefits may be rather slight in industries such as semiconductors. There is another potentially important but

---

[33]See John Barton, "Reforming the Patent System," Working Paper, Stanford Law School, 1999.

subtle disadvantage to such rent-seeking behavior on the part of public research institutions. Where the progress of public research itself depends upon reciprocal information flows between industry and public research institutions, the prospect of, say, a university's staking of exclusive rights may undermine that progress by chilling those flows. Firms may be reluctant to communicate freely about their own technology with public research institutions if they believe that technology may end up in a university patent for which they will have to pay.[34]

DISSCUSSION

Our data do not suggest that the privatization of public research flows into industry is essential to securing the commercial application of much of that information. Although we do observe a substantial impact of public research in drugs and biotechnology where patents and associated information channels are strong, we also observe a substantial impact in semiconductors where that is not the case. Moreover, across all four industries that we have examined, the most important channels from public research to industrial R&D labs are the most public ones. Moreover, our analyses suggest that even when private channels such as consulting are strong, use of public channels complement that strength.

The cross-industry differences observed in the impact of public research on industrial R&D, the importance of the different channels of communication, and the importance and role of patents in particular also offer other implications. First, public research simply matters more in some industries than others, suggesting that to the extent that Bayh-Dole and related policies have an impact, that impact will vary substantially across industries. For example, we observe that public research appears to be particularly important for three of the four industries we examined, namely drugs, biotechnology and semiconductors. So, in these settings, what is at stake may be quite large. In these three industries, we also fmd, however, important differences in the effectiveness of patents in protecting the commercialization and licensing of inventions. Thus, in drugs and biotechnology where firms report patents to be quite effective, patents on the outputs of public research that are already close to commercialization may play a role in stimulating downstream commercialization. The enforcement of exclusionary rights on more "upstream" public research outputs essential to the research of a broader community may also impose, however, considerable costs, as noted by Mazzoleni and Nelson.[35]

In contrast to the role and importance of patents in drugs and biotechnology, patents are not reported to be very effective in protecting the commercialization

[34] I thank Robert White, Director of Carnegie Mellon University's NSF Data Storage Engineering Research Center for suggesting this point.

[35] R. Mazzoleni and R. Nelson, "Economic Theories About the Benefits and Costs of Patents," *op. cit.*

of innovation in the semiconductor industry. Here, firms protect their inventions through a combination of lead time advantages, secrecy and the exploitation of complementary capabilities. Firms in semiconductors do, however, patent their inventions, but apparently a key motive, at least among larger incumbents, is to amass portfolios used to gain access to other firms' technologies via cross-licensing and to protect themselves from suits. It is not apparent how patents on public research bearing on semiconductor technology would therefore foster subsequent commercialization, which is the rationale for Bayh-Dole. If granted exclusive licenses, semiconductor firms could conceivably use patents originating from public research like their own that is to strengthen their bargaining position in the industry.[36] There is also, however, the prospect that if public research institutions attempt to enforce their property rights, they could antagonize firms and undermine the reciprocal communication flows that contribute to the research efforts of all parties.

REFERENCES

Barton, John (1999) " Reforming the Patent System," Working paper, Stanford Law School.

Cohen, W.M., Florida, R., Randazzese, L., Walsh, J. (1998), "Industry and the Academy: Uneasy Partners in the Cause of Technological Advance," in Noll, R., ed., *Challenges to Research Universities,* Brookings Institution.

Cohen, W.M., Nelson, R.R. and Walsh, J. (2000) "Protecting Their Intellectual Assets: Appropriability Conditions and Why U.S. Manufacturing Firms Patent (or Not)" National Bureau of Economic Research, Working Paper No. 7552.

Faulkner, W. and Senker, J. (1995) *Knowledge Frontiers: Public Sector Research and Industrial Innovation in Biotechnology, Engineering Ceramics, and Parallel Computing,* New York: Oxford University Press.

Gibbons, M. and Johnston, R. (1975) "The Roles of Science in Technological Innovation," *Research Policy,* vol. 3, pp. 220-242.

Hall, B. and Ham, R.M. (1998) "The Patent Paradox Revisited: Firm Strategy and Patenting in the U.S. Semiconductor Industry," NBER Conference on Patent System and Innovation, Jan. 8, 9, Santa Barbara, 1999, Working Paper.

Heller, M. and Eisenberg, R. (1998) "Can Patents Deter Innovation? The Anticommons in Biomedical Research", *Science,* Vol. 28, May 1, pp. 698-701.

Klevorick, A. K., Levin, R., Nelson, R. R. and Winter, S. (1994) " On the Sources and Significance of Interindustry Differences in Technological Opportunities," *Research Policy*, vol. 24, no. 2, pp. 195-206.

Levin, R., Klevorick, A., Nelson, R.R., and Winter, S.G. (1987) "Appropriating the Returns from Idustrial R&D", *Brookings Papers on Economic Activity,* pp. 783-820.

Mansfield, E. (1986) "Patents and Innovation: An Empirical Study," *Management Science,* 32:173-181.

[36]An example of such may be the announced pooling of 400 patent and patent applications between Stanford University and the Yamaha Corporation in the area of sound synthesis (Chronicle of Higher Education, August 7, 1998).

Mansfield, E., Schwartz, M. and Wagner, S. (1981) "Imitation Costs and Patents: An Empirical Study," *Economic Journal,* 91:907-918.

Mazzoleni, R. and Nelson, R.R. (1998), "Economic theories about the benefits and costs of patents," *Journal of Economic Issues,* 32(4):1031-1052.

Mowery, D.C., Nelson, R.R., Sampat, B.N., and Ziedonis, A.A., "The Effects of the Bayh-Dole Act on U.S. University Research and Technology Transfer: An Analysis of Data from Columbia University, the University of California and Stanford University," 1999.

Narin, F., Hamilton, Kimberly S., and Olivastro, D. (1997) "The Increasing Link between U.S. Technology and Public Science," *Research Policy*, 26(3):317-330.

Rosenberg, N. and Nelson, R.R. (1994) "American Universities and Technical Advance in Industry," *Research Policy,* 239(3):323-348.

Scherer, F.M., et al. (1959) *Patents and the Corporation.* 2nd ed. Boston, privately published.

Bioinformatics: Emerging Opportunities and Emerging Gaps[1]

Paula E. Stephan and Grant Black
Georgia State University

INTRODUCTION

A typical gene lab can produce 100 terabytes of information a year, the equivalent of 1 million encyclopedias.[2] Few biologists have the computational skills needed to fully explore such an astonishing amount of data; nor do they have the skills to explore the exploding amount of data being generated from clinical trials. The immense amount of data that are available, and the knowledge that this is but the tip of the data iceberg means that researchers must increasing-

---

[1]This paper draws on work that was prepared at the request of Paul Romer for the workshop on the Role of Human Capital in Capitalizing on Research, sponsored by the National Academy of Engineering and the National Research Council's Committee on Science, Engineering, and Public Policy, The Beckman Center, Irvine, CA, January 20-21, 1998. The paper prepared for that conference was subsequently published in *Science and Public Policy* ("Bioinformatics: Does the U.S. System Lead to Missed Opportunities in Emerging Fields? A Case Study," Dec. 1999). This paper also draws on a report prepared for the Alfred P. Sloan Foundation, "Hiring Patterns Experienced by Students Enrolled in Bioinformatics/Computational Biology Programs," May 1999. We have benefitted from the comments of participants at the workshop as well as those of Michael Teitelbaum, Mary Frank Fox, and Bill Amis. We have also benefitted from the comments of William Zumeta and Charlotte Kuh. We wish to express our appreciation to Bill Agresti, Jim Brown, Sean Eddy, Warren Ewens, Dan Gusfield, Gene Myers, Gerald Selzer, and Judy Willis for their ready willingness to speak with us while we were writing this paper. We also wish to thank all of those who responded to the survey we sent in the spring of 1999 concerning programs in bioinformatics.

[2]David Malakoff, "NIH Urged to Fund Centers to Merge Computing and Biology," *Science,* June 11, 1999, p.1742.

ly rely on an interdisciplinary approach to not only succeed in but to just proceed with research. It also means that individuals are needed who can work in the emerging field of bioinformatics, combining skills of computer science with a knowledge of biology.

This paper examines this emerging field. We begin by discussing demand for individuals who can work in the field. We then summarize the number of individuals in the pipeline who are currently being trained in the field. The indication that demand is strong and the pipeline sparsely populated leads us to ask why the response of higher education has been sluggish. We close by suggesting possible solutions to address the problem of sluggish response.

DEMAND FOR INDIVIDUALS IN BIOINFOMATICS

By all accounts the field of bioinformatics/computational biology is booming. The scientific press stresses the high salaries paid to new hires ($65,000 for persons with top master's training; $90,000 or more for Ph.D.s) and the intensity with which headhunters seek out possible candidates.[3] Universities complain that their students are "grabbed" before they are able to complete their degrees and that their faculty and students are lured to industry, creating the concern that the bioinformatics field is "eating its seed."[4]

Here we use a two-part methodology to investigate demand: we analyze position advertisements in *Science* as well as summarize data collected from a survey of programs concerning their placements of students. Figure 1 presents job openings in bioinformatics and computational biology by month for a two-year period as measured by counting position announcements in *Science*. Given the methodology, the numbers reported are a lower bound.[5] In 1996, 209 posi-

---

[3]See Eliot Marshall, "Hot Property: Biologists Who Compute," *Science*, June 21, 1996, pp. 1730-32; Eliot Marshall, "Demand Outstrips Supply," *Science*, June 21, 1996, pp. 1731; and Diane Gershon, "Bioinformatics in a Post-genomics Age," *Nature*, Vol. 389, September 25, 1997, pp. 417-18.

[4]See Eliot Marshall, "Demand Outstrips Supply," *Op cit.* and Potter Wickware, "Choices and Challenges," *Nature*, Vol. 389, September 25, 1997, pp. 420. For example, the bioinformatics staff at Johns Hopkins University's Genome Data Base fell from 35 to 20 during Fall of 1997 due to corporate recruitment. See Jocelyn Kaiser, (ed.) "Hopkins's Genetic Database to Close," *Science*, vol. 279, January 30, 1998, p. 645.

[5]*Science* and *Nature* are the two scientific journals that consistently publish employment ads related to computational biology. Our index was computed by examining job advertisements in every issue of *Science* for the years 1996 and 1997. A position was counted if the ad specifically asked for a computational biologist or a bioinformatist or the position announcement explicitly mentioned experience in computational biology or bioinformatics. Counts are lower bounds of actual position announcements in *Science* because some advertisements do not state the specific number of position openings but instead indicate more than some specified number. In such instances the lower bound was recorded. Within each calendar year every effort was made not to count repeated ads for the same position.

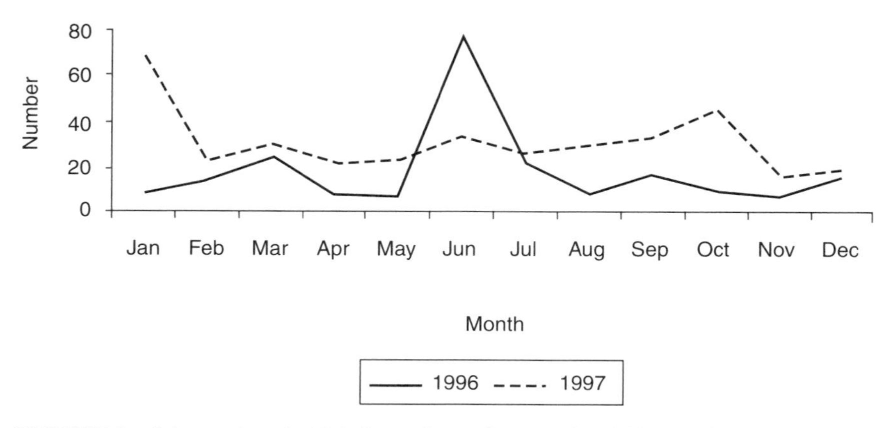

FIGURE 1 Job openings in bioinformatics and computional biology from Science ads, 1996 & 1997.

tions were advertised; in 1997 this had increased by 69 percent to 354.[6] These counts include two special advertising supplements focused on biotechnology, one in June 1996 the other in July 1997.[7] Both supplements were dominated by ads from SmithKline.[8] In a typical month (ignoring special supplements) the journal averaged 12 position announcements in 1996. This had more than doubled by 1997, rising to 25.

Table 1 organizes the information in terms of type of entity placing the ad, rather than number of position announcements. Three categories are listed: firms, universities, and other not-for-profits, including government.[9] We see that the number of entities placing ads grew from 70 to 118 between 1996 and 1997, representing a growth of 68 percent. In both years the majority—about 63 percent of entities placing ads—were firms, and the number of firms placing ads grew by 70 percent.[10] In addition to large firms, such as Bristol-Myers Squibb, Eli Lilly, SmithKline-Beecham, Pfizer, Merck, Abbott, Bayer, and Monsanto, a

[6]This finding of growth is consistent with Wendy Yee's report that from 1995-1996 the number of ads related to bioinformatices tripled. See "The Top Five Career Trends of 1996: Informatics Anything," http:\www.nextwave.org/server-java/SAM/pastloop/trend2.htm.

[7]There was also a special supplement in July 1996, apparently an addendum to the June supplement.

[8]In the 1996 supplement SmithKline said that they wanted to about double their staff of 30. In the 1997 supplement they again said they wanted to about double their staff, this time reported at 40, suggesting that SmithKline has plans to grow and is experiencing difficulty filling positions in bioinformatics/computational biology.

[9]Note that institutes that are affiliated with universities, such as Whitehead, are here counted as "other not for profit."

[10]Note: these are firm, not enterprise, counts.

TABLE 1 Number of Distinct Entities Advertising Positions in *Science*

Sector	Year 1996 (share)	Year 1997 (share)	Number of distinct position announcements in 1996 and 1997	Growth between 1996 and 1997 (%)
Firms	44 (62.8)	75 (63.6)	90	70.4
Not-for-profit universities	17 (24.3)	22 (18.6)	36	29.4
Other not for profit	9 (12.9)	21 (17.8)	27	133
Total	70	118	153	68.6

number of smaller biotech firms, such as Regeneron, Immunex, and Zenez, have position announcements. A substantial number of firms—29 to be exact—placed ads in both 1996 and 1997.

In contrast, only 36 universities ran position announcements during the period, and the growth rate for university ads was slightly less than 30 percent. Universities placing ads included the University of California at Los Angeles (UCLA), the University of California at Irvine (UC Irvine), the University of California at San Francisco (UCSF), the University of Pennsylvania, the University of Southern California (USC), and California Institute of Technology. Only three universities placed ads in both years. A number of not-for-profit entities (such as the Centers for Disease Control) also ran position announcements and the number of ads from this sector more than doubled.

Based on the position announcements, jobs in computational biology range from entry-level data analysts and programmers to senior-level scientists and research directors.[11] Lower-level positions that are more directly computer-oriented call for as little as an undergraduate science degree, and some state no degree requirements. The majority of positions call for a doctorate degree in either a science (preferably molecular biology) or computer science with considerable programming or bioinformatics experience, although a number of positions explicitly advertise for individuals with a master's degree.

The second part of the methodology to study demand involved using data collected from a survey of programs concerning the placement of students. The

[11]This analysis is based on 1996 ads only.

data are for the 12-month period March 1998 to March 1999. Of the 21 identified formal and informal programs, 16 supplied data and eight institutions/programs reported students taking jobs during the period. All but one indicated that students completed their degrees prior to employment, suggesting that earlier media reports that students are commonly recruited before completing programs do not hold for most trainees in the field.

Fifty-three individuals were reportedly placed: three at the undergraduate level, 23 at the master's level, 13 at the doctorate level and 14 with postdoctorate training. Nine students who graduated during the period chose to continue their education and training. Five of these went to a postdoc position either from a Ph.D. program or from a previous postdoc appointment. Three moved on to graduate programs, and one went to medical school. Specific placement information was ascertained for 42 of the 53 hires. Seven of the 53 placements are at an academic institution, and only one position is at a government-sponsored institution. The remaining identified placements are in the private sector and include four graduates who established their own firms.

Salary data for a subset of the hires are presented in Table 2.[12] Salary ranges are reported for 71 percent of the hires during the period. The greatest lack of information is at the postdoctorate and doctorate levels. As expected, salaries for the most part climb as the level of training rises, starting in the $40,000-$50,000 range for BAs and reaching over $100,000 for one post doc. But there are exceptions. For example, two of the three undergraduates who were placed received salaries between $50,000 and $60,000. This is higher than that earned by seven of the masters students, although ten of the 19 master's students for whom we have salary information earn more than $60,000. One masters student received a starting salary of over $100,000. Reported salaries for five hires at the doctorate level are over $70,000. One is between $80,000 and $90,000; another is over $100,000. Three postdocs received placements with a salary between $80,000 to $90,000. One was placed at a salary of over $100,000. One institution reported that one or more master's students received a signing bonus.

Our surveys of ads and programs lead us to conclude that (1) demand is strong and growing but small relative to other areas, (2) demand is driven in large part by industry, (3) salaries are high relative to other areas in the life sciences, and (4) the majority of jobs being advertised are not being filled by graduates of programs. This suggests that a number of the jobs being advertised either remain unfilled or are filled by individuals coming from outside the field. The head of Merck's computational biology program, for example, is a physicist.

---

[12]For an explanation of how the salary data were collected see Paula Stephan and Grant Black, "Bioinformatics: Does the U.S. System Lead to Missed Opportunities in Emerging Fields? A Case Study," *Science and Public Policy*, December, 1999, pp. 1-15.

TABLE 2 Salary Ranges by Training Level, January 1998 to March 1999

Training Level	Salary Range $ (number of hires)
Undergraduate	40,001-50,000 (1)
	50,001-60,000 (2)
Masters	40,001-50,000 (7)
	50,001-60,000 (2)
	60,001-70,000 (1)
	over 70,000 (8)
	over 100,000 (1)
	unknown (4)
Doctorate	60,001-70,000 (1)
	over 70,000 (5)
	80,001-90,000 (1)
	over 100,000 (1)
	unknown (5)
Postdoctorate[a]	80,001-90,000 (3)
	over 100,000 (1)
	unknown (5)

[a]Excludes placements for 1997 trainees listed in the W.M. Keck Center's 1997 *Annual Report*.

PIPELINE

Table 3 summarizes student enrollment in training programs by degree level as of March 1999. The data come from the same survey from which the salary data were collected. Of the 21 identified programs, information was gathered for 16. Information concerning type of support is also provided. We see from the table that doctorate programs dominate enrollment in terms of the number of

TABLE 3 Characteristics of Formal Training Programs as of March 1999[a]

	Undergraduate	Masters	Doctorate	Postdoctorate
Total Number of Programs	3	5	9	7
Number of Programs with Internal Support[b]	1	1	3	0
Number of Programs with External Support	2	3	8	9
Enrollments[c]	23	35	86	~25

Note: Placement information is included in the respective programs for several Ph.D. students and postdocs trained at the University of California-Santa Cruz. The nature of the reported data broadens the time period of placements, starting as early as 1995 and ending in May 1999; thus, the reporting period extends beyond the January 1998 to March 1999 period. Placement information is included for five 1997 postdocs listed in the W.M. Keck Center's 1997 *Annual Report* and, thus, in the case of the postdocs, the reporting period extends before January 1998. Due to the lack of exact counts reported, the precise completion rate cannot be determined.

[a]Baylor College of Medicine, Boston University, Northwestern University, Rutgers University, Stanford Universtiy, University of California-Santa Cruz, University of Pennsylvania, University of Washington, W.M. Keck Center for Computational Biology. The Keck Center includes Baylor College of Medicine, Rice University and the University of Houston. Note that Baylor also has a program that is independent of the Keck Center. This program is counted separately here.

[b]Based only on institutions responding to the question regarding internal sources of funding; several institutions did not respond to this question.

[c]Includes counts of students in degree programs in the Department of Computer Sciences at the University of California-Davis and the University of California-Santa Cruz; there are no formal bioinformatics/computational biology programs at UC-Davis and no formal undergraduate program at UC-Santa Cruz.

training programs and the number of students (86). The largest program is at Stanford, followed by Rutgers, the W.M. Keck Center program, and Baylor College of Medicine. The smallest enrollment in the spring of 1999 was at the bachelor's level, with the University of Pennsylvania having by far the largest program. Thirty-five students were enrolled in the various masters programs and approximately 25 in formal postdoc programs at the time of the survey. Enrollment in master's programs was expected to more than double in the fall of 1999 when new programs at George Mason University, the Georgia Institute of Technology, and Boston University came on line.

THE GAP

This strong demand in bioinformatics and the creation of new programs comes at a time when the prospects of young life scientists look less than promising. Despite the "hot" reputation of the life sciences (and in part because of its hot reputation), a large and increasing number of early-career life scientists are unable to find the type of job that will permit them to become independent researchers and establish their own lab. Sufficient concern over the career outcomes of young life scientists exists to have warranted the establishment by the National Research Council of the Committee on Dimensions, Causes, and Implications of Recent Trends in the Careers of Life Scientists. The committee issued its report in September 1998, stating that "the imbalance between the number of life-science Ph.D.'s being produced and the availability of positions that permit them to become independent investigators concerns the committee." The committee concludes that "Intense competition for jobs has created a 'crisis of expectation' among young life scientists. Much of this imbalance is the result of the professional structure of the life sciences research enterprise where "the important work of conducting experiments rests almost entirely on the shoulders of graduate students and postdoctral fellows." Recommendations of the committee included restraint of the rate of growth of the number of graduate students in the life sciences.[13]

Is it not contradictory that the committee concluded that a "crisis of expectations exists for young life scientists" at a time when demand is strong and growing in the field of bioinformatics? Why are there but nine doctoral programs in the United States in computational biology,[14] while there are approximately 194 programs in biochemistry and molecular biology and over 100 in molecular and general genetics?[15] More generally, the contrast of the two fields leads one to ask if the structure of the U.S. science enterprise leads to missed opportunities in emerging fields, particularly when the demand is heavily centered in industry.

Here we examine four interrelated explanations of this gap: The four are (1) the low incentive for individual faculty to establish such programs and attract students in the area, (2) an educational system that responds differently when demand is driven by industry as opposed to when demand is driven by universities and research labs, (3) the interdisciplinary nature of the field creates disincentives to the establishment of programs, and (4) the quick fix—turning life scientists into computational biologists—is not possible, given the skills and quantitative abilities of individuals in the life sciences, nor is the incentive present for computer scientists to opt for additional training in the life sciences.

---

[13] See National Research Council, *Trends in the Early Careers of Life Scientists*, Committee on Dimensions, Causes, and Implications of Recent Trends in the Careers of Life Scientists, Washington, D.C.: National Academy Press, 1998, p. 4.

[14] The nine programs (as of December 1997) are at the Baylor College of Medicine, the Carnegie Mellon University, George Mason University, Rice University, Rutgers University, the University of Houston, the University of Pennsylvania, the University of Pittsburgh, and Washington University.

[15] See Marvin Goldberger, Brendan Maher, and Pamela Flattau, eds. *Research Doctorate Programs in the United States: Continuity and Change*. Washington, D.C.: National Academy Press, 1995.

Lack of Incentive

The research structure that has evolved in academe in the U.S. means that faculty are extremely responsive to research funding opportunities since it is external grants that provide resources to purchase equipment and support graduate students and postdocs—the collaborators who are absolutely essential to the lab of the principal investigator (PI). Furthermore, at many medical schools in the U.S., funding not only supports the collaborators, it also supports the PI, which means that the PI can only retain his or her academic appointment as long as the PI has funding to cover the cost of the lab and the PI's salary.[16]

This suggests that an effective way to alter the educational mix of graduate students is to alter the amount of research funds directed to an area and thus provide the incentive for faculty to recruit students into the field. To what degree has this occurred in computational biology/bioinformatics.

The evidence, which is difficult to assemble due to its fragmented nature, suggests that funding agencies have only begun to do this and still not to a major extent.[17] Instead, funders have placed their targeted computational eggs in the training basket. NSF has provided training funds through its Computational Biology Activities and the Sloan and Keck foundations have targeted funds to the training of individuals in bioinformatics; non-targeted training funds have also come from NIH. While such a strategy may be best in the long run, in the short run training grant initiatives may be ignored by many faculty. This is because training grants signal collective bodies. Research grants signal individuals. It is difficult to get academic units composed of competitive PIs focused on where the next grant will come from to engage in the collective response required to succeed in creating new programs. Faculty are much more attuned to thinking about individual-investigator-initiated grants. And, in the past, little funding has been targeted by federal agencies at research in computational biology. For example, at a time when NIH supported more than 25,000 active research grants a year, only 96 R01s listed the key words "computational and biology" and only 11 R29s listed these key words. A similar statement can be made with regard to NSF CAREER grants. Of the approximately 400 active grants in 1996, only six appear to be directly related to the area of computational biology. Moreover, while many of the training grants were targeted at bioinformatics, most of these research awards were not targeted specifically to the area.[18]

---

[16]At Baylor College of Medicine, for example, 100 percent of the faculty in the biomedical department receive 80 percent of their salary from grants.

[17]For a summary of funding sources, see Table 2 in Paula Stephan and Grant Black, "Bioinformatics: Does the U.S. System Lead to Missed Opportunities in Emerging Fields?" *op. cit.*

[18]This should change in the near future if the recommendation of an NIH advisory committee made public in June 1999 are followed up. The cornerstone of the panel's recommendations is that NIH create biocomputing centers. Funding for the centers would come from a new NIH biocomputing program that would make research grants in the area (Malakoff, 1999, p. 1742).

Educational System Responds More Slowly to Demand Driven by Industry as Opposed to Demand Driven by Research Opportunities

As indicated in the introduction, demand for computational biologists is substantially driven by industry, which sees genetic data as "the major driving force" in drug discovery.[19] SmithKline Beecham is a case in point. Their June 7, 1996, full-page ad in *Science* reports that they had 30 individuals working in the area, with plans to double that number by 1997 (p. 1527). Six months later, they ran another ad in *Science*, again saying that they had plans to double their staff, this time reported to be at 40. Moreover, SmithKline has aggressively hired established researchers from academe and the non-profit research sector. In 1995 they succeeded in attracting David Searls away from the University of Pennsylvania, and shortly after Searls's arrival they hired James Fickett from the Los Alamos National Laboratory, Randall Smith from Baylor College of Medicine, and Chris Rawlings from the Imperial Cancer Research Fund in London.[20]

Does it make a difference that the demand is industry-driven, as opposed to driven by academe? We are inclined to say yes for two reasons. First, every time industry hires a faculty member it means that there is one less professor to train future computational biologists. Thus, while the practice of recruiting faculty from academe provides a ready source of knowledge, and hence spillovers from academe to industry, the practice—where replacement is difficult—impairs academe's capacity to continue the training initiatives it has already begun.[21] The Baylor program reportedly experienced difficulty when Randall Smith left to join SmithKline, and, while the program at the University of Pennsylvania survived despite Searls' departure, the remaining faculty were stretched as a result.

Second, academic departments in the life sciences are arguably not as responsive to demand driven by industry as are departments in engineering and computer sciences, which have long had a tradition of placing a sizeable number

[19] See Eliot Marshall, "Demand Outstrips Supply," *op. cit.*

[20] See Eliot Marshall, "Hot Property: Biologists Who Compute," *op. cit.*

[21] Industry arguably knows that, despite the fact that bioinformatics is likely to be a foundation for the next generation of pharmaceuticals, it is eating its own seed. This raises the question of why industry is not doing more to replenish the crop. The "winner-take-all" nature of competition in pharmaceuticals and the rapid pace of discoveries in the pharmaceutical industry undoubtedly lead industry to offer high premiums for the seed to abandon universities to take jobs in industry. But the answer as to why industry is investing so little in training future researchers in the area may rest, not on the intensity of competition, but instead on habit. The large number of research grants that have flowed into biomedical research, and the ability of researchers to support postdocs and graduate students on these grants, has made for a steady supply of individuals entering the life sciences in the past ten years or more. To the extent industry had a problem during this period it was in convincing individuals to abandon their hopes of becoming independent investigators in academe, not in locating individuals trained in the biomedical sciences. Training, except for postdoc positions within their firms, has thus not been anything that the pharmaceutical industry has felt that it needed to foster.

of their graduates in industry. Few life science Ph.D.s head directly to industry upon completing their Ph.D.s.[22] The reason stems from the fact that it is research funding—much more than the availability of jobs for graduates—that drives the size of academic programs in the life sciences. This is because research funding provides ready support for Ph.D. students in the life sciences and funding for the postdoctorate positions that recent Ph.D.s (and not-so-recent Ph.D.s) hold with such proclivity. In the biotech world of the late twentieth century, life science departments have found a ready supply of aspiring students who are willing to commit eight to ten years of their life to becoming life scientists so that they can have a shot at becoming a PI to continue working on the frontiers of knowledge. And, while it is viewed as both honorable and profitable for established faculty to work with industry, the profession would appear to still stigmatize the individual whose early career goal is to work in industry.

The Interdisciplinary Nature of Computational Biology Creates Disincentives to the Establishment of Programs

Bioinformatics requires training in computer and information science, mathematics, and the life sciences. Coordination among these three fields can often be an institutional nightmare since it involves not only cooperation across department lines but also across colleges. The department of computer sciences is often located in a college of engineering, while mathematics and life science departments are generally located in the college of arts and sciences. The situation is further complicated by the fact that universities that have medical schools often have an additional department of life sciences in the medical school.

The problems in working across department lines are difficult enough when departments are within the same college. They are compounded when departments are in different colleges or universities. For example, how are students to be advised? How are courses to be numbered and shared? How are contributions to be valued across college/university lines? And these are the simple questions. The harder questions concern which department/college will get "credit" for the new field. How will resources be shared? Who will get the new positions if individuals trained in computational biology are hired? Who will evaluate individuals' promotion and tenure?[23]

---

[22]In 1996, for example, the percentage of Ph.D. recipients with definite postgraduation plans for U.S. industry employment was 48.5 percent in engineering, 43.4 percent in computer science, and a mere 4.7 percent in the biological sciences. See National Science Foundation, *Characteristics of Recent Science and Engineering Graduates: 1995, Detailed Statistical Tables*, Arlington, VA: National Science Foundation, 1997.

[23]The fragmentation of fields within institutions not only creates problems in meeting the demand to train students in new areas. It also has negative consequences for the productivity of science and the ability of an institution to respond to changes in science over time. Studies indicate that breakthrough research significantly benefits from intense interdisciplinary activities across fields. See D.

The fields also differ in terms of career goals and opportunities for students. Michael Ashburner, director of research at the European Bioinformatics Institute, argues that more resources should be funnelled into master's programs to provide uniform, specialized training since almost all those involved in bioinformatics come from another field (Gavaghan 1997). Yet, terminal master's programs have historically been unpopular in the life sciences, in part because the ready supply of Ph.D. students and postdocs provided the needed assistance in the lab and in part because the field often stigmatized those with a terminal master's degree.[24] This stands in marked contrast to the fields of engineering and computer science where, a master's education is looked favorably upon and employment is found (and encouraged) in industry.[25]

The Lack of a Quick Fix

A plausible fix to the "shortage" of individuals in computational biology is to turn young life scientists into computational biologists—or to take those with degrees in mathematics or computer information systems and augment their skills. Indeed, without a proactive strategy, this is what an economist would predict would occur. The number of postdoctoral grants offered in the area by Sloan, NSF, and Burroughs Welcome suggests that they have adopted such a strategy.

Several reasons, however, lead us to suspect that this strategy is less effective than might originally appear at face value. First, at the doctoral level the market for individuals trained in computer science appears to be sufficiently strong to retain computer scientists in that field. Table 4 reports the 1995 median annual salary of recent Ph.D.s employed full time in the broad areas of computer and information sciences and life and related sciences. The subcategory of biological and health sciences is also included. Three cohorts are identified: 1993-94 graduates, 1990-92 graduates, and 1985-89 graduates. The large difference between salaries in computer and information sciences and salaries in life sci-

---

Hicks and J.S. Katz, "Science Policy for a Highly Collaborative Science System," *Science and Public Policy,* 1996, 23: 39-44; Rogers Hollingsworth, "Major Discoveries and Biomedical Research Organizations: Perspectives on Interdisciplinarity, Nurturing Leadership, and Integrated Structure and Culture," prepared exclusively for Interdisciplinarity Project and to be published in the University of Toronto Press; and J.S. Katz, et al. The Changing Shape of British Science, Brighton: Science Policy Research Unit at the University of Sussex, 1995.

[24] Furthermore, it is commonly believed that Ph.D.s and post docs provide new ideas to the lab and that to replace them with permanent masters-level technicians would rob the lab of this important source of ideas.

[25] Andy Brass, director of a master's-level bioinformatics course at the University of Manchester (U.K.), maintains that there is a wage premium for a masters in bioinformatics compared to molecular biology. See Helen Gavaghan, "Running to Catch Up in Europe," *Nature*, 389, September 25, 1997, pp. 420-422.

TABLE 4 Median Annual Salary of FTE Recent Ph.D. Graduates, 1995

	1993-94 Graduates	1990-92 Graduates	1985-89 Graduates
Computer and Information Sciences	$54,000	$61,000	$65,000
Life and Related Sciences	$30,400	$40,000	$52,000
Biological and Health Sciences	$30,000	$38,600	$52,000

SOURCE: NSF/SRS, *Characteristics of Doctoral Scientists and Engineers in the United States: 1995*, 1997a.

ence for those who have been out for one to two years reflects the fact that a majority of individuals in the life sciences hold postdoctoral positions upon graduating. The difference narrows as the life scientists move out of these positions, but a substantial differential of 25 percent exists for those who have been out six to ten years. This suggests that job prospects in computer science are sufficiently strong to preclude computer scientists from seeking additional formal training in the biological sciences. The quick-fix strategy is more attractive to those trained in the biological sciences where the market, as we have already indicated, is considerably weaker.

Table 4 suggests that life scientists may have the incentive to seek additional training to become computational biologists. Do they have the background *and* aptitude to transform themselves into computational biologists? The response to the University of Pennsylvania's training initiative in computational biology was remarkable. Over 200 individuals applied for the two postdoctoral positions. Yet, according to the faculty member who directs the program at the University of Pennsylvania, less than a handful qualified for the program precisely because the applicants had so little background in mathematics/statistics. There is reason to believe that this lack of quantitative background is generic to those with Ph.D.s in biology—not specific to the applicants to the University of Pennsylvania program. An examination of the requirements of five highly rated biology departments demonstrates that none have formal mathematical requirements for entry into their graduate programs; only a handful of graduate courses have a mathematics prerequisite up to introductory calculus.[26]

It is not just that life scientists lack training in math and statistics. A credible argument can be made that the typical life scientist lacks interest and excep-

[26]The five institutions reviewed are Harvard University, Johns Hopkins University, Massachusetts Institute of Technology, Stanford University, and the University of California-Berkeley. Three of the institutions state that they expect entering students to possess some mathematics knowledge, preferably at least introductory calculus.

TABLE 5 GRE Scores by Intended Field of Graduate Study, for Seniors and Nonenrolled College Graduates, 1993-96

Intended Graduate Field of Study		Mean Score	Percent of Test-takers with Score above 700	Percent of Test-takers with Score of 800
Biological Sciences	Verbal	501	3.6	0.1
	Quantitative	595	20.7	1.1
Health and Medical Sciences	Verbal	449	0.7	0.0
	Quantitative	515	5.8	0.1
Computer and Information Sciences	Verbal	483	5.4	0.2
	Quantitative	672	52.2	5.6
Mathematical Sciences	Verbal	502	6.5	0.2
	Quantitative	698	60.6	8.8

SOURCE: *1997-98 Guide to the Use of Scores*, Educational Testing Service, 1997b

tional aptitude in these areas. This is somewhat borne out by data supplied by the Graduate Records Exam. Table 5 presents data on the scores of GRE test-takers from 1993-96 by their intended field of graduate study. The data indicate that individuals intending to pursue graduate study in the biological and health sciences test substantially lower in the quantitative area than those intending to study computer and mathematical sciences.[27] While the mean quantitative score for biological sciences was 595, the score for computer sciences was 672—a difference of 77 points. Moreover, less than 21 percent of test-takers intending to enter the biological sciences achieved a score above 700 compared to 52 percent in computer and information sciences.

[27]It should be noted that this is for all test-takers *intending* to pursue graduate education, not those actually in graduate programs. Many of these test-takers will not receive admission into graduate programs, let alone leading programs in their intended field of study. A discussant suggested that the large differential may be due to the "Asian factor." Specifically, Asian students score extremely well on the quantitative portion of the test, and the fields of computer and information sciences and mathematical sciences attract a disproportionate number of Asian students compared to the life sciences. This Asian factor could lead to a lower mean test score among individuals intending to study in the life sciences compared to the other fields. This is undoubtedly true but, we suspect, does not explain away the differential. Although data are not reported for GRE scores by country of origin and intended field of study, some indication of the magnitude of the "Asian factor" is given by examining the scores by ethnicity and intended field for U.S. citizens. According to the GRE Board, the mean test score of Asian/Pacific American U.S. citizens intending to do graduate

CONCLUSION AND RECOMMNENDATIONS FOR WAYS TO ENCOURAGE THE DEVELOPMENT OF NEW PROGRAMS

This paper explores four reasons why the current educational system appears to be sluggish in responding to the increased demand for individuals trained in computational biology. The first and second reasons are interrelated. Specifically, we argue that the size and direction of Ph.D. programs in the life sciences are more responsive to signals embedded in funding opportunities for faculty research than to the signals provided by the job market for graduates. While this may appear perverse, it is the logical consequence of a research regime that places great emphasis on having doctoral students and postdoctoral students in the lab and can persist as long as there is an adequate supply of applicants. Such a supply has been forthcoming in the United States in recent years because of (1) the "hot" reputation of biotechnology; (2) the availability of immigrant scientists and (3) the ready supply of postdoctoral positions that permit graduate schools to provide placement for graduates.

The third reason for the sluggish response relates to the interdisciplinary nature of bioinformatics. Given the fields involved (mathematics, computer science, and biology), collaboration typically requires working across college lines within a university. While this is not impossible, the bureaucracy and incentive structure of academe act to discourage cooperation across disciplines. Finally, we have argued that there is no "quick fix." Individuals trained in computer science have few economic incentives to change their stripes by acquiring additional training in biology. And, if they did, the response would be far from quick since they would require a substantial amount of training in biology. In contrast, the loose labor market for young life scientists means that the incentive is there for life scientists to augment their skills and become computational biologists. But for life scientists the path may be difficult since many lack both the mathematical training and the inclination to become successful computational biologists.

---

work in the life sciences was 590, compared to 694 for Asian/Pacific American citizens intending to do graduate work in engineering and 665 for those intending to do graduate work in the physical sciences (Educational Testing Service 1997b, p. 16). Making the heroic assumption that the test scores of the U.S. Asian population reflects test scores of Asians who are noncitizens, these numbers suggest that the lower quantitative scores in the life sciences are due at least in part to the fact that Asians students who seek out training in the life sciences have lower quantitative scores than Asian students who go into engineering or the physical sciences. The same thing can be said for whites. Educational Testing reports that white U.S. citizens who intend to study in the life sciences have a mean quantitative score of 537—117 points lower than white citizens who plan to enter the physical sciences, and 152 points lower than white citizens who plan to enter engineering (p. 16). (The broad fields of engineering, life sciences, and physical sciences are used in this note because scores are not reported by ethnicity for the narrower fields given in Table 4.)

We end by proposing four possible ways to address the apparent shortage of individuals trained in bioinformatics. First, it is important to foster ways for faculty to interact across disciplines and institutional boundaries both within universities and within urban areas. Providing space for interdisciplinary efforts could encourage such interaction; the "if you build it they will come and they will talk" argument. Second, provide funds for targeted research in the area and continue to provide funds for training awards and institutional awards. While training grants can affect outcomes in the long run, there is nothing like targeted research awards to get the attention of faculty in the short run. Third, provide information to students on career outcomes in bioinformatics. This involves making faculty aware of the career outcomes of their students so that faculty can help prospective students make well-informed decisions. Finally, in an emerging field such as bioinformatics, it is important to recruit early in the pipeline in order to attract the "right" kind of mind. Perhaps it is not surprising that the University of Pennsylvania has decided that the best way to meet the demand for computational biologists is to "grow their own," offering undergraduate and master's programs in computational biology in order to attract the "right" kind of mind and integrate the curriculum at an early stage.[28]

REFERENCES

Educational Testing Service. 1997a. *1997-98 Guide to the Use of Scores*. Princeton, New Jersey: Educational Testing Service.

Educational Testing Service. 1997b. *Sex, Race, Ethnicity, and Performance on the GRE General Test*. Princeton, New Jersey: Educational Testing Service.

Gavaghan, Helen. 1997. "Running to Catch Up in Europe." *Nature*, 389:420-422.

Gershon, Diane. 1997. "Bioinformatics in a Post-genomics Age." *Nature*. 389:417-418.

Goldberger, Marvin L., Brendan A. Maher, and Pamela E. Flattau (eds.). 1995. *Research-Doctorate Programs in the United States: Continuity and Change*. Washington, D.C.: National Academy Press.

Hicks, D. and J.S. Katz. 1996. "Science Policy for a Highly Collaborative Science System." *Science and Public Policy*. 23:39-44.

Hollingsworth, Rogers. 1995. "Major Discoveries and Biomedical Research Organizations: Perspectives on Interdisciplinarity, Nurturing Leadership, and Integrated Structure and Cultures," prepared exclusively for Interdisciplinarity Project and to be published by University of Toronto Press.

Kaiser, Jocelyn (ed.). 1998. "Hopkins's Genetic Database to Close." *Science*. 279:645.

Katz, J.S., D. Hicks, M. Sharp, and B. R. Martin. 1995. *The Changing Shape of British Science*. Brighton: Science Policy Research Unit at the University of Sussex.

Malakoff, David. 1999. "NIH Urged to Fund Centers to Merge Computing and Biology." *Science*. (June 11, 1999):1742.

---

[28]Rensselaer Polytechnic Institute joined the sparse ranks of institutions offering undergraduate training, starting an undergraduate degree program in bioinformatics and molecular biology in the Fall of 1998 that is funded in large part by a $1.2 million grant from Howard Hughes Medical Institute for undergraduate education in the life sciences.

Marshall, Eliot. 1996. "Hot Property: Biologists Who Compute." *Science*. June 21: 1730-32.
Marshall, Eliot. 1996. "Demand Outstrips Supply." *Science*. June 21: 1731.
National Research Council. 1995. *Research Doctorate Programs in the United States: Continuity and Change*. M. Goldberger, B. Maher, and P. Ebert, editors. Washington, D.C.: National Academy Press.
National Research Council. 1998. *Trends in the Early Careers of Life Scientists*. Committee on Dimensions, Causes, and Implications of Recent Trends in the Careers of Life Scientists. Washington, D.C.: National Academy Press.
National Science Foundation. 1997. *Characteristics of Doctoral Scientists and Engineers in the United States: 1995, Detailed Statistical Tables*. Arlington, VA: National Science Foundation.
National Science Foundation. 1997. *Characteristics of Recent Science and Engineering Graduates: 1995, Detailed Statistical Tables*. Arlington, VA: National Science Foundation.
National Science Foundation. 1996. *Science and Engineering Doctorate Awards: 1996, Detailed Statistical Tables*. Arlington, VA: National Science Foundation.
Stephan, Paula and Grant Black. 1999. "Bioinformatics: Does the U.S. System Lead to Missed Opportunities in Emerging Fields? A Case Study." *Science and Public Policy*, (December): 1-15.
Stephan, Paula and Grant Black. 1999. "Hiring Patterns Experienced by Students Enrolled in Bionformatics/Computational Biology Programs." Report to the Alfred P. Sloan Foundation. May.
Wickware, Potter. 1997. "Choices and Challenges." *Nature*. 389:420.
Yap, Ting K., Frieder Ophir, and Robert L. Mantino. 1996. *High Performance Computational Methods for Biological Sequence Analysis*, Boston: Kluwer Academic Publishers.
Yee, Wendy. "The Top Five Career Trends of 1996: Informatics Anything," http:\www.nextwave.org/server-java/SAM/pastloop/trend2.htm.

Recent Trends in the Federal Funding of Research and Development Related to Health and Information Technology

Michael McGeary
McGeary and Smith

INTRODUCTION

This paper documents recent trends in federal funding of research related to health and information technology (IT). It shows especially the impact of the halt in growth of federal funding for several years in the mid-1990s, which resulted in a substantial shift in the composition of the federal research portfolio. In broad terms, federal investment in biomedical research has expanded relative to federal investment in most fields of the physical sciences and engineering. That shift has raised the question of whether the federal research portfolio has become "imbalanced." The paper analyzes the shifts in funding of research fields, compares them with nonfederal trends, and discusses the implications of the reductions in funding during the 1990s of many fields in the physical sciences, engineering, and geosciences.

Part II examines funding of core fields, biological and medical science in the case of health and computer science and electrical engineering in the case of IT. Part III looks at federal funding of research and development (R&D) in health and IT in a national context; that is, how federal funding relates to nonfederal funding. Part IV is an analysis of recent trends in the funding of other fields of research, including those that also contribute to progress in health and IT. In the conclusion (Part V), the question of whether recent shifts in funding— toward the biological, medical, and computer sciences and away from most fields in the physical sciences and engineering—have given rise to an unbalanced federal research portfolio is discussed. To some extent, where one stands on the imbalance question depends on where one sits, but several broader approaches to resolving the question are presented.

RECENT TRENDS IN FEDERAL FUNDING OF LIFE SCIENCES AND INFORMATION TECHNOLOGY RESEARCH

Historically, federal funding of R&D associated with advances in health and the development of computer-related products and services has been substantial and important. This is particularly the case for support of fundamental long-range research where industry investment is often missing.[1] Federal investment in the life sciences[2] and IT[3] increased many-fold from the 1950s to the 1990s. In the 1990s, however, efforts to reduce the federal budget deficit and the post-Cold War reductions in the defense budget affected federal R&D funding in terms of both amount—where funding was reduced for several years—and composition—where some research fields expanded greatly relative to others. Budget authority for federal R&D funding peaked in FY 1992 in real terms. By 1996, it had fallen by 9 percent. Real growth in the R&D budget resumed in 1997, but the federal investment in R&D did not equal the 1992 level until the FY 2000 budget. Research was less affected.[4] Federal research funding went flat in 1993 and began to increase again in 1997 (see Figure 1).

As we will see in more detail in Part IV, the pattern of funding by field in the area of science and engineering changed substantially during the period from 1993 to 1997. During this period, overall federal funding for research was flat in real terms.[5] The Administration and Congress, however, continued to increase the budget of the National Institutes of Health (NIH), which provides more than 80 percent of the federal support for the life sciences, including the biological and medical sciences.[6] At the same time, they responded to the end of the Cold

---

[1]For a recent review of the importance of the federal role in computing research and advances in computer–related technology (including networking), see National Research Council, *Funding a Revolution: Government Support for Computing Research*, Washington, D.C.: National Academy Press, 1999. For the impact of federally funded research in the pharmaceuticals and biotechnology industry, see I. Cockburn, et al. "Pharmaceuticals and Biotechnology," U.S. Industry in 2000: Studies in Competitive Performance, Washington, D.C.: National Academy Press, 1999.

[2]This includes the biological, environmental, agricultural, and medical sciences.

[3]This is composed of computer science and electrical engineering.

[4]R&D consists of basic research, applied research, and development. The focus in this paper is research (basic and applied), which can be classified by field of science and engineering. R&D is used to compare federal and nonfederal funding in Part III, because the data on funding of research by academia and industry cannot easily be broken out. Section IV looks at funding of basic research.

[5]See Stephen A. Merrill and Michael McGeary, "Who is balancing the federal research portfolio and how?" *Science* 285, (September 10, 1999): 1979, p. 1680.

[6]Unless otherwise indicated, all budget amounts and changes are expressed in constant 1999 dollars, using the fiscal year GDP deflators published in the President's 2002 budget request (OMB, 2001: Table 10.1). Except for R&D in Table II-1, the numbers are obligations rather than budget authority or outlays. Obligations are "the amounts for orders placed, contracts placed, services received, and similar transactions during a given period, regardless of when the funds were appropriated and when future payment of money is required" (NSF, 2001). In this paper, the words funding, support, and investment will be used interchangeably with the term obligations.

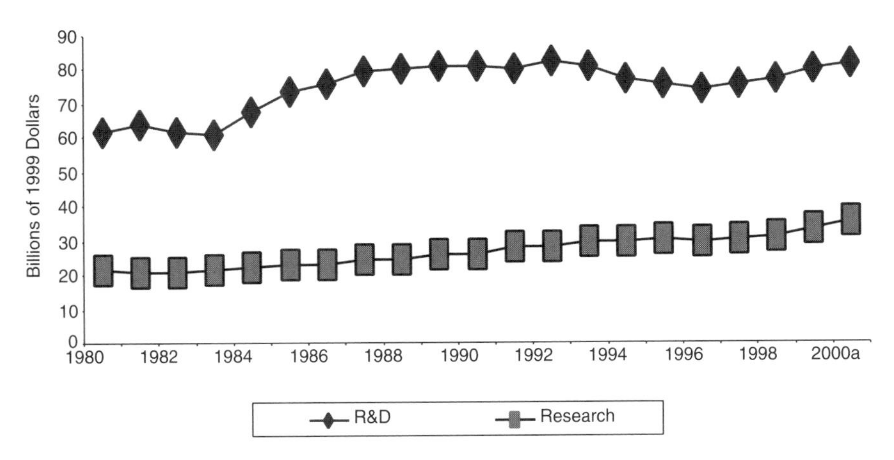

FIGURE 1 Federal funding of research and development, FY 1980–2000.
SOURCE: AAAS (2001) and NSF (2001).
[a] Preliminary

War by cutting the research budgets of the Department of Defense (DOD) and, to a lesser extent, the Department of Energy (DOE). Historically, those two agencies have provided the majority of federal funding for research in electrical engineering, mechanical engineering, materials engineering, physics, and computer science.

These trends in agency funding of research have affected, in turn, research related to life sciences and IT. Funding of life sciences research has increased while funding for most fields in engineering and the physical sciences has fallen.[7] Nearly all the departments and agencies that support R&D have increased their funding of research since 1997 (see Figure 2). As we show in Part IV, however, the reallocation of funding among fields of science and engineering that occurred in the mid-1990s has not changed.

Researchers in the biological and medical sciences have been favored by the budget success of NIH, their main source of federal support. However, support of the biological sciences and the medical sciences differs by degree (see Figure 3). Federal obligations for medical sciences increased more than that for biological sciences, especially in the first half of the decade.

The decline in DOD's budget, including its research budget after 1993, has put heavy downward pressure on the fields that rely on it for most of their

[7] Life sciences fields include biological, medical, environmental biology, and agricultural. Physical sciences include the fields of astronomy, chemistry, and physics. Engineering fields include aeronautical, astronautical, chemical, civil, electrical, mechanical, and metallurgy and materials.

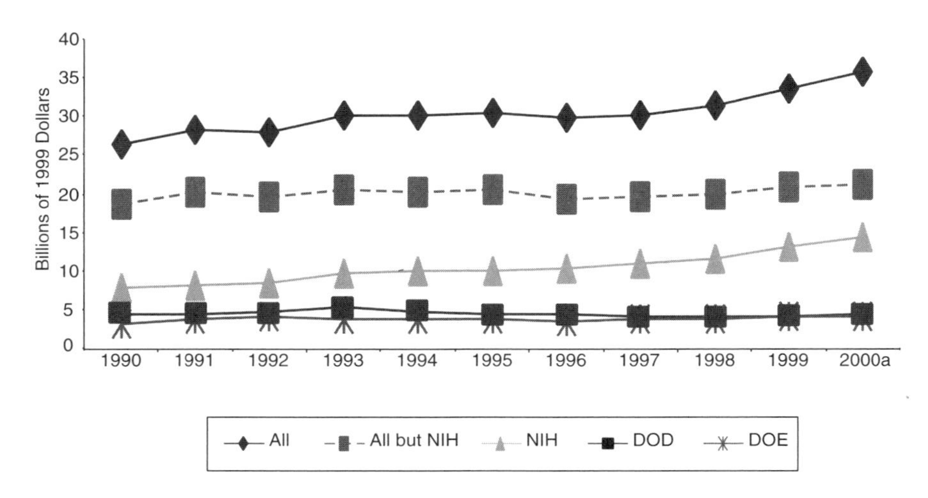

FIGURE 2 Federal funding of research, by selected agencies, FY 1990–2000.
SOURCE: Appendix Table 2.
[a] Preliminary

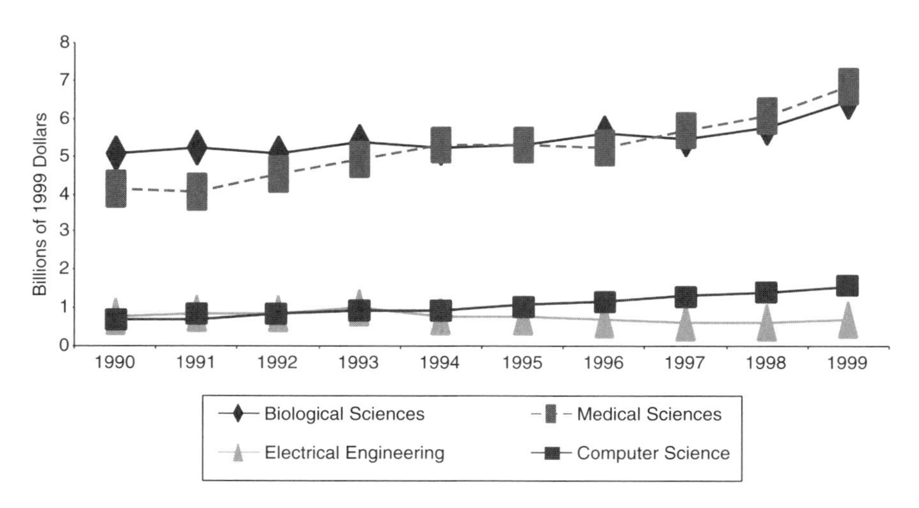

FIGURE 3 Federal funding of selected fields, FY 1990–2000.
SOURCE: Appendix Table 3.

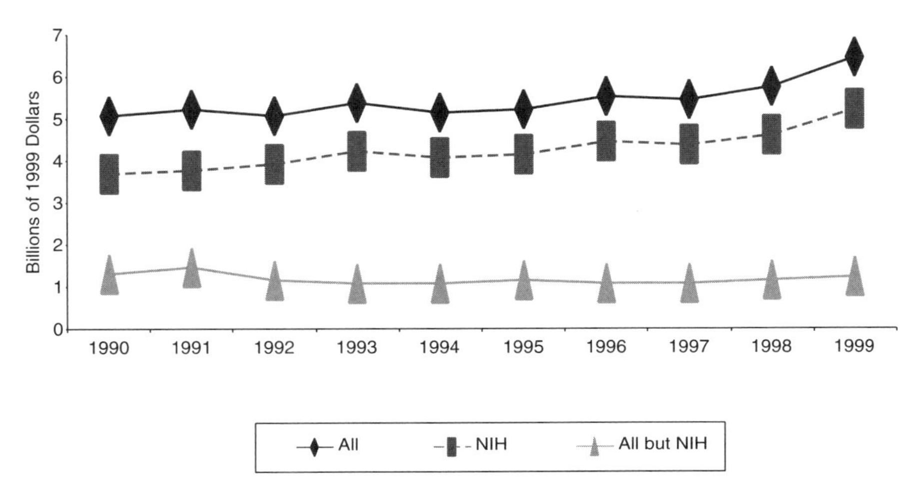

FIGURE 4 Federal funding of biological sciences research, FY 1990–1999.
SOURCE: Appendix Table 4.

federal funding. DOD funding of nearly all fields of engineering, including electrical engineering, was down substantially from 1993 to 1999. Because other agencies did not step up their support substantially, federal funding of electrical engineering declined after 1993. DOD, however, did increase its funding of particular fields of research, including computer science. Several other agencies sharply increased their investment in computers sciences research, and that field had substantially more federal support in 1999 than in 1990.

Biological Sciences

Federal obligations for biological research increased by 28 percent from 1990 to 1999 in real terms, from $5.0 billion to $6.5 billion (see Figure 4). That increase was driven by NIH's steady budget growth in the 1990s of 67 percent. NIH provides more than 80 percent of all federal funding of biological research[8] and accounted for all of the net increase in federal funding of the field from 1990 to 1999.

The emphasis on basic research and university performers is high in federally supported biological research. The ratio of basic to applied research funding remained at about 2 to 1 during the period. The share conducted at universities and colleges increased from 58 percent in 1990 to 66 percent in 1999.

[8]NSF, DOE, and DOD provide another 10 percent of the federal support.

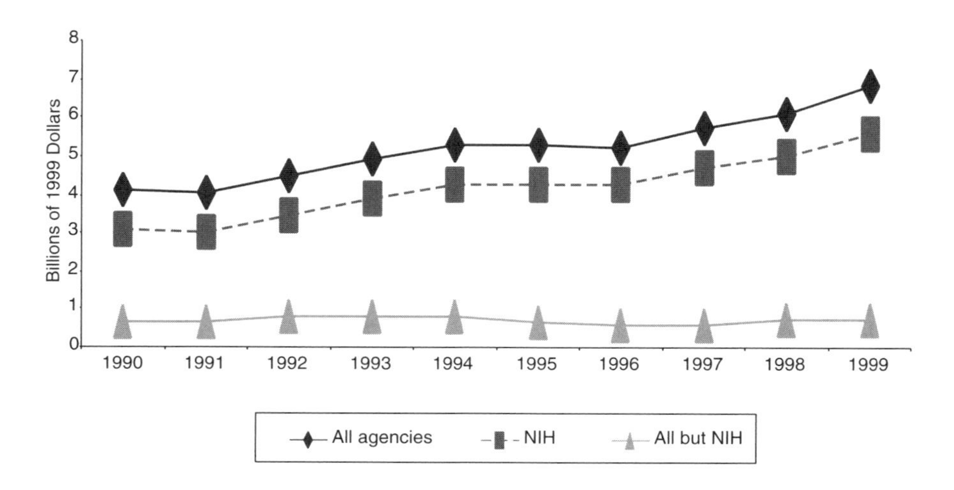

FIGURE 5 Federal funding of medical sciences research, FY 1990–1999.
SOURCE: Appendix Table 5.

Medical Sciences

Federal funding of medical sciences increased by 65 percent from 1990 to 1999, from $4.1 billion to $6.8 billion (see Figure 5). NIH accounted for 82 percent of federal funding of the medical sciences in 1999,[9] up from 75 percent in 1990, and NIH's steady budget growth during those years has permitted the agency to make such a large increase in funding.

Funds allocated to basic research and university-based research in the medical sciences are less than those allocated to biological research. Funding shifted a bit toward basic research but away from universities during the 1990s. The ratio of basic to applied research increased from 1.2 to 1 in 1990 to 1.4 to 1 in 1999. Universities and colleges received 46 percent of the funding in 1999, down from 51 percent in 1990.

Computer Science

As in biological and medical sciences research, federal support of research in computer science increased substantially during the 1990s (see Figure 6). Obligations increased by 121 percent from 1990-1999, from $685 million to $1.5 billion. In 1990, DOD provided 63 percent of the federal funding—in the

[9]Other agencies of the Department of Health and Human Services provide another 8 percent, and the Department of Veterans Affairs, DOD, and NASA account for 9 percent.

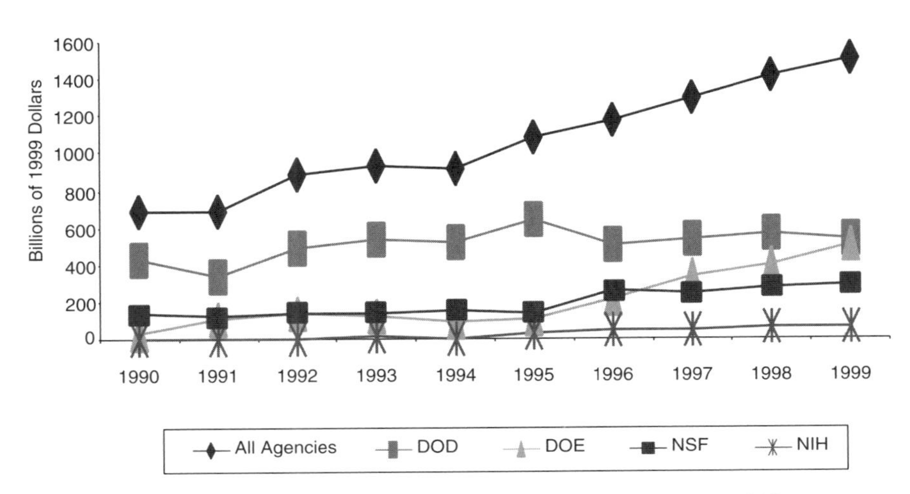

FIGURE 6 Federal funding of computers sciences research, FY 1990–1999.
SOURCE: Appendix Table 6.

amount of $433 million—with NSF a distant second at 19 percent. Despite a cut of 25 percent from 1993 to 1999 in overall research funding, DOD managed to provide $538 million for computer science in 1999, compared with $528 million in 1993. Meanwhile, other agencies substantially increased funding from 1993 to 1999—NSF by 110 percent (from $142 million to $297 million), DOE by 339 percent (from $115 million to $506 million), and NIH by 209 percent (from $20 million to $62 million). In 1999, DOD accounted for 35 percent of the federal funding, DOE for 33 percent, NSF for 20 percent, and NIH for 4 percent.

The ratio of basic to applied research in computer science was 1 to 2.5 in 1999 and this was down from 1 to 1.5 in 1990. Although federal funding of computer science research more than doubled during the 1990s, funding for basic research and university research increased much less, by 60 percent and 47 percent, respectively. The share of federal funding going to basic research fell from 40 percent in 1990 to 29 percent in 1999. In 1999, universities were receiving one-third of all federal funding for computer science, compared with half in 1990.

The trends toward more applied research and research done outside the universities are consistent with the federal cross-agency initiative on applied computing and network research that was in effect during most of the 1990s. It is also consistent with the fact that much of the DOE program is conducted in federal laboratories rather than universities. In addition to scientific opportunities and technological ripeness, the increase in federal funding of computer science research (and research in related fields) was driven by the high priority accorded it by successive administrations.

The effort began as a special interagency initiative in the FY 1992 budget. This initiative was called the High-Performance Computing and Communications (HPCC) Program. HPCC, which was initiated by the Federal Coordinating Council for Science, Engineering, and Technology under President George Bush, was later taken up by the National Science and Technology Council under President Clinton. In the FY 2002 budget, the effort is continued as the Networking and Information Technology R&D Program. Actual budget authority for the interagency initiative tracks the increase in spending on computer science research (Table 1).

TABLE 1 Funding of Interagency Networking and Computing Initiative, FY 1992–2002 (in millions of 1999 dollars)

Fiscal Year	Requested	Enacted
1992	729	749
1993	896	806
1994	1,096	1,025
1995	1,234	1,172
1996	1,255	1,073
1997	1,080	1,054
1998	1,144	
1999		1,301
2000	1,462	1,473
2001	2,315	1,853
2002	1,853	Pending

SOURCE: Annual budget requests.
NOTE: Current dollars were converted to constant 1999 dollars using GDP deflators in OMB (2001): Table 10.1.

Electrical Engineering

Federal funding of research in electrical engineering increased from $780 million in 1990 to $984 million in 1993 (26 percent), then fell to $640 million in 1997 (35 percent), before increasing again to $699 million in 1999 (see Figure 7). The field had 10 percent less funding in 1999 than in 1990 (29 percent less than its high point in 1993).[10]

In 1990, most federal funding for electrical engineering (84 percent) came from DOD. Unlike in the case of computer science, DOD did not try to sustain or increase its level of support of electrical engineering. DOD sharply reduced

[10]Historically, federal funding of electrical engineering peaked in 1987-1989 at just over $1 billion a year (in 1999 dollars).

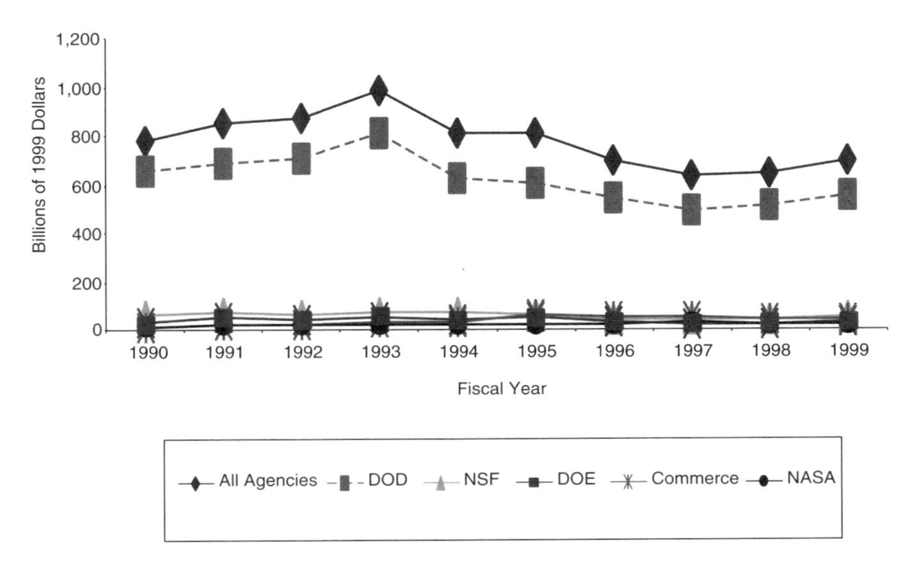

FIGURE 7 Federal funding of electrical engineering research, FY 1990–1999.
SOURCE: Appendix Table 7.

its support from 1993 to 1997. In 1999, the field had 31 percent less funding from DOD than in 1993 and 15 percent less than in 1990. Also unlike the computer science case, other agencies did not increase the amount of support enough to offset much of the DOD cut.[11] The Department of Commerce increased its support by $26 million from 1990 to 1999. DOE and NASA each provided about $5 million more in funding in 1999 than in 1990, but the 1999 levels constituted cuts from 1993 levels. NSF cut funding by $17 million from 1990 to 1999 (22 percent).

Within a shrinking budget, federal agencies increased their emphasis on basic research and research conducted at universities on electrical engineering, but most of the funding still went to applied research and research conducted in non-university settings. The ratio of basic to applied research increased from 1 to 3.4 in 1990 to 1 to 2.7 in 1999. Although federal obligations for electrical engineering declined by 10 percent from 1990 to 1999, funding for basic and university research increased, by 4 percent and 14 percent, respectively, indicating that the cuts were made in applied research and performers other than academic institutions.[12] Basic research rose from 23 percent of total federal funding

[11]Compare Appendix Table 7 with Appendix Table 6.

[12]For example, federal intramural laboratories, national laboratories, and industrial laboratories.

in 1990 to 27 percent in 1999. Universities accounted for 22 percent in 1990, compared with 28 percent in 1999.

NATIONAL TRENDS IN SUPPORT OF R&D RELATED TO LIFE SCIENCES AND IT

The impact of federal funding of research is affected by trends in funding by other institutions. These include industry, universities, and other organizations such as nonprofit research institutes, state and local governments, and foreign research institutions. The federal government was the largest source of R&D funds from World War II until 1980 and continued to be the largest source of research funds until 1995, in real terms (see Figure 8).

Industry provides 51 percent of *total* research funding and the federal government provides 34 percent, but the roles are reversed for *basic* research. The federal government provides the largest portion of basic research support, 49 percent, and industry provides 34 percent (see Figure 9).

The federal government may not need to continue funding research in areas where industry is investing its own funds. However, there may be areas that should receive more emphasis in federal research priorities despite increased investment by industry. These areas require attention because they are still too risky for industry to fund or are at a stage of research where companies cannot be

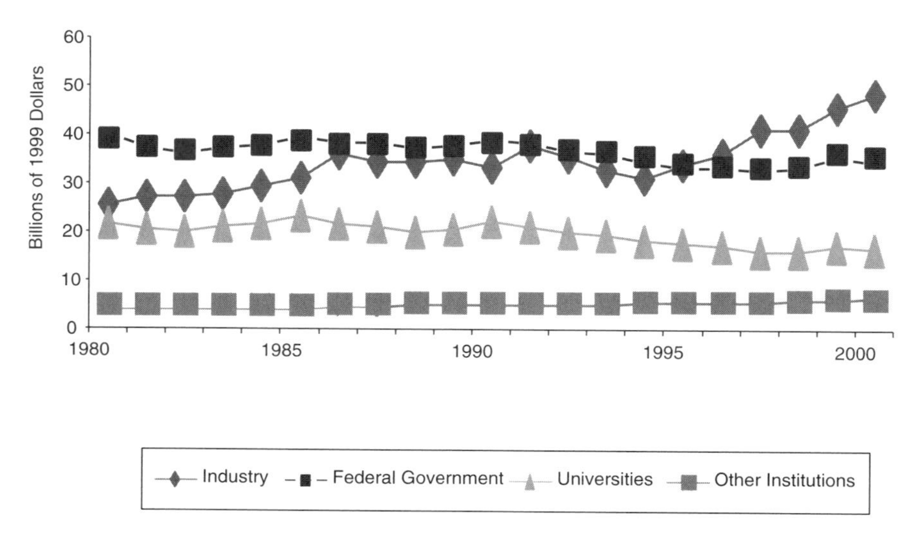

FIGURE 8 National sources of funding for science and engineering research, calendar year 1980–1999.
SOURCE: Appendix Table 8.

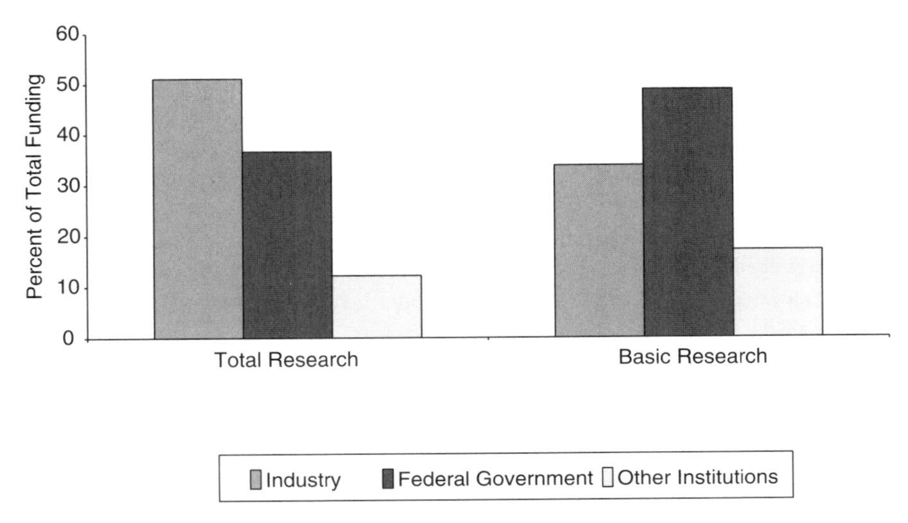

FIGURE 9 Industry and federal roles in funding basic and applied research, FY 2000.
SOURCE: Appendix Table 9.

sure of recouping their investment. Industry typically spends most of its research funding on development and short-term, applied research rather than on the long-term, fundamental research needed to generate new technologies in the future. This section looks at the relative roles of the federal, industry, academic, and nonprofit sectors in funding health and IT research.

We will approach this issue somewhat indirectly. There is no consistent database of information across sectors on the field of research that is being funded, such as biotechnology or information technology, or on the stage of research (basic or applied). The analysis in Part II used the annual NSF survey of federal funds for R&D. This survey collects information from federal agencies on the fields of research they fund and on whether the research is basic or applied. Another NSF survey—the annual survey of academic R&D expenditures—collects information from colleges and universities on how much federally and non-federally funded R&D they perform. The academic survey uses a set of fields very similar to the classification of fields used in the federal funds survey, but the inclusion of development with research makes the results non-comparable with the data collected in the federal funds survey. A third NSF survey—the annual survey of industrial R&D expenditures—classifies research of each company by a single industrial classification (SIC) code, such as "drugs and medicines" (SIC-283) and "office, computing, and accounting machines" (SIC-357), rather than by field of research, such as chemistry, computer science, and mathematics. The survey greatly expanded coverage of the service sector a few years ago, so data on new categories of R&D such as software R&D only go back a few years.

In the most recent edition of *Science and Engineering Indicators*, the National Science Board (NSB) and NSF explored the possibility of conducting "cross-sector field-of-science classification analysis".[13] Here, we will look at an updated and slightly modified version of the NSB analysis of R&D in life sciences and information technology, areas chosen by the NSB because they can be "associated with academic fields of study and with industrial end-projects that tend to be associated with those fields." The NSB graphs and text covered the period from 1985 through 1997.[14] Here, the graphs have been updated to 1998 with survey data released since NSB issued the 2000 Indicators, and they have been extended back to 1981, to give additional historical context.[15]

Life-Sciences R&D

Although company-funded R&D in the life sciences, which includes pharmaceuticals and biotechnology, has increased tremendously, federal funding for R&D in the life sciences is still larger (see Figure 10). Federal funding of life sciences R&D was $13.8 billion in 1998, compared with industry funding of $12.8 billion. Other sources of funding for university research in life sciences and bioengineering also increased, although at a much lower rate.[16] Industry and U.S. Department of Agriculture (USDA) funding related to food and agricultural biotech R&D was fairly flat.

In aggregate, these sources of R&D increased from $15.0 billion in 1981 to $36.7 billion in 1998, a real increase of 145 percent.[17] Federal funding has constituted a large but declining share of the total funding—55 percent in 1981 and 43 percent in 1998. Federal investment increased steadily, so this shift in shares resulted from the steep increase in R&D spending by pharmaceutical and biotechnology companies throughout the period.

According to NSF's survey of federal funds, half the federal funding was for basic research in 1998 ($8.0 billion of $15.8 billion). Assuming that the biopharmaceutical industry devotes the same percent of its R&D funding to basic research that drugs and medicine companies reported to NSF in 1998—17.4

---

[13]See National Science Board, *Science and Engineering Indicators*— 2000, Arlington, VA: National Science Foundation (NSB 00-01) pp. 2-34

[14]See NSB 00-01, *op. cit.:* Figures 2-22 and 2-23. The indicators also looked at chemistry-related R&D: Figure 2-21.

[15]Data from the 1999 survey of industry R&D are available but not included here. The 1999 survey used a new industry classification, and the results are therefore not strictly comparable to the pre-1999 data.

[16]Nonfederal sources of academic R&D funding include own sources (tuition, endowment), state and local government, foundations and other charitable sources, and industry. (Industry funding of academic R&D is also included in company-funded R&D and thus double counted, but the amounts are relatively small.)

[17] Total national R&D increased 90 percent during that time.

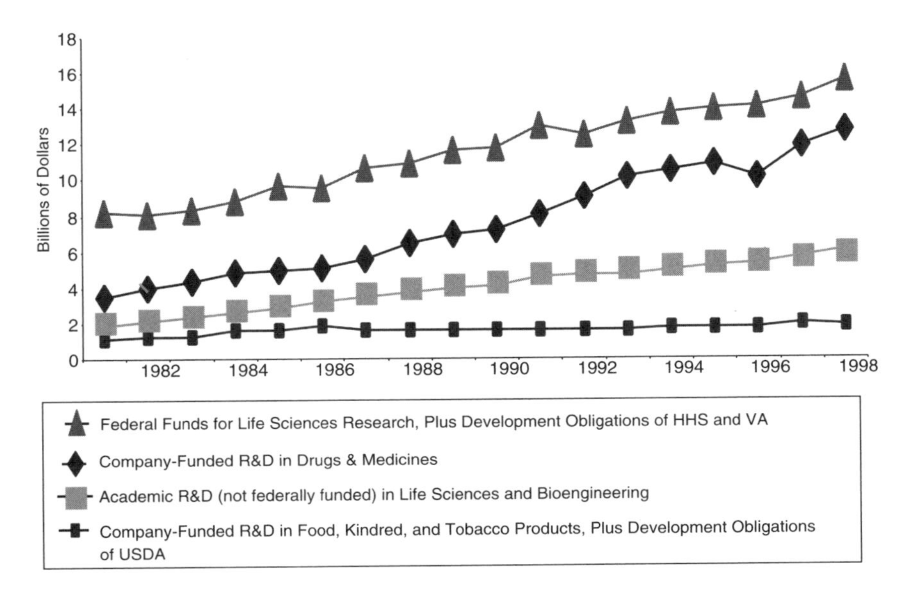

FIGURE 10 National investment in R&D related to life-sciences, 1981–1998.
SOURCE: Appendix Table 10.
NOTE: Federal R&D funds are fiscal year obligations, company R&D funds are calendar year expenditures, and academic R&D funds are fiscal year expenditures.

percent—and that one-third of nonfederal funding of university R&D is for basic research, approximately $12.5 billion, or 34 percent of the total funding of life sciences R&D, went to basic research in 1998.[18]

Information Technology R&D

Two sources of funding account for most R&D in information technology. Company-funded R&D in electrical equipment varied between $10 billion and $14 billion in the 1980s, then grew steeply in the 1990s to nearly $25 billion in 1998 (in constant 1999 dollars; see Figure 11). Company-funded R&D in office, computing, and accounting machines increased from $6 billion to $28 billion in the 1980s, then declined sharply to less than $5 billion in the mid-1990s. It jumped back up to $13 billion in 1997 before falling to $9 billion in 1998.[19]

[18]Also included is the 5.3 percent of R&D that the food industry spent on basic research in 1998. Calculated from NSF (2000):Table A-29.

[19]Flamm, in this volume, analyzes this trend in detail.

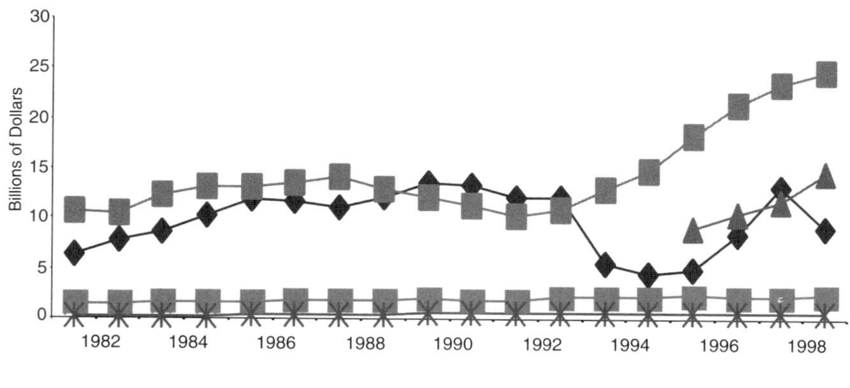

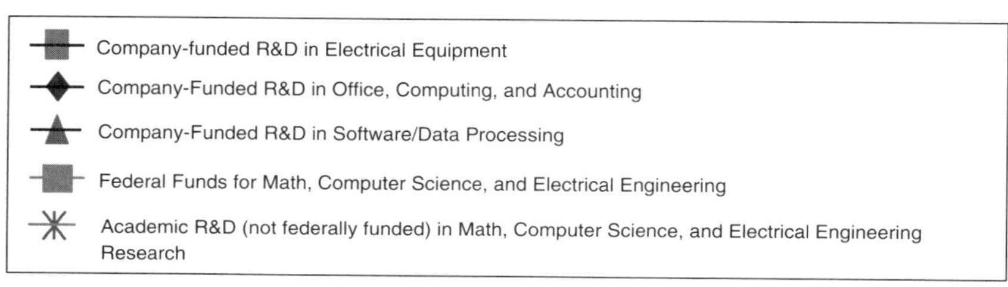

FIGURE 11 National investment in information technology R&D, 1981–1998.
SOURCE: Appendix Table 11.
NOTE: Federal R&D funds are fiscal year obligations, company R&D funds are calendar year expenditures, and academic R&D funds are fiscal year expenditures.

Federal funds for research in mathematics and computer science increased steadily over the period, from $470 million in 1981 to $1.9 billion in 1998, but federal funding of electrical engineering research, which was between $900 million and $1 billion a year in the 1980s, declined after 1993 to $650 million in 1998. Nonfederal academic R&D funding increased steadily from $80 million in 1981 to $350 million in 1998.

According to the NSF industry survey, which expanded its coverage of the service sector in 1995, R&D on software and data processing services is also large and fast-growing. It rose from $9.1 billion in 1995 to $14.5 billion in 1998.

Federal R&D funding is much smaller for IT than for life sciences (although DOD-funded development of military equipment is not included here). As approximated here, IT-related R&D totals about $40 billion, not including software R&D. Federal funding accounts for just 6 percent of the total, compared with 43 percent of life sciences R&D funding.

The matter of funding long-term, fundamental research is quite different in IT than in the life sciences. The $2.5 billion in federal funding of IT R&D includes a total of $933 million for basic research. If we assume that IT firms invest 5 percent of their R&D funding in basic research, and that half of academic R&D funding is for basic research, then national funding of basic research was $3.0 billion in 1998. That was 8 percent of the total spent nationally on R&D in IT (7 percent if software R&D is included).[20]

RECENT TRENDS IN FEDERAL FUNDING OF THE BASIC RESEARCH BASE

Merrill, McGeary, and Henderson recently updated a 1999 study by McGeary and Merrill on trends in federal funding of fields of science and engineering.[21] The first study analyzed data on actual federal research expenditures from FY 1990 through FY 1997, especially trends after 1993, when pressures from the federal budget deficit and reductions in the defense budget had stopped real growth in federal research budgets.[22] The update extends the analysis to FY 1999, the latest year for which there are data on actual expenditures.

The 1999 study tested the hypothesis that certain fields of science and engineering would do less well in budgetary terms because the agencies that provide most of their federal research funding were reducing their level of investment in research. Indeed, in 1997, although the level of federal research spending was nearly the same as it had been in 1993, a number of agencies were spending less on research than they had in 1993, including DOD (–27.5 percent), Department of the Interior (–13.3 percent), Department of Agriculture (–6.2 percent), and DOE (–5.2 percent).[23] Meanwhile, NIH's research budget was up by 11 percent. In fact, federal research spending, not counting the NIH, was down by 6 percent rather than remaining flat.

The study found, however, that the fate of individual fields did not necessarily mirror that of their chief benefactor. For example, computer science, which received the majority of its support from DOD before 1993, continued to grow strongly despite the downturn in overall DOD research funding. Conversely, bio-

---

[20]Assuming 5 percent is generous. In 1998, office, computing, and accounting machines companies reported that 3.1 percent of their R&D funding was for basic research; electrical equipment manufacturers reported 4.6 percent of their R&D went to basic research (NSF, 2000:Table A-29).

[21]See, National Research Council, *Trends in Federal Support of Research and Graduate Education.* Board on Science, Technology, and Economic Policy. Washington, D.C.: National Academy Press, 2001.

[22]See National Research Council, *Securing America's Industrial Strength.* Board on Science, Technology, and Economic Policy, Washington D.C. National Academy Press, 1999: Appendix A. For an issue-oriented summary of the study and its implications, see Stephen A. Merrill and Michael McGeary *op. cit.* 1999.

[23]See Michael McGeary and Stephen A. Merrill, "Recent Trends in Federal Spending on Scientific and Engineering Research: Impacts on Research Fields and Graduate Training," in National Research Council, *Securing America's Industrial Strength.* Washington, D.C.: National Academy Press, 1999: Table A-1.

logical sciences research, which receives most of its funding from NIH, had barely increased its funding despite the rapid increase of the overall NIH budget. Several factors might account for that discrepancy: (1) agencies with reduced research budgets did not necessarily cut all fields, or did not cut them the same; (2) other agencies stepped up their support; or (3) some combination of these factors.

Nevertheless, most fields in the physical sciences and engineering were disproportionately negatively affected because they received most of their support from DOD and DOE, and only a few have been able to increase support from other agencies. The fields with less federal funding included chemical, civil, electrical, and mechanical engineering; mathematics; physics and chemistry; and geology. Four—electrical and mechanical engineering, physics, and geology—had at least 20 percent less funding than 4 years before.

The update found that the funding situation had eased somewhat by 1999. Federal obligations for research were 12-percent greater in 1999 than in 1993. Moreover, all agencies except DOD and the Department of the Interior now had larger research budgets, although NIH accounted for most of the net growth in research funding since 1993.[24] Fewer fields had less funding than in 1993—7 of 21 compared with 11 of 21 in 1997—but 5 of them (including electrical engineering) had at least 20 percent less funding compared with 4 in 1997. Meanwhile, the number of fields with at least 20 percent more funding than in 1993 went from 1 in 1997 to 5 in 1999, including biological sciences, medical sciences, and computer science (see Figure 12).

Despite increased funding, the new pattern of allocation developed during the budget cuts of the 1993-1996 period remained. In some respects, the gap that had opened between certain fields at the top (especially biomedical and computer science) and those at the bottom (electrical and mechanical engineering, physics and chemistry, and geology) was widening.

The "Balance" Issue

The divergent trends in federal funding of fields of research in the 1990s, reinforced by the effort to double the NIH budget in 5 years beginning in FY 1999, have raised concerns among scientists and engineers, industrial leaders, and science and technology policy analysts about a possible imbalance in the federal research portfolio. Such concerns extend even to the long-term health of biomedical research itself.

Harold Varmus, formerly director of the NIH, stated his conviction that "NIH can only wage an effective war on disease only if we—as a nation and a scientific community, not just as a single agency—harness the energies of many

[24]Excluding NIH, federal obligations for research were 1.4 percent higher in 1999 than in 1993, instead of 11.7 percent, in constant dollars.

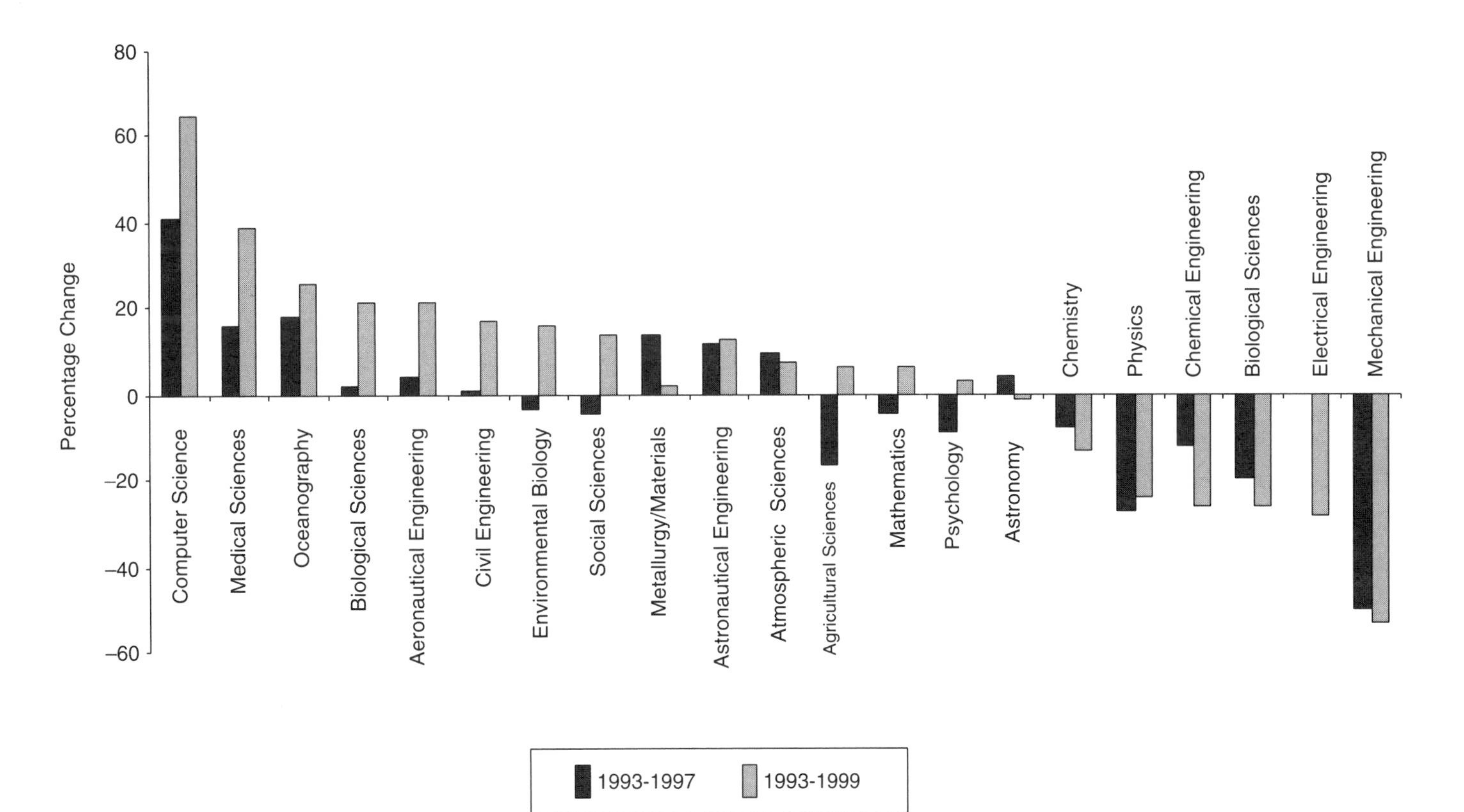

FIGURE 12 Changes in federal funding of fields of science and engineering, FY 1993–1997 and FY 1993–1999 (in constant dollars).
SOURCE: Appendix Table 12.

disciplines, not just biology and medicine. These allied disciplines range from mathematics, engineering, and computer sciences to sociology, anthropology, and behavioral sciences."[25]

The Ad Hoc Group for Medical Research Funding, which strongly supports doubling the NIH budget, has also supported more balanced funding of the range of research fields. They have noted that

> "...a strong federal research enterprise requires a balanced portfolio across the scientific and engineering disciplines. Many breakthroughs in medical research and treatment, such as magnetic resonance imaging (MRI), have come from advances in the physical sciences that were developed from basic research in physics, chemistry, and mathematics. Continued progress in medical research depends on continued advances in other areas of science and engineering."[26]

The House Committee on Science, in its annual report of views and estimates, has also expressed concern about the growth of NIH relative to other disciplines also important to scientific progress, including in biomedical fields.

> "The Committee looks forward to working with the administration and our Congressional colleagues to try to develop ways to determine whether the current portfolio is too heavily weighted toward NIH, and, if it is, to figure out what a balanced portfolio would be."[27]

Finally, in a recent editorial, Donald Kennedy, editor of *Science*, asked, "Does it really make sense for some pieces of the enterprise to be treated very well indeed and others to be held back or cut? There are good reasons for thinking it doesn't."

> "In the first place, an increasing proportion of the important problems in science are interdisciplinary in character. At *Science*, we have published contributions to nanotechnology that come from disciplines as diverse as chemistry, materials science, and electrical engineering. The climate sciences, on which we will depend in formulating international policies, draw from paleontology, oceanography, and atmospheric chemistry. The dramatic scientific gains that will flow from the sequencing of the human genome will be harvested not only by molecular biologists but also by specialists in bioinformatics, trained in such disciplines as mathematics and computer science. Nurturing fields such as these requires a balanced portfolio."[28]

---

[25]See Harold Varmus, "The Impact of Physics on Biology and Medicine." Plenary Talk, Centennial Meting of the American Physical Society, Atlanta, March 22, 1999.

[26]See Ad Hoc Group for Medical Research Funding, "Ad Hoc Group for Medical Research Funding—FY 2002 Proposal, January 2001.

[27]See House Committee on Science, "Views and Estimates of the Committee on Science for Fiscal Year 2002," March 16, 2001.

[28]See Donald Kennedy, "A Budget out of Balance," *Science* 291, 2001, p. 2337.

The current administration is less concerned about the balance issue, arguing that the research enterprise is expanding and healthy because of the increase in industrial R&D funding. At his confirmation hearing in January 2001, Mitch Daniels, Director of the Office of Management and Budget, acknowledged that the federal government has an important role to play in funding research that the private sector finds too risky or too long-term to invest in and said his office would "work to develop an appropriately balanced program of federal research." At the same hearing, Daniels' deputy director, Sean O'Keefe, spoke of the need for a balanced portfolio of research funded by both the federal government and the private sector and said that, from that perspective, the current portfolio is balanced.[29]

Trends in Federal Funding of Basic Research

Trends in basic—that is, long-term, fundamental research funding are important because industry tends to focus its funding on applied research and development. In the IT area, for example, Irving Wladawsky-Berger, General Manager of IBM's Internet Division, told the President's Information Technology Advisory Committee (PITAC) that government must take the lead in funding long-term research because industry must focus on being successful in the marketplace by developing competitive products and services and is facing increasingly short development cycles. "While the IT industry invests significantly in R&D, the bulk of the investment is product development (90 percent), and the bulk of the remaining is short-term, applied research, with only a few larger companies doing any long-term, basic research." Meanwhile, "R&D has declined as a percentage of revenues due to competitive pressures on prices and profit margins, putting further pressure on long-term research."[30]

The trends in basic research are similar to those in total research described above (see Figure 12), although federal funding of basic research was not cut as much and recovered faster (see Figure 13). In 1997, when federal obligations for basic research were nearly 3 percent more than in 1993, 6 of the 12 fields in engineering, physical sciences, and math/computer science had less funding for basic research than in 1993. And 2 of the 12 (electrical and mechanical engineering) had more than 20 percent less. In 1999, when federal basic research funding was 17 percent more than in 1993, 6 of the 12 fields still had less than in 1993. And 2 of the 12 (chemical and mechanical engineering) had more than 20 percent more. In the life sciences, 1 of the 4 fields (medical sciences) had more funding for basic research in 1997 than in 1993; in 1999, all 4 had more funding than in 1999.

---

[29]American Institute of Physics, "The Office of Management and Budget on Science and Technology," FYI, The American Institute of Physics Bulletin of Science Policy News. No. 28, March 2001.

[30]See I. Wladasky-Berger, "Information Technology: Transforming Our Society," Presentation to the President's Information Technology Advisory Committee, February 17, 1999.

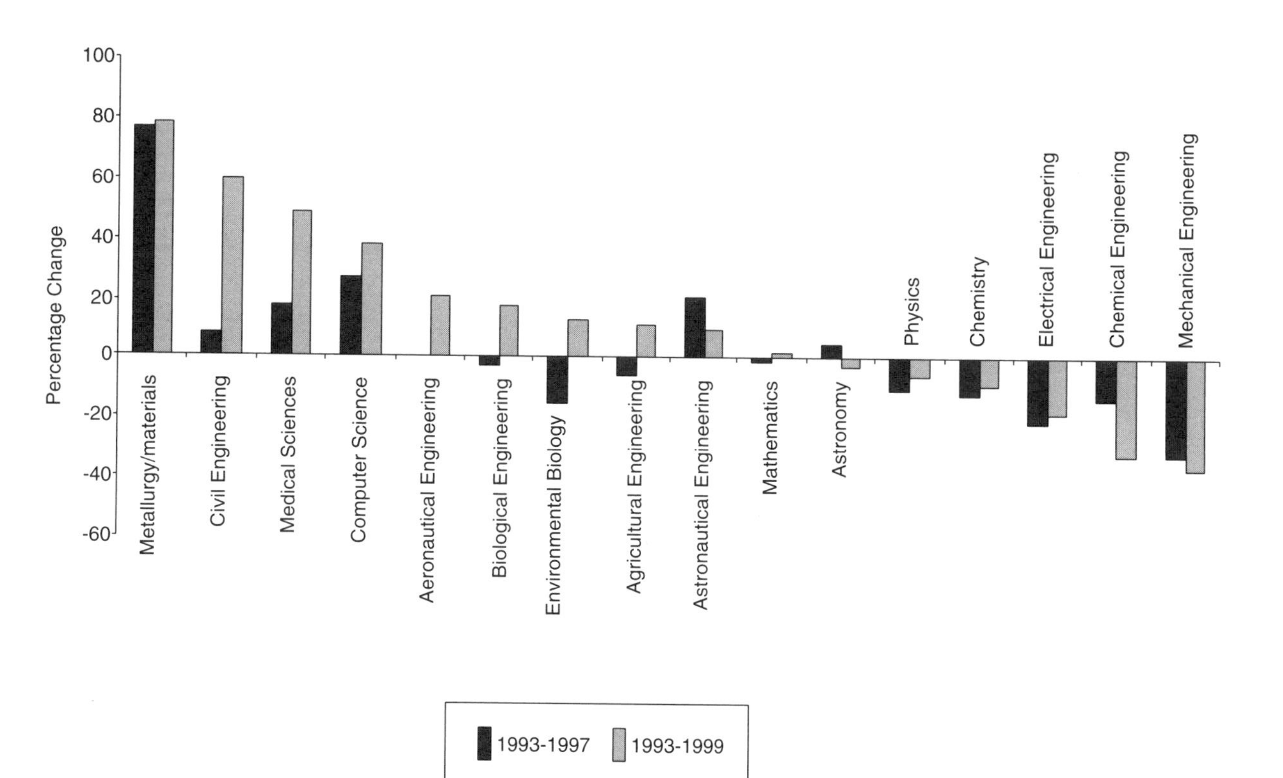

FIGURE 13 Trends in federal funding of basic research in selected fields of science and Engineering, FY 1993–1997 and FY 1993–1999 (in constant dollars).

SOURCE: Appendix Table 13.

Life Sciences

Federal support of the life sciences fields increased substantially during the 1990s. Annual funding for basic research in the biological sciences went from $3.3 billion in 1990 to $4.2 billion in 1999, a real increase of 27 percent (see Figure 14). Funding for medical sciences increased even more in the 1990s (76 percent) and was close in amount to that for the biological sciences by 1999 ($4.0 billion). The other two life sciences fields—environmental biology and agricultural science—were cut for several years in the mid-1990s and were about even with their 1990 funding levels in 1998. Each received a substantial increase in 1999, which put them ahead of 1990 by 37 percent and 13 percent, respectively.

Because 1999 was the first year of the 5-year campaign to double the NIH budget, the rate of growth for basic research in the biological and medical sciences can be expected to increase substantially from the 2.7 percent and 6.5 percent annual compound rates of growth they experienced from 1990 to 1999, respectively. From 1998 to 1999, for example, funding increased by 14.4 percent for biological sciences, and by 15.0 percent for medical sciences.

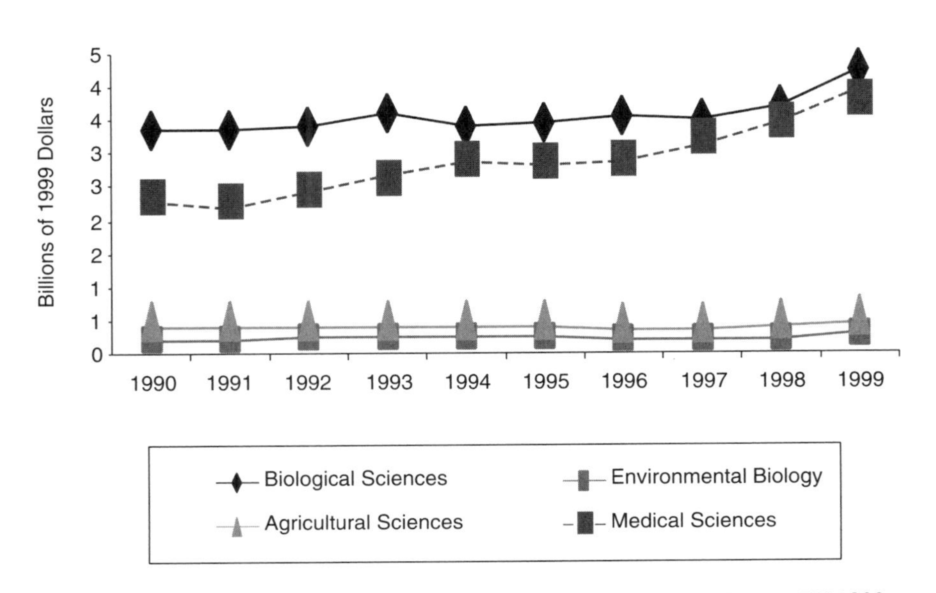

FIGURE 14 Trends in federal funding of basic research in the life sciences, FY 1990–1999.
SOURCE: Appendix Table 13.

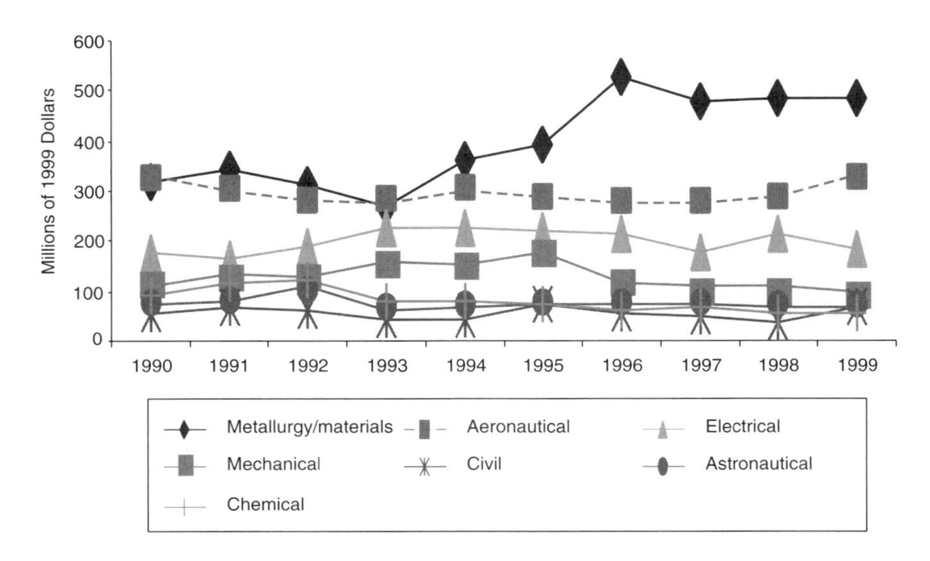

FIGURE 15 Trends in federal funding of basic research in engineering, FY 1990–1999. SOURCE: Appendix Table 13.

Engineering

Trends in federal obligations for basic research in engineering varied by field (see Figure 15). Metallurgy/materials engineering experienced substantial growth, especially from 1993 to 1996. From 1990 to 1999, funding for the field increased from $317 million to $481 million, or 52 percent.

Several other fields increased their funding for basic research for the first few years of the 1990s, then received less funding through 1999. Electrical engineering increased from $179 million in 1990 to $227 million in 1993, then declined to $186 million in 1999, 18 percent less than its high point in 1993. Chemical engineering increased from $92 million in 1990 to $120 million in 1992, dropped to $81 million in 1993, and steadily shrank to $55 million in 1999; it was down 40 percent for the decade (54 percent from its high point in 1992). Astronautical engineering rose from $76 million in 1990 to $108 million in 1992, then fell to between $60 million and $70 million from 1993 to 1999.

Funding levels for the other two fields were down for most of the decade but recovered in 1999. Aeronautical engineering had its two best years at the beginning and end. Funding for basic research in that field was $329 million in 1990 and $332 million in 1999, but it fluctuated between $275 million and $300 million during that period. Civil engineering was down from its high of $70 million in 1991 for most of the decade, jumping from $36 million in 1998 to $69 million in 1999.

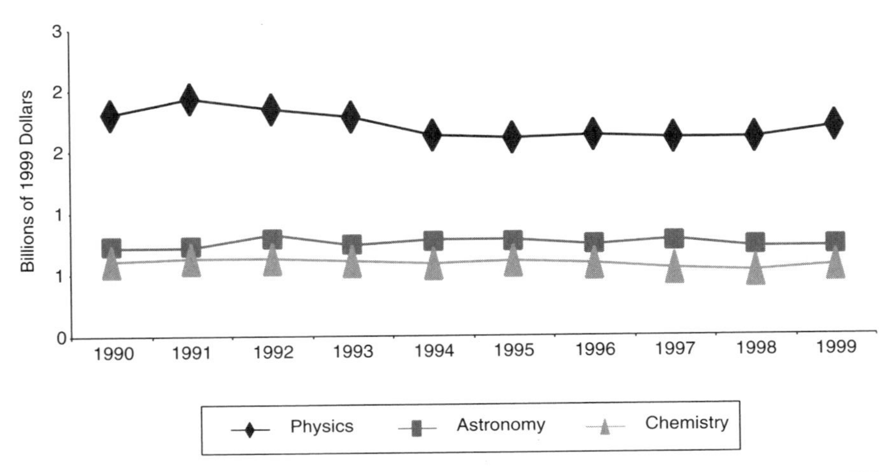

FIGURE 16 Trends in federal funding of basic research in the physical sciences, FY 1990–1999.
SOURCE: Appendix Table 13.

Physical Sciences

The general trend for funding of basic research in the physical sciences was downward in the 1990s, at least after the first year or two (see Figure 16). Funding for physics peaked in 1991 at $1.9 billion, dropped to about $1.6 billion from 1994 to 1998, and was at $1.7 billion in 1999. Physics was down by 6 percent for the decade (by 12 percent from the 1991 high point). Similarly, chemistry was funded at $612 million in 1990, $636 million in 1992, and $555 million in 1999.

Funding for astronomy also peaked early (at $834 million in 1992) but ended the decade about where it started (at $717 million compared with $707 million in 1990).

Computer Science/Mathematics

The general trend for computer science was up, except in 1994 (see Figure 17). Basic research funding was $274 million in 1990 and $483 million in 1999, an increase of 60 percent. Applied research funding was up even more—by 163 percent.

The picture for mathematics was more mixed. Funding for basic research generally increased from 1990 to 1994, then dropped for several years before recovering in 1997. The funding level was $215 million in 1990 and $271 million in 1994. In 1995 and 1996, it dropped about $100 million, then jumped to about $250 million in 1997, 1998, and 1999. Thus, funding for basic research

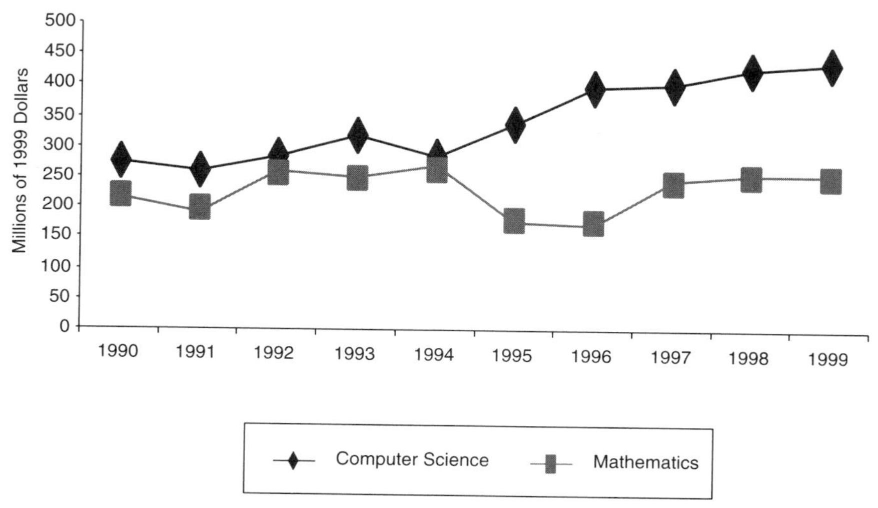

FIGURE 17 Trends in federal funding of basic research in computer science and mathematics, FY 1990–1999.
SOURCE: Appendix Table 13.

in mathematics increased by 14 percent from 1990 to 1999, but the 1999 level was 6 percent lower than it was in 1994.

CONCLUSION

Federal and nonfederal funding of research related to the core fields of health (biological sciences and medical sciences) and IT (computer science and electrical engineering) increased substantially in the 1990s. It must be kept in mind, though, that federal funding plays a much larger role in the health sector. This contributes to a much greater emphasis on basic research in the health than in the IT sector.

The growth in federal funding stems from strong congressional interest in biomedical research and the high priority that successive administrations have given to a cross-agency initiative in high-performance computing and networking dating from 1992. This growth will continue and even accelerate because of the campaign, begun in 1999, to double the NIH budget in 5 years. A similar acceleration of growth may occur in IT in response to 1998 and 1999 reports of the PITAC, which recommended a near tripling of federal investment in IT R&D by 2004.[31]

[31]See, for example, PITAC (1999). President Bush just extended PITAC until June 2003. See Executive Order of May 31, 2001, at http://www.whitehouse.gov/news/releases/2001/06/20010601-6.html.

Federal funding of research in a number of other fields—important to long-term progress in the health and IT sectors—has either not increased as much or, in the case of most fields in the physical sciences and engineering, has suffered cuts during the 1990s. This situation has given rise to a growing chorus of concerns about a possible "imbalance" in the federal portfolio of research.

Funding Trends in the Core Fields of Health and IT R&D

Federal funding has been moving into the hot fields central to health and IT R&D. For example, funding of biological sciences and medical sciences increased 28 percent and 65 percent, respectively, in real terms, from 1990 to 1999. Federal funding of computer science research increased 121 percent during the same period. Electrical engineering, however, experienced 10 percent less federal support.

Nonfederal funding patterns differ in health and IT R&D. It is impossible to know how nonfederal funding is distributed by field of research (e.g., computer science or biology), because industry R&D data are not collected by field. In this paper, we have attempted rough estimates of national R&D by including R&D funded by health and IT industries and nonfederal funds spent by academic institutions on R&D in the relevant fields, along with federal R&D funding. In health, for example, R&D funding in two industries—drugs and medicines and food—is counted, along with nonfederal expenditures on life sciences and bioengineering R&D by academic institutions. The industries representing IT are electrical equipment; office, computing, and accounting machines; and, beginning in 1995, computer and data processing services, which are included along with academic nonfederal spending on mathematics, computer science, and electrical engineering.

Nonfederal (primarily industry) funding of R&D related to health has apparently increased at a faster rate than federal funding. Federal health R&D, which is almost all research, accounted for 43 percent of the national investment in 1998 despite the strong buildup in industry funding. In IT, nonfederal funding of R&D appears to have increased at a rate slightly less than that for federal funding of computer science and electrical engineering, but federal funding still only constitutes a small part of IT R&D—less than 7 percent (5 percent if the software industry is included).[32]

Although R&D is up overall in both cases, the picture is more mixed in IT than in health R&D. The strong increase in federally funded research in computer science has been offset to some extent by the decline in funding for electrical engineering, down by 29 percent since 1993 (basic research by 18 percent).

[32]If data for software R&D funded by industry were available back to 1990, it is possible that nonfederal funding grew at a faster rate than federal funding.

Company funding of computer R&D is also down, although funding of computer peripherals, networking products, and software is probably up (these are not measured directly except, since 1995, software R&D).

The greater extent of federal support of R&D in health than in IT helps account for the larger share of basic research in health R&D. Drugs and medicines firms also devote a larger share of their R&D funding to basic research than electrical equipment and computer firms do—17 percent vs. less than 5 percent on average. As a result, basic research constitutes roughly one third of national R&D related to health, compared with less than one tenth of national R&D related to IT.

Trends in Federal Funding of Other Fields Contributing to Progress in Health and IT

Most other fields of research, including those that have been or promise to be important in health and IT in the long run, have received a shrinking share of federal investment during most of the 1990s. In some cases, they are receiving less in absolute terms than earlier in the decade. Fields increasing at a rate less than the biological, medical, and computer sciences since 1993 include environmental biology, agriculture, materials engineering, mathematics, atmospheric science, psychology, and the social sciences. Fields whose level of funding has fallen absolutely since 1993 include chemical engineering, electrical engineering, chemistry, physics, and geology (all but chemistry by 25 percent or more).

Most of the shift in the composition of federal funding of research occurred during the period from 1993, when federal funding stopped growing, and 1997, when growth resumed. The increases in biomedical and computer research were deliberate efforts on the part of Congress and the Administration, respectively. The cuts, however, were not the result of a high-level plan or policy. They were the byproduct of priority setting by the departments in light of their individual missions. Many of the fields with reduced funding (physics, chemical engineering, and electrical engineering) were primarily supported by DOD and DOE in the early 1990s, agencies that cut their research budgets substantially in the 1993-1996 period (by 22 and 8 percent, respectively).[33]

The shift in federal funding among fields was not a planned one in response to a formal assessment of national needs or research opportunities. It was a "resultant," that is, it is what happened after each federal agency and its committees in Congress set priorities without regard to national-level considerations. But what levels are optimal for each agency, taken one by one, may not be

[33]Similarly, geology, another field that experienced greatly reduced funding, is primarily funded by the Department of the Interior (U.S. Geological Survey), which cut its research budget by 12 percent from 1993 to 1996.

optimal overall, especially as the rationale for research in the physical sciences and engineering shifts from Cold War exigencies to economic growth and competitiveness. Industry's need for a flow of research results in the physical sciences and engineering, and for physical scientists and engineers trained in academic institutions conducting the research, may exceed the future needs DOD has for research and talent in those areas, but there is no mechanism for making the adjustment.

Balance in the Federal Portfolio

The federal budget process should provide adequate and steady support for a diverse portfolio of research activities that reflects scientific opportunities and national needs and also provides for a healthy infrastructure of facilities and talented, well-trained personnel. At present, however, there is no well-accepted methodology for setting priorities across fields of science and engineering. It is difficult to achieve agreement within a field on the appropriate balance among funding current research projects, a state-of-the-art infrastructure of research facilities and equipment, and the training of new researchers for the future. The absence of clear guidance in the form of policy priorities, or of a process for analyzing the overall federal research budget, makes it difficult to balance funding among complementary fields of research. In such circumstances, one person's idea of research portfolio imbalance can well be another's notion of what constitutes a healthy investment in a "hot" area of research—though both may honestly seek to meet major national goals through their respective proposals.

In the case at hand, some core fields in health and IT research are expanding rapidly and are accounting for most of the increase in federal funding in recent years (three fields—the biological, medical, and computer sciences—received more than two-thirds of the net increase in federal funding of research from 1990 to 1999). And that trend has no doubt continued since 1999, the last year for which there are data on actual funding by field, and will continue at least through 2003, the final year of the 5-year effort to double the NIH budget. But is this disparity in rates of growth among fields' funding leading to an imbalance among fields?

There are several approaches to answering this question. One approach is to treat research funding as an investment portfolio in which uncertainty and risk are minimized through diversification.[34] This approach is based on the uncertainty inherent in research.[35] It is not possible to predict where and when break-

[34]See M. McGeary and P.M. Smith, "The R&D Portfolio: A Concept for Allocating Science and Technology Funds," *Science*, 274, 1996, pp. 1484-1485.

[35]See R.E. Gomery, "An Unpredictability Principle for Basic Research," in A.H. Teich, S.D. Nelson, and C. McEnaney, eds., *AAAS Science and Technology Policy Yearbook: 1995*. Washington, D.C.: American Association for the Advancement of Science.

throughs will occur, and the payoffs may come years later, perhaps in another field. Who knew in 1945, when nuclear magnetic resonance was discovered in condensed matter, that that basic research finding would lead to the development of magnetic resonance imaging (MRI) technology for improved health care and research 30 years later?[36] This approach would ask if all fields were being adequately funded, including those that are not necessarily identified with popular goals of research, because advances in any field may lead to a breakthrough in another. The increase in interdisciplinary research adds to the prudence of investing in a broad range of fields.

A second approach is to see where the United States stands in comparison with research in other countries in each field and make adjustments to ensure not only that the nation is preeminent in certain fields and among the world leaders in the other fields.[37] International benchmarking avoids the problem of setting priorities across fields. This approach would ask if the reduced funding in fields such as physics, chemistry, geology, and some fields of engineering is putting the United States way behind the performance of other countries in those fields.

A third approach is to examine carefully the degree to which progress in high-priority fields might depend on advances in other fields. In this case, because of the sequencing of the human and other genomes, advances in computing are clearly going to contribute to health research and enable new applications; at the same time, computing advances may well be based on biological systems (see workshop proceedings in this volume). But advances in both areas—IT and health—have and almost certainly will continue to rely on contributions from other fields, including chemistry, physics, mathematics, various engineering fields, and the social and behavioral sciences. In this approach, for example, the federal research portfolio is out of balance to the extent that progress in health research depends on fields other than the biological and medical sciences that are not receiving funding commensurate with their role.

This situation came about because decision-making on health research historically has assumed that funding for the other important disciplines would be provided by other agencies (for example, chemistry, physics, and chemical engineering). To a large extent it was, until the end of the Cold War, reinforced by the federal budget deficit, caused DOD and DOE to reduce their research budgets.

The trend toward interdisciplinary work makes it more prudent to invest in a broad portfolio of research and not concentrate funding in a few fields. The editorial in *Science* by Donald Kennedy quoted earlier takes this approach in noting that progress in a number of hot fields—for example, nanotechnology,

---

[36]See National Research Council, *Trends in Federal Support of Research and Graduate Education*, Washington, D.C.: National Academy Press, 2001.

[37]See National Research Council, *Science, Technology, and the Federal Government: National Goals for a New Era*, Committee on Science, Engineering, and Public Policy, Washington, D.C.: National Academy Press, 1993.

global climate change research, genomics, and energy supply—depend on advances in a number of other fields. He concludes that capturing interdisciplinary complementarities requires a balanced portfolio.

A fourth and perhaps simplest way to approach the issue is not to ask if biomedical research or any other field is receiving too large a share of federal research funding. Instead, ask if there are fields that are underfunded with respect to scientific opportunities, including their contributions to other fields, especially fields central to the achievement of national goals such as better health, economic competitiveness, and environmental protection. International comparisons are useful here. This approach is also consistent with the reality that balance in the federal research portfolio cannot easily be achieved by reallocating funding from one agency to another. The R&D budget, including funding for research, is never considered as an integrated whole, especially in Congress. Each R&D program competes with other programs in its agency and the other agencies under the jurisdiction of its appropriations subcommittee, not with other R&D budgets in other appropriations subcommittees. Thus the argument shifts from saying, say, NIH gets too much to saying, say, physics does not get enough, in part because of its role in improving health among many other worthy applications such as computing and other information technologies.

Each of the approaches to assessing the federal research portfolio outlined above would examine the magnitude of the shift among fields that took place in the mid-1990s, focus on fields that stopped growing or began shrinking, examine their importance to the nation's leadership in science and technology, and ask if the decentralized decision-making that led to no or negative funding growth makes sense from an overall point of view. The answer may well be yes, given changes in scientific opportunity or national goals or both, but currently there is no mechanism for addressing the question in the first place. In the early 1970s, when DOD, NASA, and the Atomic Energy Commission cut research funding and federal support of the physical sciences and engineering dropped sharply, NSF received a 20 percent increase, half designated for support of important research being dropped by the mission agencies.[38] In the late 1970s, the Office of Management and Budget held funding in reserve for R&D program expansions and additions deemed advisable from the President's point of view after review of the agency requests (Smith, 2001).[39] There has been no similar plan to adjust the impact of shifts in agency priorities. Proponents of diminishing fields must make their best case to agencies and congressional committees with juris-

[38]See M. Lomask, *A Minor Miracle: The Informal History of the National Science Foundation*, NSF 76-18, Washington, D.C.: National Science Foundation, 1976, pp. 239-240.

[39]See P.M. Smith, "Science and Technology in the Carter Presidency," Paper presented at the Office of Science, Technology, and Public Policy 25th Anniversary Symposium, Massachusetts Institute of Technology, Cambridge, MA, May 1.

diction over their programs except perhaps to convince the House and Senate budget committees to give larger allocations to the relevant appropriations subcommittees.

REFERENCES

Ad Hoc Group for Medical Research Funding. 2001. "Ad Hoc Group for Medical Research Funding – FY 2002 Proposal," January. At: http://www.aamc.org/research/adhocgp/fy2002.doc.

AIP (American Institute of Physics). 2001. "The Office of Management and Budget on Science and Technology." *FYI, The American Institute of Physics Bulletin of Science Policy News*. No. 28(March 9).

Cockburn, I, R. Henderson, L. Orsenigo, and G. P. Pisano. 1999. "Pharmaceuticals and Biotechnology," pp. 363-398 in National Research Council, *U.S. Industry in 2000: Studies in Competitive Performance*. Washington, D.C.: National Academy Press.

Gomery, R. E. 1995. "An Unpredictability Principle for Basic Research," pp. 5-17 in A. H. Teich, S. D. Nelson, and C. McEnaney, eds., *AAAS Science and Technology Policy Yearbook: 1995*. Washington, D.C.: American Association for the Advancement of Science.

House Committee on Science. 2001. "Views and Estimates of the Committee on Science for Fiscal Year 2002." March 16.

Kennedy, D. 2001. "A Budget Out of Balance." *Science*. 291(March 23): 2337.

Lomask, M. 1976. *A Minor Miracle: An Informal History of the National Science Foundation*. NSF 76-18. Washington, D.C.: National Science Foundation.

Merrill, S.A., and M. McGeary. 1999. "Who's Balancing the Federal Research Portfolio and How?" *Science*. 285(September 10):1679-1680.

McGeary, M., and S. A. Merrill. 1999. "Recent Trends in Federal Spending on Scientific and Engineering Research: Impacts on Research Fields and Graduate Training," Appendix A in National Research Council, *Securing America's Industrial Strength*. Washington, D.C.: National Academy Press.

McGeary, M., and P. M. Smith. 1996. "The R&D Portfolio: A Concept for Allocating Science and Technology Funds." *Science*. 274(29 November):1484-1485.

National Academy of Sciences. 2001. *A Life-Saving Window on the Mind and Body: The Development of Magnetic Resonance Imaging*. Washington, D.C.: National Academy of Sciences, March 2001. At: www/beyonddiscovery.org/beyond/BeyondDiscovery.nsf/files/PDF MRI.pdf/$file/MRI PDF.pdf.

National Research Council. 1993. *Science, Technology, and the Federal Government: National Goals for a New Era*. Committee on Science, Engineering, and Public Policy. Washington, D.C.: National Academy Press

National Research Council. 1999. *Funding a Revolution: Government Support for Computing Research*. Computer Science and Telecommunications Board. Washington, D.C.: National Academy Press

National Research Council. 1999. *Securing America's Industrial Strength*. Board on Science, Technology, and Economic Policy. Washington, D.C.: National Academy Press

National Research Council. 2001. *Trends in Federal Support of Research and Graduate Education*. Washington, D.C.: National Academy Press

National Science Board. 2000. *Science and Engineering Indicators—2000*. Arlington, VA: National Science Foundation (NSB 00-01).

National Science Foundation. 2000. *Research and Development in Industry: 1998*. Arlington, VA: National Science Foundation (NSF 01-305).

Office of Management and Budget. 2001. *Historical Tables, Budget of the United States Government, Fiscal Year 2002*. Washington, D.C.: U.S. Government Printing Office.

President's Information Technology Advisory Committee (PITAC). 1999. *Information Technology Research: Investing in Our Future*. Report to the President. Arlington, VA: National Coordinating Office for Computing, Information, and Communications.

Smith, P. M.. 2001. "Science and Technology in the Carter Presidency." Paper presented at the Office of Science and Technology Policy 25th Anniversary Symposium, Massachusetts Institute of Technology, Cambridge MA, May 1. Forthcoming in symposium volume. Cambridge, MA: MIT Press.

Varmus, H. 1999. "The Impact of Physics on Biology and Medicine." Plenary Talk, Centennial Meeting of the American Physical Society, Atlanta, March 22. At: www.mskcc.org/medical_professionals/president_s_pages/speeches/the_impact_of_physics_on_biology_and_medicine.html.

Wladawsky-Berger, I. 1999. "Information Technology: Transforming Our Society," presentation at the seventh meeting of the President's Information Technology Advisory Committee, February 17. See "Transformations" PDF at: www.ccic.gov/ac/agenda-17feb99.html.

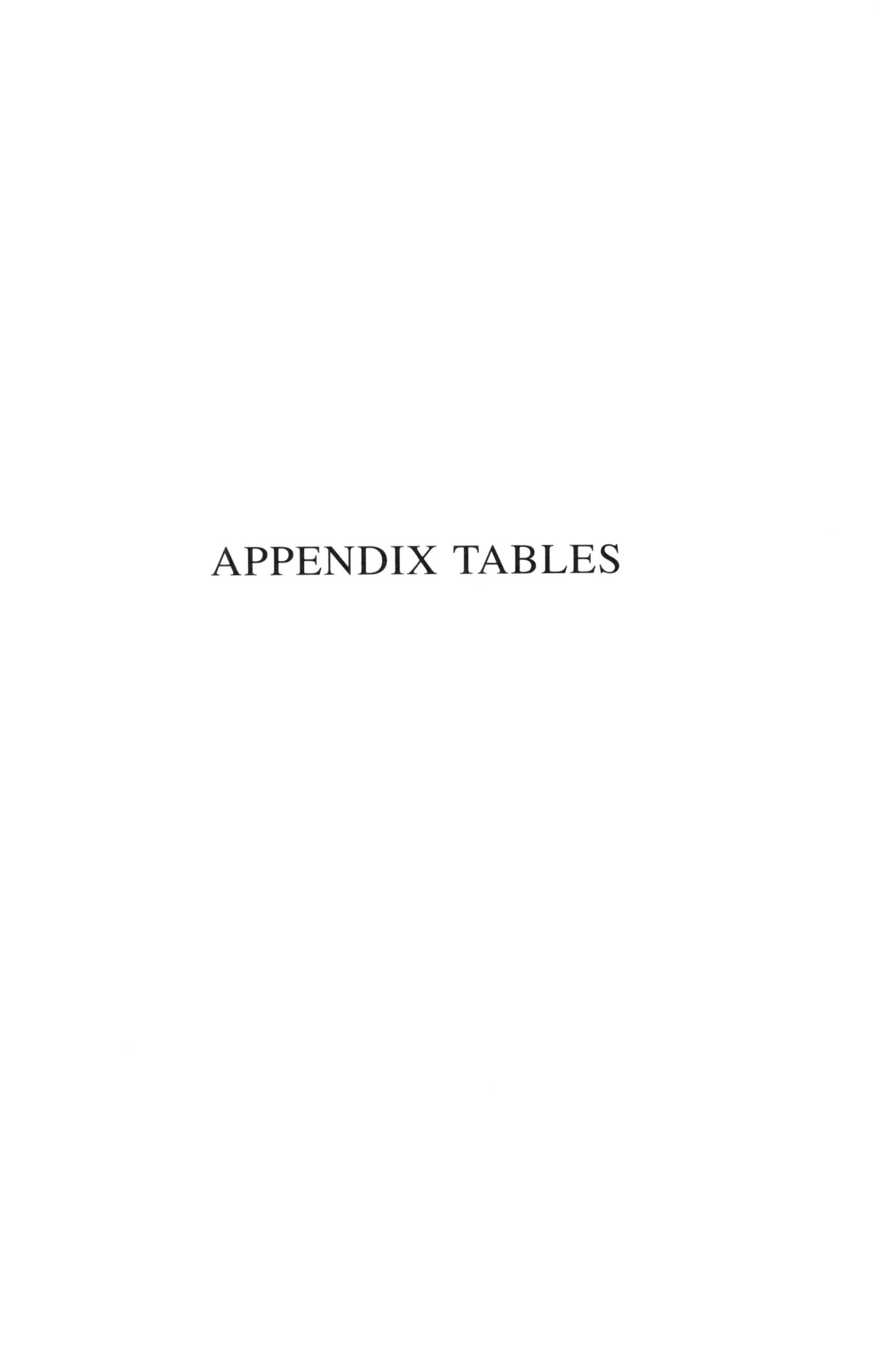

APPENDIX TABLES

Appendix Tables

TABLE 1 Federal funding of R&D and Research (in constant 1999 dollars)

	1999 Dollars:		GDP	Current Dollars:	
	R&D	Research	Deflator	R&D	Research
1980	61.880	21.717	0.5340	33.044	11.597
	63.721	20.849	0.5858	37.327	12.213
1982	61.803	20.775	0.6268	38.738	13.022
	61.192	21.783	0.6544	40.042	14.254
1984	67.808	22.077	0.6785	46.007	14.979
	73.810	23.017	0.7009	51.734	16.133
1986	76.096	22.990	0.7178	54.621	16.502
	79.515	24.326	0.7375	58.645	17.941
1988	80.132	24.486	0.7617	61.033	18.650
	81.276	26.255	0.7909	64.284	20.766
1990	80.963	26.346	0.8207	66.446	21.622
	80.628	28.110	0.8526	68.747	23.968
1992	82.222	27.991	0.8750	71.941	24.491
	81.406	30.017	0.8959	72.928	26.891
1994	77.658	29.950	0.9152	71.074	27.411
	75.876	30.409	0.9351	70.948	28.434
1996	74.659	29.631	0.9537	71.206	28.260
	76.037	30.201	0.9723	73.934	29.366
1998	77.007	31.356	0.9862	75.942	30.922
	80.171	33.528	1.0000	80.171	33.528
2000[a]	82.186	35.643	1.0193	83.769	36.330

SOURCES: For R&D: http://www.aaas.org/spp/dspp/rd/hist02p.pdf For research: Tables C-95, C-95a, C-98, C-98a in NSF, *Federal Funds for Research and Development, Fiscal Years 1999, 2000, and 2001*. Volume 49. For deflator, Table 10.1 in OMB, *Historical Tables, Budget of the United States Government, Fiscal Year 2002*. Washington, D.C.: U.S. Government Printing Office, 2001.

[a]Preliminary

TABLE 2 Federal Funding of Research, by Selected Agencies, FY 1990–2000 (in billions of 1999 dollars)

	1990	1991	1992	1993	1994	1995	1996	1997	1998	1999	2000[a]
All	26.3	28.1	28.0	30.0	30.0	30.4	29.6	30.2	31.4	33.5	35.6
All but NIH	18.6	20.2	19.5	20.4	20.0	20.4	19.3	19.4	19.9	20.7	21.2
NIH	7.7	8.0	8.5	9.6	9.9	10.0	10.4	10.8	11.4	12.9	14.4
DOD	4.3	4.4	4.7	5.3	4.6	4.5	4.2	3.9	4.0	4.1	4.4
DOE	3.1	3.8	3.9	3.8	3.6	3.7	3.5	3.7	3.8	3.9	4.1

SOURCE: National Science Foundation/SRS, *Survey of Federal Funds for Research and Development: Fiscal Years 1999, 2000, and 2001*.

NOTE: Constant dollar conversions were made using the fiscal year GDP deflators in OMB, Historical Tables, Budget of the United States Government, Fiscal Year 2002. Washington, D.C.: U.S. Government Printing Office, 2001, Table 10.1. They are: FY 1990=.8207, FY 1991=.8526, FY 1992=.8750, FY 1993=.8959, FY 1994=.9152, FY 1995=.9351, FY 1996=.9537, FY 1997=.9723, FY 1998=.9862, FY 1999=1.0000, FY 2000=1.0204.

[a]Preliminary

TABLE 3 Federal Obligations for Research in Selected Fields, FY 1990-1999 (in millions of 1999 dollars)

											Change, FY90-99		Change, FY93-99	
	1990	1991	1992	1993	1994	1995	1996	1997	1998	1999	Amount	Percent	Amount	Percent
Biological sciences	5.1	5.2	5.1	5.3	5.2	5.3	5.6	5.5	5.8	6.5	1.4	28.3%	1.1	21.2%
Medical sciences	4.1	4.1	4.5	4.9	5.3	5.3	5.2	5.7	6.1	6.8	2.7	65.0%	1.9	38.3%
Electrical engineering	0.8	0.9	0.9	1.0	0.8	0.8	0.7	0.6	0.6	0.7	–0.1	–10.4%	–0.3	–29.0%
Computer science	0.7	0.7	0.9	0.9	0.9	1.1	1.2	1.3	1.4	1.5	0.8	121.3%	0.6	64.4%

SOURCE: National Science Foundation/SRS, *Survey of Federal Funds for Research and Development: Fiscal Years 1999, 2000, and 2001.*

NOTE: Constant dollar conversions were made using the GDP deflators provided by OMB in January 2001: FY 1990=.8207, FY 1991=.8526, FY 1992=.8750, FY 1993=.8959, FY 1994=.9152, FY 1995=.9351, FY 1996=.9537, FY 1997=.9723, FY 1998=.9862, FY 1999=1.0000.

TABLE 4a Federal Obligations for Biological Sciences Research, by Agency, FY 1990–1999 (in millions of 1999 dollars)[a]

										Change, FY90-99		Change, FY93-99		
	1990	1991	1992	1993	1994	1995	1996	1997	1998	1999	Amount	Percent	Amount	Percent
USDA	141.8	166.7	155.9	161.1	145.5	139.2	125.0	150.7	130.6	152.7	10.8	7.6%	–8.4	–5.2
Commerce	0.0	12.0	26.3	18.6	24.7	40.5	61.0	25.6	36.0	37.5	37.5	NA	18.9	102.0
DOD	71.6	77.6	82.2	96.6	60.1	64.9	63.6	67.8	139.2	161.9	90.4	126.3%	65.4	67.7
DOE	194.3	178.0	186.3	201.7	184.8	212.7	196.3	167.3	191.8	188.8	–5.5	–2.8%	–12.9	–6.4
DHHS	4,143.7	4,274.3	4,074.1	4,404.0	4,245.2	4,284.6	4,626.6	4,555.8	4,792.5	5,358.3	1,214.6	29.3%	954.3	21.7
NIH	*3,730.7*	*3,749.7*	*3,931.6*	*4,252.9*	*4,076.6*	*4,121.5*	*4,460.2*	*4,392.7*	*4,647.6*	*5,227.5*	1,496.8	40.1%	*974.5*	*22.9*
Interior	63.6	55.8	55.8	52.8	89.0	81.4	73.4	78.1	62.8	82.0	18.4	29.0%	29.2	55.4
EPA	89.9	109.7	119.9	0.0	0.0	0.0	0.0	0.0	0.0	52.2	–37.6	–41.9%	52.2	0.0
NASA	58.6	72.6	63.5	113.9	120.7	124.5	113.7	116.4	128.6	125.5	66.8	113.9%	11.5	10.1
NSF	226.5	217.4	249.7	239.3	252.9	258.3	244.3	247.2	285.9	302.2	75.7	33.4%	62.9	26.3
All Others	60.2	64.3	55.0	58.6	62.7	61.2	63.9	62.9	19.0	16.6	–43.5	–72.3%	–42.0	–71.6
TOTAL	5,050.1	5,228.5	5,068.6	5,346.5	5,185.5	5,267.2	5,567.7	5,471.8	5,786.4	6,477.6	1,427.5	28.3%	1,131.2	21.2
All but NIH	1,319.4	1,478.8	1,137.0	1,093.5	1,108.9	1,145.7	1,107.5	1,079.1	1,138.8	1,250.2	–69.3	–5.3%	156.6	14.3

SOURCE: National Science Foundation/SRS, *Survey of Federal Funds for Research and Development: Fiscal Years 1999, 2000, and 2001.*

NOTE: Constant dollar conversions were made using the GDP deflators provided by OMB in January 2001: FY 1990=.8207, FY 1991=.8526, FY 1992=.8750, FY 1993=.8959, FY 1994=.9152, FY 1995=.9351, FY 1996=.9537, FY 1997=.9723, FY 1998=.9862, FY 1999=1.0000.

[a] Biological sciences includes anatomy, biochemistry, biology, biometry and biostatistics, biophysics, botany, cell biology, entomology and parasitology, genetics, microbiology, neuroscience (biological), nutrition, physiology, zoology, and other biological disciplines. It excludes environmental and agricultural disciplines.

TABLE 4b Federal Obligations for Biological Sciences Research, FY 1990–1999 (in millions of current dollars)[a]

	1990	1991	1992	1993	1994	1995	1996	1997	1998	1999
USDA	116.4	142.1	136.4	144.3	133.1	130.2	119.2	146.5	128.8	152.7
Commerce	0.0	10.3	23.0	16.6	22.6	37.9	58.2	24.9	35.5	37.5
DOD	58.7	66.2	71.9	86.5	55.0	60.7	60.7	65.9	137.3	161.9
DOE	159.4	151.7	163.0	180.7	169.2	198.9	187.2	162.7	189.1	188.8
DHHS	3,400.7	3,644.3	3,564.9	3,945.5	3,885.2	4,006.5	4,412.4	4,429.6	4,726.4	5,358.3
NIH	*3,061.8*	*3,197.0*	*3,440.2*	*3,810.2*	*3,730.9*	*3,854.0*	*4,253.7*	*4,271.0*	*4,583.4*	*5,227.5*
Interior	52.2	47.6	48.8	47.3	81.5	76.1	70.0	75.9	61.9	82.0
EPA	73.7	93.6	104.9	0.0	0.0	0.0	0.0	0.0	0.0	52.2
NASA	48.1	61.9	55.5	102.1	110.4	116.4	108.4	113.1	126.8	125.5
NSF	185.9	185.3	218.5	214.4	231.4	241.5	233.0	240.4	282.0	302.2
All Others	49.4	54.9	48.1	52.5	57.4	57.2	60.9	61.1	18.7	16.6
TOTAL	4,144.6	4,457.8	4,435.1	4,789.9	4,745.8	4,925.4	5,309.9	5,320.2	5,706.6	6,477.6
All but NIH	1,083	1,261	995	980	1,015	1,071	1,056	1,049	1,123	1,250

SOURCE: National Science Foundation/SRS, Survey of Federal Funds for Research and Development: Fiscal Years 1999, 2000, and 2001.

[a] Biological sciences includes anatomy, biochemistry, biology, biometry and biostatistics, biophysics, botany, cell biology, entomology and parasitology, genetics, microbiology, neuroscience (biological), nutrition, physiology, zoology, and other biological disciplines. It excludes environmental and agricultural disciplines.

TABLE 5a Federal Obligations for Medical Sciences Research, by Agency, FY 1990-1999 (in millions of 1999 dollars)[a]

										Change, FY90-99		Change, FY93-99		
	1990	1991	1992	1993	1994	1995	1996	1997	1998	1999	Amount	Percent	Amount	Percent
USDA	29.9	25.7	33.4	32.9	32.6	29.5	26.0	24.0	24.6	24.7	–5.2	–17.4%	–8.2	–24.8%
Commerce	1.7	3.0	1.9	4.8	5.2	20.6	17.6	22.2	11.5	8.2	6.5	NA	3.4	69.1%
DOD	202.3	210.6	252.5	248.4	248.9	219.8	162.0	176.3	258.5	224.4	22.1	10.9%	–24.0	–9.7%
DOE	44.9	135.6	141.8	54.6	77.4	50.7	47.4	53.8	54.4	53.0	8.1	17.9%	-1.6	–2.9%
DHHS	3,494.9	3,366.0	3,658.8	4,104.9	4,530.5	4,656.0	4,617.2	5,077.3	5,358.5	6,091.4	2,596.5	74.3%	1,986.5	48.4%
NIH	*3,077.5*	*2,983.5*	*3,449.8*	*3,876.4*	*4,250.4*	*4,253.7*	*4,254.3*	*4,685.1*	*4,957.0*	*5,575.5*	*2,498.1*	*81.2%*	*1,699.1*	*43.8%*
Interior	0.0	0.0	0.0	0.0	0.0	0.0	0.0	0.0	0.0	0.0	0.0	0.0%	0.0	0.0%
EPA	23.0	13.2	15.3	19.1	20.1	22.5	20.9	22.1	25.4	0.3	–22.8	–98.9%	–18.9	0.0%
NASA	63.5	83.4	72.2	116.1	123.0	126.8	113.0	124.8	123.3	118.2	54.8	86.3%	2.1	1.8%
NSF	0.0	0.0	0.0	0.0	0.0	0.0	0.0	0.0	0.0	0.0	0.0	0.0%	0.0	0.0%
All Others	261.1	221.1	292.5	335.4	275.0	191.3	210.7	190.0	224.0	281.0	19.9	7.6%	–54.4	–16.2%
TOTAL	4,121.3	4,058.5	4,468.5	4,916.2	5,312.6	5,317.3	5,214.8	5,690.4	6,080.2	6,801.2	2,679.9	65.0%	1,885.0	38.3%
All but NIH	626.4	692.5	809.7	811.3	782.2	661.3	597.6	613.0	721.8	709.8	83.4	13.3%	–101.5	–12.5%

SOURCE: National Science Foundation/SRS, *Survey of Federal Funds for Research and Development: Fiscal Years 1999, 2000, and 2001.*

NOTE: Constant dollar conversions were made using the GDP deflators provided by OMB in January 2001: FY 1990=.8207, FY 1991=.8526, FY 1992=.8750, FY 1993=.8959, FY 1994=.9152, FY 1995=.9351, FY 1996=.9537, FY 1997=.9723, FY 1998=.9862, FY 1999=1.0000.

[a] Medical sciences includes dentistry, internal medicine, neurology, obstetrics and gynecology, ophthalmology, otolaryngology, pathology, pediatrics, pharmacology, pharmacy, preventive medicine, psychiatry, radiology, surgery, veterinary medicine, and other medical disciplines.

TABLE 5b Federal Obligations for Medical Sciences Research, by Agency, FY 1990–1999 (in current dollars)[a]

	1990	1991	1992	1993	1994	1995	1996	1997	1998	1999
USDA	24.5	21.9	29.2	29.4	29.8	27.6	24.8	23.3	24.3	24.7
Commerce	1.4	2.6	1.7	4.3	4.8	19.3	16.8	21.6	11.3	8.2
DOD	166.1	179.6	220.9	222.6	227.8	205.6	154.5	171.4	255.0	224.4
DOE	36.9	115.6	124.1	48.9	70.8	47.4	45.2	52.3	53.7	53.0
DHHS	2,868.3	2,869.9	3,201.5	3,677.5	4,146.3	4,353.8	4,403.4	4,936.7	5,284.5	6,091.4
NIH	*2,525.7*	*2,543.7*	*3,018.6*	*3,472.9*	*3,890.0*	*3,977.6*	*4,057.4*	*4,555.4*	*4,888.6*	*5,575.5*
Interior	0.0	0.0	0.0	0.0	0.0	0.0	0.0	0.0	0.0	0.0
EPA	18.9	11.3	13.4	17.1	18.4	21.1	20.0	21.5	25.0	0.3
NASA	52.1	71.1	63.2	104.0	112.6	118.6	107.8	121.3	121.6	118.2
NSF	0.0	0.0	0.0	0.0	0.0	0.0	0.0	0.0	0.0	0.0
All Others	214.3	188.5	256.0	300.5	251.7	178.8	200.9	184.7	220.9	281.0
TOTAL	3,382.4	3,460.3	3,909.9	4,404.4	4,862.1	4,972.2	4,973.4	5,532.7	5,996.3	6,801.2
	514	590	708	727	716	618	570			

SOURCE: National Science Foundation/SRS, *Survey of Federal Funds for Research and Development: Fiscal Years 1999, 2000, and 2001.*
[a] Medical sciences includes dentistry, internal medicine, neurology, obstetrics and gynecology, ophthalmology, otolaryngology, pathology, pediatrics, pharmacology, pharmacy, preventive medicine, psychiatry, radiology, surgery, veterinary medicine, and other medical disciplines.

TABLE 6a Federal Obligations for Computer Sciences Research, by Agency, FY 1990–1999 (in millions of 1999 dollars)

										Change, FY90-99		Change, FY93-99		
	1990	1991	1992	1993	1994	1995	1996	1997	1998	1999	Amount	Percent	Amount	Percent
USDA	1.7	1.7	2.9	2.6	2.8	2.7	2.2	0.0	0.0	0.0	−1.7	−100.0	−2.6	−100.0%
Commerce	6.1	15.8	22.3	26.0	34.9	89.6	60.4	61.8	58.4	58.6	52.6	NA	32.6	125.3%
DOD	432.8	338.4	488.8	528.0	518.6	636.4	506.8	538.5	569.5	538.0	105.2	24.3	10.0	1.9%
DOE	27.8	102.4	132.9	115.4	97.9	103.9	210.8	327.6	401.3	506.3	478.5	1721.7	390.9	338.7%
DHHS	1.6	1.1	1.8	23.6	14.5	37.3	56.3	60.2	67.3	75.1	73.5	4646.0	51.5	217.8%
NIH	*0.2*	*0.0*	*1.8*	*20.1*	*0.9*	*25.9*	*46.0*	*52.0*	*55.0*	*61.9*	*61.7*	*27213.9*	*41.8*	*208.5%*
Interior	12.4	14.4	11.8	12.1	15.2	14.7	14.8	15.8	6.1	4.5	−7.9	∠64.0	−7.6	−63.2%
EPA	5.2	5.3	6.6	0.0	0.0	0.0	0.0	0.0	0.0	6.4	1.3	25.0	6.4	0.0%
NASA	55.1	59.9	50.1	26.6	28.2	29.1	27.6	28.4	25.6	25.2	−29.9	−54.3	−1.5	−5.6%
NSF	132.9	124.6	140.8	141.8	159.0	132.7	265.6	248.4	270.6	297.1	164.1	123.5	155.3	109.5%
All Others	9.5	22.6	24.1	46.0	35.9	37.9	29.8	20.0	19.5	4.9	−4.6	−47.9	−41.0	−89.2%
TOTAL	684.9	686.1	882.0	922.1	907.1	1,084.2	1,174.2	1,300.7	1,418.3	1,516.1	831.1	121.3	594.0	64.4%

SOURCE: National Science Foundation/SRS, *Survey of Federal Funds for Research and Development: Fiscal Years 1999, 2000, and 2001.*

NOTE: Constant dollar conversions were made using the GDP deflators provided by OMB in January 2001: FY 1990=.8207, FY 1991=.8526, FY 1992=.8750, FY 1993=.8959, FY 1994=.9152, FY 1995=.9351, FY 1996=.9537, FY 1997=.9723, FY 1998=.9862, FY 1999=1.0000.

TABLE 6b Federal Obligations for Computer Sciences Research, by Agency, FY 1990–1999 (in current dollars)

	1990	1991	1992	1993	1994	1995	1996	1997	1998	1999
USDA	1.4	1.4	2.5	2.3	2.6	2.5	2.1	0.0	0.0	0.0
Commerce	5.0	13.5	19.5	23.3	32.0	83.8	57.6	60.1	57.6	58.6
DOD	355.2	288.6	427.7	473.0	474.6	595.1	483.3	523.5	561.7	538.0
DOE	22.8	87.3	116.3	103.4	89.6	97.2	201.0	318.6	395.8	506.3
DHHS	1.3	0.9	1.6	21.2	13.3	34.8	53.7	58.5	66.3	75.1
NIH	*0.2*	*0.0*	*1.6*	*18.0*	*0.9*	*24.2*	*43.8*	*50.6*	*54.3*	*61.9*
Interior	10.1	12.2	10.3	10.8	13.9	13.8	14.1	15.4	6.0	4.5
EPA	4.2	4.5	5.8	0.0	0.0	0.0	0.0	0.0	0.0	6.4
NASA	45.2	51.1	43.8	23.9	25.8	27.2	26.3	27.6	25.3	25.2
NSF	109.1	106.2	123.2	127.0	145.5	124.1	253.3	241.5	266.8	297.1
All Others	7.8	19.2	21.1	41.2	32.9	35.4	28.5	19.5	19.3	4.9
TOTAL	562.1	584.9	771.8	826.1	830.2	1,013.9	1,119.8	1,264.7	1,398.7	1,516.1

SOURCE: National Science Foundation/SRS, *Survey of Federal Funds for Research and Development: Fiscal Years 1999, 2000, and 2001.*

TABLE 7a Federal Obligations for Electrical Engineering Research, by Agency, FY 1990–1999 (in millions of 1999 dollars)

											Change, FY90-99		Change, FY93-99	
	1990	1991	1992	1993	1994	1995	1996	1997	1998	1999	Amount	Percent	Amount	Percent
USDA	0.5	0.7	0.8	0.8	0.5	0.5	0.4	0.6	0.6	0.6	0.1	23.6%	-0.2	-23.5%
Commerce	15.1	17.9	23.8	26.8	31.3	60.3	51.6	51.6	42.1	41.5	26.4	NA	14.7	55.0%
DOD	652.0	685.4	702.8	807.4	628.2	601.3	548.5	487.8	514.3	555.6	-96.4	-14.8%	-251.8	-31.2%
DOE	25.7	50.3	44.3	48.1	46.1	54.9	26.0	25.0	25.6	30.8	5.1	19.8%	-17.3	-35.9%
DHHS	0.0	0.0	0.0	0.0	0.0	0.0	0.0	0.0	0.0	0.0	0.0	0.0%	0.0	0.0%
NIH	*0.0*	*0.0*	*0.0*	*0.0*	*0.0*	*0.0*	*0.0*	*0.0*	*0.0*	*0.0*	*0.0*	*0.0%*	*0.0*	*0.0%*
Interior	0.9	1.1	0.9	0.2	0.1	0.1	0.1	0.1	0.2	0.6	−0.2	−26.5%	0.4	192.2%
EPA	0.5	1.0	1.0	0.0	0.0	0.0	0.0	0.0	0.0	0.3	−0.3	−46.4%	0.3	0.0%
NASA	14.1	19.5	16.9	23.4	24.8	25.5	25.5	25.7	20.4	19.7	5.6	40.0%	−3.7	−15.8%
NSF	60.2	67.9	60.9	66.9	72.7	61.0	41.0	43.4	40.8	46.8	−13.4	−22.2%	−20.1	−30.0%
All Others	10.9	12.1	16.1	10.3	6.8	5.1	9.1	5.7	4.2	2.7	−8.2	−74.9%	−7.6	−73.4%
TOTAL	779.9	856.1	867.4	983.9	810.5	808.8	702.2	639.9	648.3	698.7	−81.2	−10.4%	−285.2	−29.0%

SOURCE: National Science Foundation/SRS, *Survey of Federal Funds for Research and Development: Fiscal Years 1999, 2000, and 2001.*

NOTE: Constant dollar conversions were made using the GDP deflators provided by OMB in January 2001: FY 1990=.8207, FY 1991=.8526, FY 1992=.8750, FY 1993=.8959, FY 1994=.9152, FY 1995=.9351, FY 1996=.9537, FY 1997=.9723, FY 1998=.9862, FY 1999=1.0000.

TABLE 7b Federal Obligations for Electrical Engineering Research, by Agency, FY 1990-1999 (in current dollars)

	1990	1991	1992	1993	1994	1995	1996	1997	1998	1999
USDA	1.4	1.4	2.5	2.3	2.6	2.5	2.1	0.0	0.0	0.0
USDA	0.4	0.6	0.7	0.7	0.4	0.5	0.3	0.6	0.6	0.6
Commerce	12.4	15.3	20.8	24.0	28.7	56.3	49.2	50.2	41.6	41.5
DOD	535.1	584.4	615.0	723.4	574.9	562.3	523.1	474.3	507.2	555.6
DOE	21.1	42.9	38.7	43.1	42.2	51.4	24.8	24.3	25.2	30.8
DHHS	0.0	0.0	0.0	0.0	0.0	0.0	0.0	0.0	0.0	0.0
NIH	*0.0*	*0.0*	*0.0*	*0.0*	*0.0*	*0.0*	*0.0*	*0.0*	*0.0*	*0.0*
Interior	0.7	1.0	0.8	0.2	0.1	0.1	0.1	0.1	0.2	0.6
EPA	0.5	0.9	0.8	0.0	0.0	0.0	0.0	0.0	0.0	0.3
NASA	11.5	16.6	14.8	21.0	22.7	23.9	24.4	25.0	20.1	19.7
NSF	49.4	57.9	53.3	60.0	66.5	57.0	39.1	42.2	40.2	46.8
All Others	9.0	10.3	14.1	9.2	6.2	4.8	8.7	5.5	4.2	2.7
TOTAL	640.0	729.9	759.0	881.5	741.8	756.3	669.7	622.2	639.3	698.7

SOURCE: National Science Foundation/SRS, *Survey of Federal Funds for Research and Development: Fiscal Years 1999, 2000, and 2001.*

TABLE 8 National Trends in Research Funding, FY 1980–2000 (in billions of 1999 dollars)

	1999 Dollars:					Basic research in current dollars:	
	Industry	Federal Government	Universities	Other Institutions	GDP Deflator	Industry 83	Fed 82
1980	25.7	40.2	21.9	4.0	0.5444	2,255	10,913
	27.3	37.0	20.4	3.7	0.5953	2,566	10,868
	27.2	36.0	20.2	3.7	0.6323	2,798	11,070
	27.9	37.4	21.3	3.7	0.6574	3,116	11,736
	29.5	37.7	21.5	3.8	0.6819	3,617	12,350
1985	31.1	39.5	23.2	4.0	0.7034	3,956	12,951
	36.1	38.4	21.6	4.3	0.7188	5,484	13,747
	34.5	38.2	21.2	4.6	0.7405	5,519	14,385
	34.5	37.3	20.0	4.7	0.7656	5,329	15,115
	35.0	38.0	20.4	4.9	0.7948	5,635	16,105
1990	33.1	39.0	22.0	5.0	0.8257	5,444	16,296
	37.9	38.6	21.4	5.1	0.8558	7,983	17,114
	35.6	36.6	19.7	5.2	0.8766	7,541	17,132
	32.8	36.1	19.4	5.2	0.8977	7,587	17,450
	31.0	34.7	18.4	5.4	0.9164	7,627	17,395
1995	33.9	33.3	17.5	5.5	0.9363	6,844	17,272
	36.0	33.6	17.5	5.7	0.9545	8,306	18,082
	41.2	32.6	16.4	5.8	0.9731	10,136	18,726
	41.4	33.4	16.3	6.0	0.9852	12,979	19,970
	45.4	35.3	17.5	6.3	1.0000	13,999	21,132
2000	48.2	34.6	16.6	6.4	1.0204	15,174	21,804
105.9	45.6%	32.7%	15.7%	6.1%			

SOURCE: Table B-6 in *National Patterns of R&D Resources: 2000 Data Update* (NSF 01-310). See <http://www.nsf.gov/sbe/srs/nsf01309/start.htm>

NOTE: Includes basic and applied research funding. FY 2000 data are preliminary. Universities includes colleges. Other institutions includes nonprofits, state and local governments, and foreign institutions.

			Applied research in current dollars:				
U&C 71	Othr 84	Othr 69	Industry 125	Fed 124	U&C 113	Othr 126	Othr 111
954	811	538	11,742	10,988	540	523	304
987	821	542	13,677	11,173	581	532	319
1,081	893	556	14,419	11,709	607	554	313
1,185	970	575	15,245	12,825	643	582	312
1,281	1,044	610	16,500	13,390	688	614	328
1,460	1,136	698	17,900	14,836	743	640	355
1,676	1,243	804	20,451	13,838	824	677	395
1,804	1,356	849	19,993	13,911	912	736	429
1,907	1,471	879	21,117	13,422	1,020	807	470
2,058	1,593	919	22,160	14,116	1,121	867	501
2,230	1,718	979	21,881	15,920	1,192	911	523
2,372	1,852	1,017	24,476	15,928	1,216	945	522
2,398	1,987	1,025	23,668	14,909	1,219	991	521
2,411	2,112	1,012	21,847	14,968	1,255	1,046	527
2,505	2,249	1,032	20,826	14,369	1,308	1,093	539
2,557	2,370	1,089	24,891	13,870	1,336	1,126	569
2,738	2,538	1,148	26,074	13,964	1,388	1,162	582
2,936	2,611	1,167	29,971	13,008	1,490	1,247	592
3,076	2,801	1,179	27,763	12,957	1,595	1,350	
3,265	2,974	1,223	31,352	14,199	1,676	1,427	628
3,435	3,130	1,264	34,047	13,525	1,762	1,500	648

TABLE 9 Sources of Funding for Basic and Applied Research, FY 2000 (percentages)

	Industry	Federal Government	Other Institutions	Total
Total Research	51.1	36.7	12.2	100.0
Basic Research	33.9	48.7	17.5	100.1

SOURCE: NSF, National Patterns of Research and Development Resources: 2000 Data Update (NSF 01-309).

TABLE 10a National Investment in Life Sciences-related R&D, 1981–1998 (in millions of 1999 dollars)

	Total	Company-funded R&D in drugs and medicines	Company-funded R&D in food, kindred, & tobacco products	Federal obligations for research in life sciences	DSS obligations for development
1981	15,004	3,467	1,068	7,452	700
1982	15,603	3,911	1,229	7,504	501
1983	16,690	4,405	1,253	7,876	479
1984	18,050	4,854	1,585	8,265	515
1985	19,227	4,949	1,615	9,046	574
1986	19,848	5,088	1,781	8,993	584
1987	21,498	5,530	1,626	9,914	731
1988	22,719	6,400	1,532	10,090	802
1989	24,295	6,935	1,565	10,688	907
1990	24,784	7,166	1,511	10,694	1,028
1991	27,393	8,118	1,492	11,243	1,705
1992	28,090	9,051	1,581	11,305	1,164
1993	30,027	10,173	1,498	12,000	1,267
1994	31,084	10,503	1,611	12,313	1,385
1995	32,162	10,896	1,673	12,615	1,454
1996	31,649	10,235	1,639	12,639	1,469
1997	34,491	11,906	1,836	13,011	1,596
1998	36,680	12,755	1,737	13,762	1,903
1981-1998	144.5%	267.9%	62.6%	84.7%	171.7%
1990-1998	48.0%	78.0%	14.9%	28.7%	85.1%
1994-1998	18.0%	21.4%	7.8%	11.8%	37.4%

SOURCES: National Science Foundation/SRS, *Research and Development in Industry: 1998* (NSF 01-305). See http://www.nsf.gov/sbe/srs/nsf01305/htmstart.htm. National Science Foundation/SRS, *Federal Funds for Research and Development, Fiscal Years 1999, 2000, and 2001, Vol. 49.*

National Science Foundation/SRS, *Academic Research and Development Expenditures: Fiscal Year 1999 [Early Release Tables].* See http://www.nsf.gov/sbe/srs/srs01407/start.htm.

VA obligations for development	USDA obligations for development	AcademicR&D (not federally funded) in life sciences & bio-engineering/bio-medical engineering	As % of GDP	As % of Total R&D
29	55	2,232	0.285%	12.36%
22	49	2,387	0.303%	12.20%
23	46	2,607	0.310%	12.18%
26	45	2,758	0.313%	12.03%
26	45	2,971	0.321%	11.78%
22	45	3,336	0.320%	11.86%
26	39	3,633	0.336%	12.60%
25	40	3,830	0.341%	12.99%
26	45	4,128	0.352%	13.61%
27	57	4,301	0.353%	13.46%
27	71	4,737	0.392%	14.57%
26	75	4,888	0.390%	14.89%
32	85	4,973	0.406%	16.27%
27	84	5,160	0.404%	16.83%
20	87	5,418	0.407%	16.40%
22	84	5,562	0.387%	15.31%
23	102	6,017	0.404%	15.80%
8	109	6,407	0.411%	15.93%
−71.6%	95.9%	187.0%		
−69.5%	90.8%	49.0%		
−70.2%	29.3%	24.2%		

TABLE 10b National Investment in Life Sciences-related R&D, 1981–1998 (in current dollars)

	Total	Company-funded R&D in drugs and medicines	Company-funded R&D in food, kindred, & tobacco products	Federal obligations for research in life sciences	DSS obligations for development*
1981	8,932	2,064	636	4,436	417
1982	9,866	2,473	777	4,745	317
1983	10,972	2,896	824	5,178	315
1984	12,308	3,310	1,081	5,636	351
1985	13,524	3,481	1,136	6,363	404
1986	14,267	3,657	1,280	6,464	420
1987	15,919	4,095	1,204	7,341	541
1988	17,394	4,900	1,173	7,725	614
1989	19,310	5,512	1,244	8,495	721
1990	20,464	5,917	1,248	8,830	849
1991	23,443	6,947	1,277	9,622	1,459
1992	24,624	7,934	1,386	9,910	1,020
1993	26,955	9,132	1,345	10,772	1,137
1994	28,485	9,625	1,476	11,284	1,269
1995	30,113	10,202	1,566	11,811	1,361
1996	30,209	9,769	1,564	12,064	1,402
1997	33,563	11,586	1,787	12,661	1,553
1998	36,137	12,566	1,711	13,558	1,875
1999				15,422	2,184
2000 (est.)				17,422	2,494
2001 (est.)					2,660

SOURCE: Company-funded R&D=NSF/SRS, *Research and Development in Industry: 1998.* NSF 01-305.

Academic R&D (not federally funded) For 1992-1999: NSF/SRS, *Academic Research and Development Expenditures, Fiscal 1999 [Early Release Tables],* Table B-5. For 1985-1991=Calculated from Appendix Tables 6-6 and 6-7 in NSB, *Science and Engineering Indicators 2000.* For 1981-1985=WebCASPAR.

Federal obligations for computer science research and electrical engineering research=NSF/SRS, *Federal funds for Research and Development, Fiscal Years 1999, 2000, and 2001,* Tables C-107 and C-107a.

DHHS, USDA, and VA obligations for development: NSF/SRS, *Federal funds for Research and Development, Fiscal Years 1999, 2000, and 2001,* Tables C-101 and C-101a.

* DHHS obligations for development exclude the social services agencies (Administration on Aging and Administration on Children and Families and the Social Security Administration when it was part of DHHS). NIH accounts for most of the rest (86 percent in 1999).

VA obligations for development	USDA obligations for development	Academic R&D (not life sciences & bioengineering /biomedical engineering)	GDP	Total R&D
17	33	1,329	3,131,000	72,267
14	31	1,509	3,259,000	80,848
15	30	1,714	3,535,000	90,075
18	31	1,881	3,933,000	102,344
18	32	2,090	4,213,000	114,778
16	32	2,398	4,453,000	120,337
19	29	2,690	4,743,000	126,299
19	31	2,932	5,108,000	133,930
21	36	3,281	5,489,000	141,914
22	47	3,551	5,803,000	152,051
23	61	4,054	5,986,000	160,914
23	66	4,285	6,319,000	165,358
29	76	4,464	6,642,000	165,714
25	77	4,729	7,054,000	169,214
19	81	5,073	7,401,000	183,611
21	80	5,309	7,813,000	197,330
22	99	5,855	8,318,000	212,379
8	107	6,312	8,790,000	226,872
7	126	6,736	9,299,000	244,143
7	131			
7	141			

TABLE 11a National Investment in Life Sciences-Related R&D, 1981–1998 (in millions of 1999 dollars)

	Total without software	Total with software	Company-funded R&D in office, computing, & accouting machines	Company-funded R&D in electrical equipment	Company-funded R&D in computer and data processing services	Federal obligations for research in math & computer science
1981	18,811	18,811	6,462	10,766	0	469
1982	20,093	20,093	7,819	10,568	0	554
1983	22,775	22,775	8,570	12,409	0	637
1984	25,364	25,364	10,282	13,253	0	645
1985	27,176	27,176	11,968	13,180	0	817
1986	27,402	27,402	11,658	13,588	0	856
1987	27,533	27,533	11,064	14,111	0	866
1988	27,485	27,485	12,209	13,029	0	840
1989	28,048	28,048	13,494	12,047	0	925
1990	26,886	26,886	13,307	11,223	0	1,019
1991	25,011	25,011	12,175	10,359	0	1,056
1992	25,716	25,716	12,108	10,856	0	1,323
1993	21,387	21,387	5,477	13,013	0	1,365
1994	22,015	22,015	4,450	14,772	0	1,421
1995	26,318	35,445	5,019	18,221	9,126	1,686
1996	32,778	43,281	8,520	21,326	10,503	1,647
1997	39,504	51,135	13,140	23,376	11,631	1,718
1998	36,970	51,482	9,024	24,744	14,512	1,865
1981-1998	96.5%	173.7%	39.6%	129.8%	NA	297.8%
1990-1998	37.5%	91.5%	–32.2%	120.5%	NA	83.1%
1994-1998	67.9%	133.8%	102.8%	67.5%	NA	31.2%

SOURCES: National Science Foundation/SRS, *Research and Development in Industry: 1998* (NSF 01-305). See http://www.nsf.gov/sbe/srs/nsf01305/htmstart.htm.

National Science Foundation/SRS, *Federal Funds for Research and Development, Fiscal Years 1999, 2000, and 2001, Vol. 49.*

National Science Foundation/SRS, *Academic Research and Development Expenditures: Fiscal Year 1999 [Early Release Tables].* See http://www.nsf.gov/sbe/srs/srs01407/start.htm.

Academic R&D in computer science & math (not federally funded)	Federal obligations for electrical engineering	Academic R&D in electrical engineering (not federally funded)	Without software		With software	
			As % of GDP	As % of Total R&D	As % of GDP	As % of Total R&D
99	936	79	0.358%	15.50%	0.358%	15.50%
106	968	79	0.390%	15.71%	0.390%	15.71%
117	936	105	0.424%	16.62%	0.424%	16.62%
135	924	126	0.440%	16.90%	0.440%	16.90%
165	891	155	0.454%	16.65%	0.454%	16.65%
175	938	187	0.442%	16.37%	0.442%	16.37%
216	1,061	214	0.430%	16.14%	0.430%	16.14%
220	955	233	0.412%	15.71%	0.412%	15.71%
260	1,061	262	0.406%	15.71%	0.406%	15.71%
282	775	280	0.383%	14.60%	0.383%	14.60%
284	853	284	0.358%	13.30%	0.358%	13.30%
274	866	290	0.357%	13.63%	0.357%	13.63%
282	983	267	0.289%	11.59%	0.289%	11.59%
285	810	278	0.286%	11.92%	0.286%	11.92%
292	807	294	0.333%	13.42%	0.448%	18.07%
281	702	303	0.400%	15.86%	0.529%	20.94%
298	639	332	0.462%	18.10%	0.598%	23.43%
334	649	355	0.414%	16.05%	0.577%	22.36%
236.9%						
18.3%						
17.3%						

TABLE 11b National Investment in Information Technology-related R&D, 1981-1998 (in millions of current dollars)

	Total without software	Total with software	Company-funded R&D in office, computing, & accouting machines	Company-funded R&D in electrical equipment	Company-funded R&D in computer and data processing services
1981	11,198	11,198	3,847	6,409	
1982	12,705	12,705	4,944	6,682	
1983	14,972	14,972	5,634	8,158	
1984	17,296	17,296	7,011	9,037	
1985	19,116	19,116	8,418	9,271	
1986	19,697	19,697	8,380	9,767	
1987	20,388	20,388	8,193	10,449	
1988	21,043	21,043	9,347	9,975	
1989	22,293	22,293	10,725	9,575	
1990	22,200	22,200	10,988	9,267	
1991	21,404	21,404	10,419	8,865	
1992	22,543	22,543	10,614	9,516	
1993	19,199	19,199	4,917	11,682	
1994	20,175	20,175	4,078	13,537	
1995	24,642	33,187	4,699	17,060	8,545
1996	31,287	41,312	8,132	20,356	10,025
1997	38,441	49,759	12,787	22,747	11,318
1998	36,423	50,720	8,890	24,378	14,297
1999					
2000 (est.)					
2001 (est.)					

SOURCE: Company-funded R&D = NSF/SRS, *Research and Development in Industry: 1998*. NSF 01-305.

Aacademic R&D (not federally funded).

For 1992-1999: NSF/SRS, *Academic Research and Development Expenditures, = Fiscal 1999 [Early Release Tables]*, Table B-5. For 1985-1991: Calculated from Appendix Tables 6-6 and 6-7 in NSB, *Science and Engineering Indicators 2000*. For 1981-1985: WebCASPAR.

Federal obligations for computer science research and electrical engineering research = NSF/SRS, *Federal Funds for Research and Development, Fiscal Years 1999, 2000, and 2001*, Tables C-107 and C-107a.

Federal obligations for research in math & computer science	Academic R&D in computer science & math (not federally funded)	Federal obligations for electrical engineering	Academic R&D in electrical engineering (not federally funded)	GDP	Total R&D
279	59	557	47	3,131,000	72,267
350	67	612	50	3,259,000	80,848
419	77	615	69	3,535,000	90,075
440	92	630	86	3,933,000	102,344
575	116	627	109	4,213,000	114,778
615	126	674	135	4,453,000	120,337
641	160	786	159	4,743,000	126,299
643	168	731	179	5,108,000	133,930
735	206	843	208	5,489,000	141,914
841	233	640	231	5,803,000	152,051
904	243	730	243	5,986,000	160,914
1,160	240	759	254	6,319,000	165,358
1,225	253	882	240	6,642,000	165,714
1,302	261	742	255	7,054,000	169,214
1,579	273	756	275	7,401,000	183,611
1,572	268	670	289	7,813,000	197,330
1,672	290	622	323	8,318,000	212,379
1,837	329	639	350	8,790,000	226,872
1,837	382	699	368	9,299,000	244,143
2,008				9,963,000	264,622
2,131					

TABLE 12 Real Changes in Federal Obligations for Research, by Field, FY 1993–1997 and FY 1993–1999 (in constant dollars)*

	1993–1997	1993–1999
Computer science	41.1%	64.4%
Medical sciences	15.7%	38.3%
Oceanography	17.8%	25.9%
Biological sciences	2.3%	21.2%
Aeronautical engineering	4.5%	20.9%
Civil engineering	0.8%	16.8%
Environmental biology	–3.5%	16.0%
Social sciences	–4.9%	13.5%
Metallurgy/materials	14.0%	1.5%
Astronautical engineering	11.2%	12.6%
Atmospheric sciences	9.0%	7.1%
Agricultural Sciences	–17.1%	6.7%
Mathematics	–4.4%	6.4%
Psychology	–8.7%	2.9%
Astronomy	4.0%	–1.1%
Chemistry	–7.6%	–13.4%
Physics	–27.8%	–24.6%
Chemical engineering	–11.8%	–25.9%
Geological sciences	–20.1%	–25.9%
Electrical engineering	–35.0%	–29.0%
Mechanical engineering	–49.8%	–53.9%

SOURCE: National Science Foundation/SRS, *Survey of Federal Funds for Research and Development, Fiscal Years 1999, 2000, and 2001.*

* CAVEAT: Percentage changes in some fine fields should be used with caution because, due to changes in NSF reporting procedures in 1996, the amounts reported by NSF in 1999 for some fields are not comparable to those reported in 1993. Mechanical engineering is most affected by the change. NSF also changed its classification of funding among fields in the physical and environmental sciences.

TABLE 13 Real Changes in Federal Obligations for Basic Research in Selected Fields, FY 1993–1997 and FY 1993–1999 (in constant dollars)*

	1993–1997	1993–1999
Metallurgy/materials	76.6%	78.6%
Civil engineering	8.2%	59.5%
Medical sciences	17.5%	48.9%
Computer science	26.2%	38.1%
Aeronautical engineering	0.7%	20.7%
Biological sciences	–3.0%	17.8%
Environmental biology	–14.9%	12.5%
Agricultural sciences	–-6.3%	11.1%
Astronautical engineering	20.0%	10.6%
Mathematics	–1.0%	2.6%
Astronomy	4.7%	–3.0%
Physics	–10.1%	–5.2%
Chemistry	–12.2%	–8.6%
Electrical engineering	–21.6%	–18.1%
Chemical engineering	–13.9%	–32.1%
Mechanical engineering	–32.3%	–37.4%

SOURCE: National Science Foundation/SRS, *Survey of Federal Funds for Research and Development, Fiscal Years 1999, 2000, and 2001.*

* CAVEAT: Percentage changes in some fine fields should be used with caution because, due to changes in NSF reporting procedures in 1996, the amounts reported by NSF in 1999 for some fields are not comparable to those reported in 1993. Mechanical engineering is most affected by the change. NSF also changed its classification of funding among fields in the physical and environmental sciences.

VI

ANNEX

Annex A:

Biographies of Contributors*

GRANT BLACK

Grant Black is a Research Associate at the Andrew Young School of Policy Studies, Georgia State University, where he is involved in research related to the economics of science. He has contributed to research funded by the Alfred P. Sloan Foundation, the Andrew W. Mellon Foundation, the National Science Foundation, and the United States Agency for International Development. He has also participated in national and international conferences on science and technology policy issues.

Black's research interests center on the economics of science, particularly the role of knowledge and its movement in the economy and the careers of scientists. Recent research has examined the importance of the local knowledge infrastructure on small-firm innovation, patterns of research collaboration, and individual patenting behavior. Other research has focused on the impact of immigration on scientific labor markets; women and minorities in the sciences; and educational training and labor market outcomes in the emerging field of bioinformatics. Black also has interests in the Small Business Innovation Research Program, the largest federal R&D program targeting small high-tech businesses.

Black received a B.S. and an M.A. in economics from the University of Missouri, St. Louis. He has taught economics at the University of Missouri, St.

*As of July 2001.

Louis, and Georgia State University since 1994. He will serve as a visiting scholar in the economics department at the University of Pretoria, South Africa, in 2002.

WESLEY COHEN

Wesley Cohen (Ph.D., Economics, Yale University, 1981) is Professor of Economics and Social Science in the Department of Social and Decision Science at Carnegie Mellon University (CMU) and is a Research Associate of the National Bureau of Economic Research. He also holds faculty appointments in CMU's Department of Engineering and Public Policy and its Heinz School of Policy and Management.

Focusing on the economics of technological change, Cohen's research examines the links between firm size, market structure and innovation, firms' abilities to exploit outside knowledge, the determinants of innovative activity across industries and firms, the knowledge flows affecting innovation, the means that firms use to protect their intellectual property, and the links between university research and industrial R&D, among other related subjects. Recently, he coordinated a major comparative survey research study in the U.S. and Japan on the nature and determinants of industrial R&D, and is currently engaged in a multi-year, NSF-funded research project on patenting and its impact on innovation.

He has published in numerous scholarly journals, including the *American Economic Review*, the *Economic Journal*, *Review of Economics and Statistics*, the *Journal of Industrial Economics*, the *Administrative Science Quarterly*, *Management Science* and the *Strategic Management Journal*, and served for five years as a Main Editor for *Research Policy*. He is also currently serving on the National Academies' Committee on Intellectual Property Rights in the Knowledge-Based Economy.

He has taught courses on the economics of technological change, the economics of entrepreneurship, industrial organization economics, policy analysis and organizational behavior.

KENNETH FLAMM

Kenneth Flamm is Dean Rusk Chair in International Affairs at the Lyndon B. Johnson School of Public Affairs at the University of Texas at Austin.

From 1993 to 1995, Flamm served as Principal Deputy Assistant Secretary of Defense for Economic Security and Special Assistant to the Deputy Secretary of Defense for Dual Use Technology Policy. He was awarded the Department's Distinguished Public Service Medal in 1995 by Defense Secretary William J. Perry. Prior to his service at the Defense Department, he spent eleven years as a Senior Fellow in the Foreign Policy Studies Program at The Brookings Institution.

Flamm has been a professor of economics at the Instituto Tecnológico Autónomo de México in Mexico City, the University of Massachusetts, and George Washington University. He has also been an adviser to the Director General of Income Policy in the Mexican Ministry of Finance and a consultant to the Organization for Economic Cooperation and Development, the World Bank, the National Academy of Sciences, the Latin American Economic System, the U.S. Department of Defense, the U.S. Department of Justice, the U.S Agency for International Development, and the Office of Technology Assessment of the U.S. Congress.

Among Dr. Flamm's publications are *Mismanaged Trade? Strategic Policy and the Semiconductor Industry* (1996), *Changing the Rules: Technological Change, International Competition, and Regulation in Communications* (ed., with Robert Crandell, 1989), *Creating the Computer* (1988), and *Targeting the Computer* (1987). He is currently completing an analytical study of the post-Cold War defense industrial base.

Kenneth Flamm, an expert on international trade and the high technology industry and member of the National Research Council's Committee on Government-Industry Partnerships for the Development of New Technologies, teaches classes in micro-economic theory, international trade, and defense economics. He received a bachelor's degree from Stanford University and a Ph.D. in economics from the Massachusetts Institute of Technology.

MICHAEL MCGEARY

Michael McGeary is a political scientist specializing in science, health, and technology policy analysis and writing. He works as an independent consultant to government agencies, foundations, and nonprofit organizations on issues related to science and technology. His areas of expertise include funding of research and development; research priority setting, funding, and management of biomedical research at NIH; graduate education, training, and employment of scientists and engineers; merit review systems at NSF and NIH; and the role of research in industrial innovation.

McGeary was on the staff of the National Academy of Sciences (NAS) from 1981 until 1995, where he was staff director for more than a dozen major reports. Since leaving the NAS, as a consultant, he has helped draft several NAS reports, including *Allocating Federal Funds for Science and Technology*, trends in the Federal Science and Technology (FS&T) budget (1996-1998), and *Trends in Federal Support of Research and Graduate Education*. He has also worked on health research issues. As a consultant, he helped draft the Institute of Medicine report *Scientific Opportunities and Public Needs: Improving Priority Setting and Public Input at the National Institutes of Health* (1998) and analyzed the implementation of the War on Cancer for the President's Cancer Panel (1999).

Between graduate study at the Massachusetts Institute of Technology (1972-1976) and coming to the NAS, Mr. McGeary taught political science and urban studies at Wellesley College (1976-78) and worked on studies of presidential management at the National Academy of Public Administration (1978-1980). His undergraduate degree is from Harvard College.

PAULA STEPHAN

Paula Stephan is a Professor of Economics, Andrew Young School of Policy Studies, Georgia State University, and served as the founding associate dean of the school from 1996-2001. Her research interests focus on the careers of scientists and engineers and the process by which knowledge moves across institutional boundaries in the economy. Stephan's research has been supported by the Alfred P. Sloan Foundation, the Andrew Mellon Foundation, the Exxon Education Foundation, the National Science Foundation, the North Atlantic Treaty Organization and the U.S. Department of Labor. She has served on several National Research Council committees including the Committee on Dimensions, Causes, and Implications of Recent Trends in the Careers of Life Scientists, Committee on Methods of Forecasting Demand and Supply of Doctoral Scientists and Engineers, and the Committee to Assess the Portfolio of the Science Resources Studies Division of NSF. She is a regular participant in the National Bureau of Economic Research's meetings in Higher Education and has testified before the U.S. House Subcommittee on Basic Science. She currently is serving a three year term as a member of CEOSE, the National Science Foundation's Committee on Equal Opportunity in Science and Engineering.

Dr. Stephan graduated from Grinnell College (Phi Beta Kappa) with a B.A. in Economics and earned both her M.A. and Ph.D. in Economics from the University of Michigan. She has published numerous articles in journals such as *The American Economic Review, Science, The Journal of Economic Literature, Economic Inquiry* and *Social Studies of Science*. Stephan coauthored with Sharon Levin *Striking the Mother Lode in Science*, published by Oxford University Press, 1992. The book was reviewed in *Science, Chemical and Engineering News, Journal of Economic Literature, The Southern Economic Journal* and *The Journal of Higher Education*. Her research on the careers of scientists has been the focus of articles in *The Economist, Science* and *The Scientist*. Stephan is a frequent presenter at meetings such as the American Economic Association, the American Association for the Advancement of Science, and the Society for the Social Studies of Science. Stephan reviews regularly for the National Science Foundation and a number of academic journals including *The American Economic Review, The American Sociological Review, Economic Inquiry, The Journal of Political Economy,* and *The Journal of Human Resources*.

Dr. Stephan has lectured extensively in Europe. She was a visiting scholar at the Wissenschaftszentrum Berlin für Sozialforschung, Berlin, Germany, intermittently during the period 1992-1995.

JOHN WALSH

John Walsh is an Associate Professor of Sociology at the University of Illinois at Chicago. Dr. Walsh's current research focuses on industrial R&D in the U.S. and Japan, including the impact of patents and patent policy on innovation, and the relations between universities and industrial research.

He has been published in numerous journals, including the *Journal of the American Society for Information Sciences*, *Public Opinion Quarterly*, *Work and Occupations*, *Social Studies of Science* and *Communications of the ACM*. Dr. Walsh authored *Supermarkets Transformed: Understanding Organizational and Technological Innovation* and coauthored *Mapping Crime in its Community Setting*. Dr. Walsh has contributed book reviews for publications such as the *American Journal of Sociology*, *Contemporary Sociology*, and *Management Learning*.

Dr. Walsh received his Ph.D. in Sociology from Northwestern University.

Annex B:

Participants List*
25 April 2000 Conference

Stanley Abramowitz
National Institute of Science & Technology

Zoltan Acs
University of Baltimore

Ted Agres
Washington Times

Jane Alexander*
Defense Advanced Research Projects Agency

J.M. Alliare

Pablo Amor
Delegation of the European Commission

Kiyoshi Ando
Nikkei

Robert Archibald
The College of William and Mary

David Audretsch
Indiana University

Gary Bachula
U.S. Department of Commerce

Wendy Baldwin
National Institutes of Health

Michael Baum
NIST

Ed Behrens
Procter and Gamble

Arpad Bergh
Optoelectronics Industry Development Association

Grant Black
Georgia State University

* Speakers

Robert Blackburn*
Chiron

David Blumenthal
Harvard Medical School/
Massachusetts General Hospital

Congressman Sherwood Boehlert*
U.S. House of Representatives

Mark Boguski*
National Center for Biotechnology
Information

John K. Boidock
Texas Instruments Incorporated

Jeff Bond
BMDO/Department of Defense

Shannon Bond
Government Accounting Office

William Bonvillian*
Office of Senator Lieberman

Michael Borrus*
Petkevich & Partners

Robert Boyd
Knight Ridder

Richard Bradshaw
Department of Enegry

Jeffrey Brancato
Office of the Executive Vice Provost
Massachusetts Institute of
Technology

Karen Brown
National Institute of Standards and
Technology

John Burgess
The Washington Post

William Camp
Sandia National Laboratories

Peter Cahill
BRTRC, Inc.

Elias Carayannis
ISTP

Kelly Carnes
U.S. Department of Commerce

Marvin Cassman*
National Institutes of Health

Mike Champness
Business-Higher Education Forum

Y.T. Chien
National Science Foundation

McAlister Clabaugh
National Research Council

Mel Ciment
National Science Foundation

Iain Cockburn
Boston University

Timothy Coffey*
Naval Research Laboratory

Wes Cohen*
Carnegie Mellon University

Sara Comley
International Observer, Press

Rita Colwell
National Science Foundation

Ereceline Companyo
Organization for Economic Cooperation and Development

B. Anne Craib
Dewey Ballantine

Jack Crowley

Michael Czinkota
Georgetown University

Stephen Dahms*
San Diego State University

K.C. Das
Office of the Secretary of Technology, Commonwealth of Virginia

Mike Davey
Congressional Research Service

Lance Davis
National Academy of Engineering

Will Davis
OECD Washington Center

Adriaan M. de Graaf
National Science Foundation

Brian Delroy
Embassy of Australia

Pierre Desrochers
Johns Hopkins University

Gerald Dinneen
National Research Council

Robert Eagan
Sandia National Laboratories

Chris Edwards
Joint Economic Committee

Mitch Eggers
Genometrix, Inc.

Rebecca Eisenberg
University of Michigan Law School

Stephan Esquires
Defense Advanced Research Projects Agency

Stephen Eule
House Science Committee

Fouad Ezra
Procter & Gamble

Maryann Feldman*
Johns Hopkins University

Frank Fernandez
Defense Advanced Research Projects Agency

David Festa
U.S. Department of Commerce

Kevin Finneran
Issues in Science and Technology

Eric A. Fischer
The Library of Congress

Kenneth Flamm*
Lyndon B. Johnson School of Public Affairs

Alexander Flax
National Academy of Engineering

Sam Fuller
Analog Devices

Cita Furlani
Advanced Technology Program

R. Michael Gadbaw
International Law and Policy

Paul G. Gaffney
Office of Naval Research

Lori Garver
National Aeronautics and Space Administration

James F. Gibbons
Paul G. Allen Center for Integrated Learning

Dan Goldin*
National Aeronautics and Space Administration

Jorge A. Goldstein
Sterne, Kessler, Goldstein & Fox

David Goldston*
Office of Congressman Boehlert

Jo Anne Goodnight
National Institutes of Health

Jeffrey L. Gren
Medical Equipment and Instrumentation

Margaret Grucza
Industrial Research Institute

Victoria D. Hadfield
Semiconductor Equipment and Materials International

Serge Hagege
Embassy of France

Lee Halcomb
National Aeronautics and Space Administration

Bronwyn Hall
National Bureau of Economic Research

Rebecca Henderson
Massachusetts Institute of Technology

Derek Hill
National Science Foundation

Alice Hogan
National Security and International Affairs

Paul M. Horn*
IBM Corporation

John B. Horrigan
National Research Council

Thomas Howell, Esq.
Dewey Ballantine

Kent Hughes
The Woodrow Wilson Center

William James
Research & Development

John Jankowski
National Science Foundation

Kenan Patrick Jarboe

Dale Jorgenson*
Harvard University

Tom Kalil*
National Economic Council

Christine Kelley
National Institutes of Health

Maryellen Kelley
National Institute of Standards and Technology

Kathleen Kingscott
International Business Machines

Karen Koppeschaar
Embassy of the Netherlands

Jeffrey D. Kueter
National Coalition for Advanced Manufacturing

Scott Kulicke
Kulicke and Soffa Industries, Inc.

Patrice Laget
Delegation of the European Commission

Ralph Landau
Stanford University

Jean Francois Large
Embassy of France

Roif Lehming
National Science Foundation

Richard Levin
Yale University

Rachel E. Levinson
Office of Science and Technology Policy

Harris Liebergot
National Institute of Science and Technology

Michael Lieberman
National Aeronautics and Space Administration

William Long
Business Performance Research Associates

Janet Lynch
General Electric

Tom Mays
Morrison and Foerster, LLP

Anne-Marie Mazza
National Research Council

Clark McFadden*
Dewey Ballantine

Robert McGuckin
Economic Research

Steve Merrill
National Research Council

Ernest Moniz
Department of Energy

Youhyoun Moon
Korean Embassy

Duncan Moore
White House Office
of Science and Technology Policy

Gordon Moore*
Intel Corp.

William Morin
R. Wayne Sayer and Associates

Mark Myers
Xerox Corporation

Kesh S. Narayanan
National Science Foundation

Karah Nazor
National Research Council

Richard Nelson
Columbia University

Robert Norwood
National Aeronautics and Space
Administration

Richard Nunno
Congressional Research Service

John Oldfield
Conference Board

Scott Pace
RAND

Erik Pages
NCDE

Edward Penhoet*
University of California, Berkeley

Barry Press
Washington CORE

Susan Pucie
National Institutes of Health

Samuel M. Rankin III
American Mathematical Society

Alan Rapoport
National Science Foundation

Diane Raynes
Government Accounting Office

Lawrence M. Rausch
National Science Foundation

Proctor Reid
National Academy of Engineering

Greg Reyes*
Schering-Plough

Josephine Robinson
Joint Economic Committee

Philippa Rogers
British Embassy

Alton D. Romig*
Science, Technology, and
Components

Sally Rood
Federal Laboratory Consortium

Peter Rooney
Council on Competitiveness

Leon Rosenberg*
Princeton University

Richard Rosenbloom
Harvard Business School

Deborah Rudolph
Institute of Electrical and Electronic Engineers

Richard Russell
House Committee on Science

R. Wayne Sayer
R. Wayne Sayer & Associates

Jeffrey Schloss
National Human Genome Research Institute

Craig Schultz
National Research Council

Alan Sears
Defense Advanced Research Projects Agency

Arun Seraphin
Office of Senator Joseph Lieberman

Jerry Sheehan
National Research Council

Richard Sheehan

Claudine Simson
Nortel Networks

Larry Smarr
National Center for Supercomputing Applications

William Spencer*
SEMATECH

Todd Spener
Charter Financial

Richard Spivack
National Institute of Standards and Technology

Kathryn E. Stein
Office of Therapeutics Research and Review

Paula E. Stephan*
Georgia State University

Gary W. Strong
Information Technology Office

Michael Steurewalt
National Science Foundation

Debbie Stine
National Research Council

Richard Swaja
National Institutes of Health

Gregory Tassey
National Institute of Standards and Technology

David Tennenhouse
Defense Advanced Research Projects Agency

Richard Thayer
Telecommunications and Technologies International

Roland Tibbetts

Phillipe Tondeur
National Science Foundation

Alan Tonelson
U.S. Business and Industrial Council Educational Foundation

Charles Trimble*
Trimble Navigation

Eric Truett
National Research Council

Robert Tuch
German-American Academic Council

James Turner
House Committee on Science

Paul Uhlir
National Academy of Sciences

Debra Van Opstal
Council on Competitiveness

Samuel Venneri
National Aeronautics and Space Administration

Nicholas Vonortas
George Washington University

Caroline Wagner
RAND

John P. Walker
Axys Pharmaceuticals, Inc.

Andrew Wang
National Institute of Standards and Technology

Kevin Wheeler
Committee on Small Business

Charles W. Wessner
Board on Science, Technology, and Economic Policy

James Wilson
House Science Commitee

Sandra Wilson
OECD Washington Center

Patrick Windham*
Windham Associates

Raymond Wolf
National Science Foundation

Isabel Wolte
Embassy of Austria

Benjamin Wu
House Subcommittee on Technology

William A. Wulf
National Academy of Engineering

Ed Zadjura
General Accounting Office

Annex C:

Bibliography

Ad Hoc Group for Medical Research Funding. 2001. "Ad Hoc Group for Medical Research Funding – FY 2002 Proposal." January. http://www.aamc.org/research/adhocgp/fy2002.doc.

American Institute of Physics. 2001. "The Office of Management and Budget on Science and Technology." *FYI, The American Institute of Physics Bulletin of Science Policy News.* No. 28 (March 9).

Audretsch, David B. and Roy Thurik. 1999. *Innovation, Industry, Evolution, and Employment.* Cambridge, UK: Cambridge University Press.

Barton, John. 1999. "Reforming the Patent System." Working Paper, Stanford Law School.

Bingham, Richard. 1998. *Industrial Policy American Style: From Hamilton to HDTV.* New York: M.E. Sharpe.

Brander, J. A. and B. J. Spencer. 1983. "International R&D Rivalry and Industrial Strategy," *Review of Economic Studies.* 50 (4).

Branscomb, L. M. and J. Keller, editors. 1998. *Investing in Innovation: Creating a Research and Innovation Policy.* Cambridge, MA: MIT Press.

Stokes, Donald. 1997. *Pasteur's Quadrant. Washington, D.C.* Brookings Institute Press.

Coburn, C. and D. Berglund. 1995. *Partnerships: A Compendium of State and Federal Cooperative Technology Programs,* Columbus, OH: Battelle Press.

Cockburn, I., R. Henderson, L. Orsenigo, and G. P. Pisano. 1999. "Pharmaceuticals and Biotechnology," pp. 363-398 in National Research Council, *U.S. Industry in 2000: Studies in Competitive Performance.* Washington, D.C.: National Academy Press.

Cohen, W. M., R. Florida, L. Randazze, and J. Walsh. 1998. "Industry and the Academy: Uneasy Partners in the Cause of Technological Advance" in Noll R., editor, *Challenges to Research Universities.* Washington, D.C.: The Brookings Institution.

Cohen, W. M., R. R. Nelson, and J. Walsh. 2000. *Protecting their Intellectual Assets: Appropriability Conditions and Why U.S. Manufacturing Firms Patent (or not).* Washington, D.C.: National Bureau of Economic Research Working Paper No. 7552.

Congressional Budget Office. 1999. *Current Investments in Innovation in the Information Technology Sector: Statistical Background.* Washington, D.C.: CBO (April).

Council of Economic Advisors. 1995. *Supporting Research and Development to Promote Economic Growth: The Federal Government's Role.* Washington, D.C.: Executive Office of the President.

Council of Economic Advisors. 2001. *Economic Report to the President.* Washington, D.C.: Executive Office of the President (H.Doc 107-2).

Devine, Kate. 2000. "Cell Signaling Alliance Gets Under Way." *The Scientist.* 14(20):1.

DeVol, Ross C. et al. 1999. *America's High-Tech Economy: Growth, Development, and Risks for Metropolitan Areas.* Santa Monica, CA: Milken Institute. July 13. http://www.milken-inst.org.

Dixit, A. K. and A. S. Kyle. 1985. "The Use of Protection and subsidies for Entry Promotion and Deterrence," *American Economic Review.* 75(1).

Educational Testing Service. 1997. *1997-98 Guide to the Use of Scores*, New Jersey: Educational Testing Service.

Educational Testing Service. 1997. *Sex, Race, Ethnicity, and Performance on the GRE General Test.* Educational Testing Service.

Faulkner, W. and J. Senker. 1995. *Knowledge Frontiers: Public Sector Research and Industrial Innovation in Biotechnology, Engineering Ceramics, and Parallel Computing.* New York: Oxford University Press.

Flamm, Kenneth. 1987. *Targeting the Computer.* Washington, D.C.: The Brookings Institution.

Flamm, Kenneth. 1998. *Creating the Computer.* Washington, D.C.: The Brookings Institution.

Flat Panel Display Task Force. 1994. *Building U.S. Capabilities in Flat Panel Displays: Final Report.* Washington, D.C.: Department of Defense (October).

Gavaghan, Helen. 1997. "Running to Catch Up in Europe." *Nature.* 389:420-22.

Gershon, Diane. 1997. "Bioinformatics in a Post-genomics Age." *Nature.* 389:417-18.

Gibbons, M. and R. Johnston. 1975. "The Roles of Science in Technological Innovation." *Research Policy.* 3:220-242.

Goldberger, Marvin L., Brendan A. Maher, and Pamela E. Flattau, eds. 1995. *Research-Doctorate Programs in the United States: Continuity and Change.* Washington, D.C.: National Academy Press.

Gomery, R. E. 1995. "An Unpredictability Principle for Basic Research," pp. 5-17 in A. H. Teich, S. D. Nelson, and C. McEnaney, eds., *AAAS Science and Technology Policy Yearbook: 1995.* Washington, D.C.: American Association for the Advancement of Science.

Gompers, P. and J. Lerner. 2000. *The Venture Capital Cycle.* Cambridge, MA: The MIT Press.

Gordon, Robert J. 2000. "Does the "New Economy" Measure up to the Great Inventions of the Past?" *Journal of Economic Perspectives* 14(4) Fall.

Grossman, Gene and Elhannan Helpman. 1993. *Innovation and Growth in the Global Economy*, Cambridge, MA: MIT Press.

Hall, B. and R. M. Ham. 1998. "The Patent Paradox Revisited: Firm Strategy and Patenting in the U.S. Semiconductor Industry," NBER Conference on Patent System and Innovation, January 8-9, 1999. Working Paper.

Henderson, Rebecca, Luigi Orsenigo, and Gary P. Pisano. 1999. "The Pharmaceutical Industry and the Revolution in Molecular Biology: Interactions Among Scientific, Institutional, and Organizational Change," in David C. Mowery and Richard R. Nelson Sources of Industrial Leadership: Studies of Seven Industries. New York: Cambridge University Press. pp. 267-311.

Heller, M. and R. Eisenberg. 1998. "Can Patents Deter Innovation? The Anticommons in Biomedical Research." *Science.* 28:698-701.

Hicks, D. and J. S. Katz. 1996. "Science Policy for a Highly Collaborative Science System." *Science and Public Policy.* 23:39-44.

Hollingsworth, Rogers. 2000. "Major Discoveries and Biomedical Research Organizations: Perspectives on Interdisciplinarity, Nurturing Leadership, and Integrated Structure and Cultures," in Weingart and Stehr, eds. *Practising Interdisciplinarity.* Toronto: University of Toronto Press..

Horrigan, J. 1999. "Cooperating Competitors: A Comparison of MCC and SEMATECH." Mongraph. Washington, D.C.: National Research Council.

Jorgenson, D. and Kevin J. Stiroh. 2000. "Raising the Speed Limit: U.S. Economic Growth in the Information Age." *Brookings Papers on Economic Activity* 1. Washington, D.C.: The Brookings Institution.

Kaiser, Jocelyn, editor. "Hopkins's Genetic Database to Close." *Science*. 279:645.

Katz, J.S., D. Hicks, M. Sharp, and B. R. Martin. 1995. *The Changing Shape of British Science*. Brighton: Science Policy Research Unit at the University of Sussex.

Kennedy, D. 2001. "A Budget Out of Balance." *Science*. 291:2337.

Kinney, Martin, editor. 2000. *Understanding Silicon Valley: The Anatomy of an Entrepreneurial Regime*. Stanford, CA: Stanford University Press.

Kleinman, Daniel Lee. 1995. *Politics on the Endless Frontier: Postwar Research Policy in the United States*. Durham, NC: Duke University Press.

Klevorick, A. K., R. Levin, R. R. Nelson, and S. G. Winter. 1994. "On the Sources and Significance of Interindustry Differences in Technological Opportunities." *Research Policy*. 24(2):195-206.

Koopman, Georg and Hans-Eckart Scharer, eds. 1996. The Economics of High-Technology Competition and Cooperation in Global Markets. Baden-Baden, Germany: HWWA (Institute for Economic Research).

Krugman, Paul. 1991. *Geography and Trade*, Cambridge, MA: MIT Press.

Krugman, P. and M. Obsfeldt. 1994. *International Economics: Theory and Policy*. 3d Edition, New York: Addison-Wesley Publishing Company.

Krugman, Paul. 1994. *Peddling Prosperity*. New York: W.W. Norton Press.

Krugman, Paul. 1990. *Rethinking International Trade*. Cambridge, MA: MIT Press.

Larson, Charles F. 2000. "The Boom in Industry Research." *Issues in Science and Technology*. 16(4):27.

Levin, R., A. Klevorick, R. R. Nelson, and S. G. Winter. 1987. "Appropriating the Returns from Industrial R&D." Brookings Papers on Economic Activity. pp. 783-820.

Lomask, M. 1976. *A Minor Miracle: An Informal History of the National Science Foundation*. NSF 76-18. Washington, D.C.: National Science Foundation.

Long, William F. 1999. *Advanced Technology Program: Performance of Completed Projects—Status Report Number 1*. NIST Special Publication 950-1. March.

MacIlwain, Colin. 2001. "Bush Favours Research at Pentagon and NIH." *Nature*. April 12.

Malakoff, David. 1999. "NIH Urged to Fund Centers to Merge Computing and Biology." *Science* (1742).

Mansfield, E. 1986. "Patents and Innovation: An Empirical Study." *Management Science*. 32:173-181.

Mansfield, E. 1991. "Academic Research and Industrial Innovation." *Research Policy*. 20(1).

Mansfield, E., M. Schwartz, and S. Wagner. 1981. "Imitation Costs and Patents: An Empirical Study." *Economic Journal*. 91:907-918.

Malakoff, David. 1999. "NIH Urged to Fund Centers to Merge Computing and Biology." *Science*. (June):1742.

Marshall, Eliot. 1996. "Hot Property: Biologists Who Compute." *Science*, (June 21):1730-32.

Marshall, Eliot. 1996. "Demand Outstrips Supply." *Science*, (June 21):1731.

Mazzoleni, R., and R. R. Nelson. 1998. "Economic Theories about the Benefits and Costs of Patents." *Journal of Economic Issues*. 32(4):1031-1052.

McGeary, M., and P. M. Smith. 1996. "The R&D Portfolio: A Concept for Allocating Science and Technology Funds," *Science*. 274:1484-1485.

Merrill, Stephen A. and Michael McGeary. 1999. "Who's Balancing the Federal Research Portfolio and How?" *Science*. 285:1679-1680.

Mowery, David C. 1998. "Collaborative R&D: How Effective is it?" *Issues in Science and Technology*. 15(1):37-44.

Mowery, David and Richard R. Nelson, eds. 1999. *Sources of Industrial Leadership: Studies of Seven Industries*. New York: Cambridge University Press.

Mowery, David C. 2000. *Using Cooperative Research and Development Agreements as S&T Indicators: What do We Have and What Would We Like?* Presentation to National Science Foundation conference, *Workshop on Strategic Research Partnerships*. October 13, 2000. Publication of proceedings pending.

Mowery, David C. and Richard R. Nelson, eds. 1999. *Sources of Industrial Leadership: Studies of Seven Industries*. New York: Cambridge University Press.

Mowery, David C., Rosemarie Ziedonis, and Greg Linden. 2001. "National Technology Policy in Global Markets." in *Innovation Policy in the Knowledge-Based Economy*, Maryann P. Feldman and Albert N. Link, eds.. Boston, Dordrecht and London: Kluwer Academic Publishers.

Nadiri, Ishaq. 1993. *Innovations and Technological Spillovers*. NBER Working Paper No. 4423.

Narin, F., Kimberly S. Hamilton, and D. Olivastro. 1997. "The Increasing Link Between U.S. Technology and Public Science." *Research Policy*, 26(3):317-330.

National Academy of Sciences. 2001. *A Life-Saving Window on the Mind and Body: The Development of Magnetic Resonance Imaging*. Washington, D.C.: National Academy of Sciences, March 2001. At: www/beyonddiscovery.org/beyond/BeyondDiscovery.nsf/files/PDF MRI.pdf/$file/MRI PDF.pdf.

National Research Council. 1986. *The Positive Sum Strategy*. Ralph Landau and Nathan Rosenberg, eds. Washington, D.C.: National Academy Press.

National Research Council. 1993. *Science, Technology, and the Federal Government,* Committee on Science, Engineering, and Public Policy. Washington, D.C.: National Academy Press.

National Research Council. 1995. *Allocating Federal Funds for Science and Technology*. Washington, D.C.: National Academy Press.

National Research Council. 1995. *Research Doctorate Programs in the United States: Continuity and Change*. M. Goldberger, B. Maher, and P. Ebert, editors. Washington, D.C.: National Academy Press.

National Research Council. 1996. *Conflict and Cooperation in National Competition for High Technology Industry,* Washington, D.C.: National Academy Press.

National Research Council. 1998. *Trends in the Early Careers of Life Scientists*. Washington, D.C.: National Academy Press.

National Research Council. 1999. *The Advanced Technology Program: Challenges and Opportunities*. Charles W. Wessner, ed. Washington, D.C.: National Academy Press.

National Research Council. 1999. *Capitalizing on Investments in Science and Technology*. Washington, D.C.: National Academy Press.

National Research Council. 1999. *Funding a Revolution: Government Support for Computing Research*. Washington, D.C.: National Academy Press.

National Research Council. 1999. *Industry-Laboratory Partnerships: A Review of the Sandia Science and Technology Park Initiative*. Charles W. Wessner, ed. Washington, D.C.: National Academy Press.

National Research Council. 1999. *Information Technology: Frontiers for a New Millennium*, Washington, D.C.: Government Printing Office (February).

National Research Council. 1999. *New Vistas in Transatlantic Science and Technology Cooperation*. Charles W. Wessner, ed. Washington, D.C.: National Academy Press.

National Research Council. 1999. *Securing America's Industrial Strength*. Washington, D.C.: National Academy Press.

National Research Council. 1999. *The Small Business Innovation Research Program: Challenges and Opportunities*. Charles W. Wessner, ed. Washington, D.C.: National Academy Press.

National Research Council. 2000. *Bioinformatics: Converting Data to Knowledge*. Washington, D.C.: National Academy Press.

National Research Council. 2000. *The Dynamics of Long-Term Growth: Gaining and Losing Advantage in the Chemical Industry*. Washington, D.C.: National Academy Press.

National Research Council. 2000. *Experiments in International Benchmarking of U.S. Research Fields*. Washington, D.C.: National Academy Press.
National Research Council. 2000. *Making IT Better: Expanding Information Technology Research to Meet Society's Needs*. Washington, D.C.: National Academy Press
National Research Council. 2000. *The Small Business Innovation Research Program: An Assessment of the Department of Defense Fast Track Initiative*. Charles W. Wessner, ed. Washington, D.C.: National Academy Press.
National Research Council. 2001. *The Advanced Technology Program: Assessing Outcomes*. Charles W. Wessner, ed. Washington, D.C.: National Academy Press.
National Research Council. 2001. *A Review of the New Initiatives at the NASA Ames Research Center*. Charles W. Wessner, ed. Washington, D.C.: National Academy Press.
National Research Council. 2001. *Observations on the President's Fiscal Year 2002 Federal Science and Technology Budget*. Washington, D.C.: National Academy Press.
National Research Council. 2001. *Trends in Federal Support of Research and Graduate Education*. Washington, D.C.: National Academy Press.
National Science Board. 2000. *Science and Engineering Indicators—2000*. Arlington, VA.: National Science Foundation (NSB 00-01).
National Science Foundation. 1997. *Characteristics of Doctoral Scientists and Engineers in the United States: 1995, Detailed Statistical Tables*. Division of Science Resources Studies. Washington, D.C.: National Science Foundation.
National Science Foundation. 1997. *Characteristics of Recent Science and Engineering Graduates: 1995, Detailed Statistical Tables*. Division of Science Resources Studies. Arlington, VA: National Science Foundation.
National Science Foundation. 1997. *Science and Engineering Doctorate Awards: 1996, Detailed Statistical Tables*. Division of Science Resources Studies. Arlington, VA: National Science Foundation.
National Science Foundation. 2000. *Research and Development in Industry: 1998*. Arlington, VA: National Science Foundation (NSF 01-305).
National Science and Technology Council, Subcommittee on Computing, Information, and Communications R&D, Committee on Technology. 1999. "Information Technology Research: Investing in Our Future." Washington, D.C.: Government Printing Office.
Office of Management and Budget. 2001. *Historical Tables, Budget of the United States Government, Fiscal Year 2002*. Washington, D.C.: Government Printing Office.
Okimoto, Daniel I. 1989. *Between MITI and the Market: Japanese Industrial Policy for High Technology*, Stanford: Stanford University Press.
Ophuls, William. 1997. *Ecology and the Politics of Scarcity: Prologue to a Political Theory of the Steady State*, San Francisco: Freeman.
"Post-Genomic Cultures." 2001. *Nature*. (February 1).
President's Information Technology Advisory Committee. 1999. *Information Technology Research: Investing in Our Future*. Report to the President. Arlington, VA.: National Coordinating Office for Computing, Information, and Communications.
Rausch, Lawerence M. 1995. *Asia's New High-Tech Competitors*. Arlington, VA: National Science Foundation (NSF 95-309).
Romer, Paul. 1990. "Endogenous Technological Change," *Journal of Political Economy*. 98(5):71-102.
Rosenberg, N. and R. R. Nelson. 1994. "American Universities and Technical Advance in Industry." *Research Policy*. 23(3):323-348.
Rosenbloom, Richard and William Spencer. 1996. *Engines of Innovation: U.S. Industrial Research at the End of an Era*. Boston: Harvard Business Press.
Rudolph, Frederick B. and Larry V. McIntire, eds. 1996. *Biotechnology: Science, Engineering, and Ethical Challenges for the 21st Century*. Washington, D.C.: Joseph Henry Press.

Russo, Eugen. 2000. "Stepping Up Mouse Sequencing." *The Scientist.* 14(22):12.
Russo, Eugen and Paul Smaglik. 1999. "Single Nucleotide Polymorphisms: Big Pharma Hedges its Bets." *The Scientist.* 13(15):1.
Saxeninan, Analee. 1994. *Regional Advantage: Culture and Competition in Silicon Valley and Route 128.* Cambridge, MA: Harvard University Press.
Scherer, F. M. et al. 1959. Patents and the Corporation. 2nd Edition. Boston, privately published.
Smith, P. M.. 2001. "Science and Technology in the Carter Presidency." Paper presented at the Office of Science and Technology Policy 25th Anniversary Symposium, Massachusetts Institute of Technology. Cambridge, MA: MIT Press.
Stephan, Paula and Grant Black. 1999. "Bionformatics: Does the U.S. System Lead to Missed Opportunities in Emerging Fields? A Case Study." *Science and Public Policy,* (December):1-15.
Stephan, Paula and Grant Black. 1999. "Hiring Patterns Experienced by Students Enrolled in Bionformatics/Computational Biology Programs." Report to the Alfred P. Sloan Foundation. May.
Stokes, Donald. 1997. *Pasteur's Quadrant.* Washington, D.C.: The Brookings Institution Press.
Tyson, Laura. 1992. *Who's Bashing Whom? Trade Conflict in High Technology Industries.* Washington, D.C.: Institute for International Economics.
Varmus, Harold. 1996. "The Impact of Physics on Biology and Medicine." Plenary talk, Centennial Meeting of the American Physical Society, Atlanta.
Varmus, Harold. 2000. "Squeeze on Science." *Washington Post.* October 4.
Vogel, David. 1996. *Kindred Strangers: The Uneasy Relationship Between Politics and Business in America.* Princeton: Princeton University Press.
Wladawsky-Berger, I. 1999. "Information Technology: Transforming Our Society," presentation at the seventh meeting of the President's Information Technology Advisory Committee, February 17. See "Transformations" PDF at: www.ccic.gov/ac/agenda-17feb99.html.
Wickware, Potter. 1997. "Choices and Challenges." *Nature.* 389:420.
Wilkinson, Sophie. 2000. "Big Science Takes on Cellular Signaling," *Chemical & Engineering News.* October 16.
Yap, Ting K., Frieder Ophir, and Robert L. Mantino. 1996. *High Performance Computational Methods for Biological Sequence Analysis.* Boston: Kluwer Academic Publishers.
Yee, Wendy. "The Top Five Career Trends of 1996: Informatics Anything." http:\\www.nextwave.org/server-java/SAM/pastloop/trend2.htm.
Zachary, G. Paschal. 1997. *Endless Frontier: Vannevar Bush, Engineer of the American Century.* New York: The Free Press.